MATHEMATICS FOR MATHS AND SCIENCE STUDENTS

By

Peter Martin Jones BSc (Hons) BA (Hons)

CONTENTS

MODULE A1

ALGEBRA 1 – GENERAL ALGEBRA

Revision Notes

BASIC ALGEBRAIC OPERATIONS

Addition

We know that when we add two numbers together we get the same result whichever way round we write them:

E.g. $3 + 2 = 5$ and $2 + 3 = 5$

It doesn't matter how many numbers there are; we can add them together in any order and get the same result.

The same rule applies to symbolic algebraic terms like a, b, r, s, t etc

i.e. $a + b + r = b + a + r = r + a + b$ etc

However, we can only add like terms together:

$ab + 3ab = 4ab$

e.g. we cannot add together $ab + abc$ to obtain a single combined term.

Multiplication

Exactly the same principle applies to the operation of multiplication

$a \times b = b \times a$ etc.

This property of addition and multiplication is called **commutative**

However it is usual write such products in the form $ab = ba$

Additionally $5x(x + 3) = 5x(x) + 5x(3) = 5x^2 + 15x$

Bracketed terms are treated in the same way:

$(x + 4) \times (x - 3) = x \times x - 3 \times x + 4 \times x - 12$ First multiply the RH Bracket terms by x and then by 4

Again in this case, it is usual to write such expressions as:

$$(x + 4)(x - 3) = x^2 - 3x + 4x - 12$$
$$= x^2 + x - 12 \qquad \text{Because we can add } -3x \text{ and } 4x \text{ together}$$

Subtraction

When we look at subtraction, however, we can clearly see it is not commutative since

$3 - 2 = 1$ but $2 - 3 = -1$.

Again we can only subtract like terms from each other:

$$3p - p = 2p$$
$$p - 2p = -p$$
$$5ab - ba = 4ab \qquad \text{Remember } ba = ab$$

Division

We also know that $6 \div 2 = 3$

This is the same as $\dfrac{3 \times \cancel{2}}{\cancel{2}} = 3 \times 1 = 3$

Because we can divide the top 2 by the lower 2 which is equal to 1

Similarly we can carry out the same operation with symbols:

$$\frac{abp}{ap} = b \qquad \text{Because we can divide the top and bottom by } p$$

$$\frac{ax^2bq^3}{bxqr} = \frac{a \times x \times x \times b \times q \times q \times q}{bxqr} = \frac{a \times x \times q \times q}{r} \ or \ \frac{axq^2}{r} \qquad \text{Dividing top and bottm by } b, x \text{ and } q$$

$$\frac{1}{p}(3p + 4p^2) \equiv \frac{3p + 4p^2}{p} \equiv \frac{3p}{p} + \frac{4p^2}{p} \equiv 3 + 4p$$

Coefficients

The numbers in front of a symbol are called **coefficients**

Hence the term $4ab$ has a coefficient of 4

Expressions

All of the above symbolic examples are called **expressions**

e.g $\dfrac{abp}{ap}$ and $\dfrac{ax^2bq^3}{bxqr}$ are expressions

EQUATIONS

Equations differ from expressions in that expressions are linked by an equals (=) sign:

$2x = 4$ is an equation with the solution $x = 2$ \quad Dividing both sides by 2 gives $x = 2$

$ab - 4c = 2ab$ is also an equation as is:

$$x^2 + 4x + 3 = 0$$

$ab^2 = ab$ is also an equation

If we divide both sides by ab we get $b = 1$

Manipulating Equations

It is essential that the student gains complete familiarity and confidence with manipulating all kinds of algebraic expressions and equations.

Consider simplifying the equation:

$$2x^2 + 2x + 4 = x^2 + 3x$$

What we need to do is to collect all like terms together remembering that each time we move a quantity from one side of the equals sign to the other we change its sign.

$$2x^2 + 2x + 1 = x^2 + 4x$$
$$2x^2 - x^2 + 2x - 4x + 1 = 0 \qquad \text{Moving the RH terms to the LHS}$$
$$x^2 - 2x + 1 = 0 \qquad \text{and changing the signs}$$

In many cases we can solve the equation to obtain an answer for the unknown **variable**, in this case, x

The answer to the above equation is $x = 1$ but some equations may have several solutions.

To verify that $x = 1$ is a solution to the equation $x^2 - 2x + 1 = 0$, simply replace x with 1:

$$(1)^2 - 2(1) + 1 = 1 - 2 + 1 = 0$$

Whenever a solution is obtained it is advisable to check it like this to make sure it actually is a solution and that we have not made a mistake or that the solution is incorrect.

If we have an equation like $\dfrac{ax}{b} + c = 4$ and we want to express b in terms of a, c and x we proceed to try and rearrange the equation to get b on the LHS of the = sign so that we finish up with $b =$ some expression containing a, c and x .e.g

$$ax + bc = 4$$

Hence

$$bc = 4 - ax \qquad \text{Subtracting } ax \text{ from both sides of the equation}$$

To isolate b, we divide throughout by c to get

$$b = \frac{4}{c} - \frac{ax}{c}$$
$$Or$$
$$b = \frac{4 - ax}{c}$$

There are some golden rules when working with equations:

1. If you want to multiply or divide a term by something, you must do it to all terms.

So if you want to multiply/divide a term by say, 3 – you must also multiply/divide **all** terms by 3.

2. Whatever you do to the LHS of an equation you must also do to the RHS of the equation and vise-versa.

So if you want to multiply/divide the complete LHS by $2a$, you must also multiply/divide the complete RHS by $2a$

3. If you wish to add or subtract an amount from one side of the equation, you must do it to the other.

4. However, we can multiply or divide any fractional term individually providing we do the same to both the numerator and denominator. E.g

$$\frac{4a}{2b} = \frac{8a}{4a} \qquad \text{Multiplying top and bottom by } 2$$

Or

$$\frac{2pq}{r^2} = \frac{6apq}{3ar^2} \qquad \text{Multiplying top and bottom by } 3a$$

FACTORISING

If we look at the equation $x^2 - 2x - 3 = 0$ and want to find out the value of x, we can **factorise** the LHS to help us.

What we want to do is to rearrange the LHS as a product of two **factors**:

In this case $(x - 3)(x + 1) = 0$

If we multiply these two factors together we simply follow the rule for multiplication

$$(x - 3)(x + 1) = x^2 + x - 3x - 3$$
$$= x^2 - 2x - 3$$

So we have $(x - 3)(x + 1) = 0$

If two factors multiplied together results in 0 (zero) then one of them must be 0 (zero)

i.e. $x - 3 = 0 \ or \ x + 1 = 0$

If $x - 3 = 0$ then $x = 3$

If $x + 1 = 0$ then $x = -1$

Check that these are both solutions to $x^2 - 2x - 3 = 0$ (they are!)

The problem is that we need to know what to put in the brackets so lets take a look at how we came to the result that:

$$(x - 3)(x + 1) = x^2 - 2x - 3$$

First, we know that the term not involving x (3) has factors of 1 and 3 and the x^2 term has a coefficient of 1. Hence we know the factors must be

$$(x \pm 1)(x \pm 3)$$

This is because the x terms in each bracket multiply together to give x^2 and the 1 and 3 multiply together to give 3

A little trial and error then provides the final answer. The only possibilities are:

$$(x+1)(x+3)$$
$$(x+1)(x-3)$$
$$(x-1)(x+3)$$
$$(x-1)(x-3)$$

Only one of these products will be equal to the original equation $x^2 - 2x - 3$
Each possibility has to be worked out until you obtain the correct result.

You will gain experience as you practice factorising and with many problems you will be able to write down the factors straight away.

Factorising less straightforward equations follows the same procedure:

Solve the equation $4x^2 + 2x - 10 = 2$

First we rearrange this equation to give
$$4x^2 + 2x - 12 = 0 \qquad \text{By subtracting 2 from each side}$$
Or
$$2x^2 + x - 6 = 0 \qquad \text{Dividing throughout by 2}$$

In this case we know we must have:

$$(2x \pm 6)(x \pm 1)$$
$$(2x \pm 3)(x \pm 2)$$

Working out all of these possibilities as before we get:

$$(2x - 3)(x + 2) = 0$$

Hence
$$2x - 3 = 0 \Rightarrow x = \frac{3}{2}$$
or
$$x + 2 = 0 \Rightarrow x = -2$$

THE FACTOR THEOREM

It becomes more difficult to factorise expressions like:

$$x^3 + 6x^2 + 11x + 6$$

But we can make certain immediate observations that can help.

To obtain the term involving x^3 we will need up to three factors

The last term (6) has factors of 1,2 and 3

Hence for this expression the possible factors are

$(x \pm 1), (x \pm 2)$ or $(x \pm 3)$

However, to avoid the tedium of trying out all the possibilities we can make use of the **Factor Theorem** which states:

If $(x - a)$ is a factor of an expression containing powers of x, then substituting for $x = a$ in that expression gives a result of zero (0).

Conversely if such a substitution does not result in the expression reducing to zero then $(x - a)$ is not a factor of that expression.

Hence in our case if $(x - 1)$ is a factor of $x^3 + 6x^2 + 11x + 6$ then we can substitute $x = 1$ in it to check:

$(1)^3 + 6(1)^2 + 11(1) + 6 = 1 + 6 + 11 + 6 = 24$ So $(x - 1)$ is not a factor

Trying $(x + 1)$ we substitute $x = -1$ and check again:

$(-1)^3 + 6(-1)^2 + 11(-1) + 6 = -1 + 6 - 11 + 6 = 0$. Hence $(x + 1)$ is a factor

Testing for the other factors in the same way reveals the other factors are:

$(x + 2)$ and $(x + 3)$

Hence $(x + 1)(x + 2)(x + 3) = x^3 + 6x^2 + 11x + 6$

If we consider the expression $x^3 + x^2 - 12x$

We can immediately see that it has an obvious factor (x) so we can write:

$x^3 + x^2 - 12x = x(x^2 + x - 12)$

Now we can look at the easier option of finding factors for $(x^2 + x - 12)$

Which we can see by inspection must be $(x + 4)(x - 3)$

Hence $x^3 + x^2 - 12x = x(x + 4)(x - 3)$

Always look for an obvious common factor like this first which makes finding the remaining factors easier.

PASCALS TRIANGLE

When we have to expand expressions like $(x + y)^5$, it can become tedious and error prone to do this manually.

Look at question 5 to see how using this technique makes the expansion much easier.

ALGEBRA 1 WORKED SOLUTIONS

1.1 Simplify the following expressions

(i) $4 \times 3x = 12x$

(ii) $4a \times 3b = 12ab$

(iii) $4a \times 2abc = 8a^2bc$

(iv) $ac(6ab)^2 = ac \times 36a^2b^2 = 36a^3b^2c$

(v) $ax^2(3 + ac + x^2) = 3ax^2 + a^2c + x^4$

(vi) $6x^2 + bx \div x = 6x + b$

(vii) $4ab^2 + a^2bc + abc^2 \div ab = 4b + ac + c^2$

(viii) $5x^2 + 2x + 3 - 3x + x^2 - 4 = 6x^2 - x - 1$

(ix) $ap^2 - 2pq + q^2 - pq + 2q^2 + 2ap^2 = 3ap^2 - 3pq + 3p^2$

(x) $a^2b^2x - ab^2x + b^2a - 3ab + 2ab^2x - 2ab^2 = a^2b^2x + ab^2x + b^2a - 3ab$

1.2 Expand and simplify the following expressions

(i) $(x + 2)(x + 3) = x^2 + 3x + 2x + 6 = x^2 + 5x + 6$ Expanding Brackets

(ii) $(2t + 3)(t - 2) = 2t^2 - 4t + 3t - 6 = 2t^2 - t - 6$ Expanding Brackets

(iii) $(ab - c)(a + 3) = a^2b + 3ab - ac - 3c$ Expanding Brackets

(iv) $(2p - 3)(p - 4)^2$

$= (2p - 3)(p^2 - 8p + 16) = 2p^3 - 16p^2 + 32p - 3p^2 - 24p - 48$

$= 2p^3 - 19p^2 + 8p - 48$ Expanding brackets & Simplifying

(v) $(a^2b - 3)(4 + a) = 4a^2b - a^3b - 12 - 3a$ or $4a^2b - a^3b - 3a - 12$ Expanding Brackets

(vi) $(x + 3)(x + 3) = (x + 3)^2$ Perfect square

$\qquad\qquad = x^2 + 6x + 9$ Expanding Bracket

(vii) $-3x(x + 1)^2 = -3x(x^2 + 2x + 1)$

$\qquad\qquad = -3x^3 - 6x^2 - 3x$ Expanding Brackets & Mutiplying by $-3x$

1.3 Expand, simplify and fully factorise $-2x(x - 3)(x + 5) + x^2 + 3x + 7$

$= -2x(x^2 + 5x - 3x - 15) + x^2 + 3x + 7$ Expanding bracket

$= -2x^3 - 10x^2 + 6x + 30x + x^2 + 3x + 7$ Multiplying bracketed term by -2x

$= -2x^3 - 3x^2 + 33x + 7$ Simplifying

$= -2(x - 7)(x^2 + 5x + 1)$ Factorising

1.4   Expand $(x+y)^3$

$$= (x+y)(x+y)(x+y)$$
$$= (x+y)(x^2+2xy+y^2) \qquad \text{Expanding last two brackets}$$
$$= x^3 + 2x^2y + xy^2 + x^2y + 2xy^2 + y^3 \qquad \text{Multiplying result by first bracket}$$
$$= x^3 + 3x^2y + 3xy^2 + y^3 \qquad \text{Simplifying}$$

1.5   Expand $(x+y)^5$

It starts to become tedious to multiply $(x+y)$ together 5 times so instead we use Pascals triangle.

```
              1       1
          1       2       1
      1       3       3       1
    1     4       6       4       1
  1     5     10      10      5       1
```

This gives the following result:

$$(x+y)^5 = x^5 + 5x^4y + 10x^3y^2 + 10x^2y^3 + 5xy^4 + y^5$$

Note that the sum of the powers of x and y must equal 5 for each term in this case.

1.6   Factorise the following expressions

(i) $x^2 + 3x + 2 = (x+1)(x+2)$ By inspection

(ii) $x^2 + 2x - 8 = (x-2)(x+4)$ By inspection

(iii) $3x^2 - 7x - 6 = (3x+2)(x-3)$ By inspection

(iv) $12p^2 + 6pq - 6q^2$

$$= 6(2p^2 + pq - q^2) \qquad \text{Taking out common factor of 6 first allowing}$$
$$= 6(2p-q)(p+q) \qquad \text{expression to be factorised more easily}$$

(v) $a^2b^2 + 13ab + 42 = (ab+6)(ab+7)$ By inspection

(vi) $\pi R^2 - \pi r^2 R = \pi R(R - r^2)$ By inspection

1.7   Factorise $x^3 - 6x^2 + 11x - 6$

It starts to become difficult to factorise expressions like this simply by inspection. We really need a more systematic approach such as that provided by use of the Factor Theorem

As the factors of 6 are 1, 2 and 3, the possible factors of $x^3 - 6x^2 + 11x - 6$ *are* $(x \pm 1), (x \pm 2)$ *and* $(x \pm 3)$ Trying each of these in turn gives:

$x = -1 \Rightarrow x^3 - 6x^2 + 11x - 6 = (-1)^3 - 6(-1)^2 + 11(-1) - 6 \neq 0$ Hence $(x+1)$ is not a factor

$x = +1 \Rightarrow x^3 - 6x^2 + 11x - 6 = (1)^3 - 6(1)^2 + 11(1) - 6 = 0$ Hence $(x-1)$ is a factor

$x = -2 \Rightarrow x^3 - 6x^2 + 11x - 6 = (-2)^3 - 6(-2)^2 + 11(-2) - 6 \neq 0$ Hence $(x+2)$ is not a factor

$x = +2 \Rightarrow x^3 - 6x^2 + 11x - 6 = (2)^3 - 6(2)^2 + 11(1) - 6 = 0$ Hence $(x-2)$ is a factor

$x = -3 \Rightarrow x^3 - 6x^2 + 11x - 6 = (-3)^3 - 6(-3)^2 + 11(-3) - 6 \neq 0$ Hence $(x+3)$ is not a factor

$x = +1 \Rightarrow x^3 - 6x^2 + 11x - 6 = (3)^3 - 6(3)^2 + 11(3) - 6 = 0$ Hence $(x-3)$ is a factor

Hence $x^3 - 6x^2 + 11x - 6 = (x-1)(x-2)(x-3)$

1.8 Factorise $x^4 - 2x^2 - 13x^3 + 38x - 24$ using the Factor Theorem

As the factors of 24 are 1, 2, 3, 4, 6, 8, 12, the possible factors are
$(x \pm 1), (x \pm 2), (x \pm 3), (x \pm 4), (x \pm 6), (x \pm 8)$ and $(x - 12)$

Trying each of these in turn in $x^4 - 2x^2 - 13x^3 + 38x - 24$ as before shows:

$x = +1 \Rightarrow x^4 - 2x^2 - 13x^3 + 38x - 24 = 0 \Rightarrow (x-1)$ is a factor

$x = -1 \Rightarrow x^4 - 2x^2 - 13x^3 + 38x - 24 \neq 0 \Rightarrow (x+1)$ is not a factor

$x = +2 \Rightarrow x^4 - 2x^2 - 13x^3 + 38x - 24 = 0 \Rightarrow (x-2)$ is a factor

$x = -2 \Rightarrow x^4 - 2x^2 - 13x^3 + 38x - 24 \neq 0 \Rightarrow (x+2)$ is not a factor

$x = +3 \Rightarrow x^4 - 2x^2 - 13x^3 + 38x - 24 = 0 \Rightarrow (x-3)$ is a factor

$x = -3 \Rightarrow x^4 - 2x^2 - 13x^3 + 38x - 24 \neq 0 \Rightarrow (x+3)$ is not a factor

$x = +4 \Rightarrow x^4 - 2x^2 - 13x^3 + 38x - 24 \neq 0 \Rightarrow (x-4)$ is not a factor

$x = -4 \Rightarrow x^4 - 2x^2 - 13x^3 + 38x - 24 = 0 \Rightarrow (x+4)$ is a factor   Stop at this point

$x = +6 \Rightarrow x^4 - 2x^2 - 13x^3 + 38x - 24 \neq 0 \Rightarrow (x-6)$ is not a factor

$x = +8 \Rightarrow x^4 - 2x^2 - 13x^3 + 38x - 24 \neq 0 \Rightarrow (x-8)$ is not a factor

$x = -8 \Rightarrow x^4 - 2x^2 - 13x^3 + 38x - 24 \neq 0 \Rightarrow (x+8)$ is not a factor

Hence $x^4 - 2x^2 - 13x^3 + 38x - 24 = (x-1)(x-2)(x-3)(x+4)$

Note that once we have found the first four factors we could stop at this point since the product of four factors containing x gives a term in x^4 which is the highest power of x in the expression. We have simply carried on testing the remaining possibilities to demonstrate that they are not factors.

This shows we should always test with the *lowest* numerical possibilities for factors first and carry on testing until the number of factors found is equal to the highest power of x in the expression. However, in some cases the number of factors found may be less than the highest power of x in an expression.
e.g.: $x^3 + 4x^2 + 4x + 1 = (x+1)(x^2 + 3x + 1)$

1.9 If $(x - 4)$ is a factor of $x^3 + px + 16$. Find the value of p

Since $(x - 4)$ is a factor of this expression we set $x^3 + px + 16 = 0$ and substitute $x = 4$ in this expression as follows:

If $(x - 4)$ is a factor of $x^3 + px + 16$ then:

$$(4)^3 + 4p + 16 = 0$$

$$\Rightarrow 64 + 4p + 16 = 0$$

$$\Rightarrow 4p = -80$$

$$\Rightarrow p = -20$$

Hence the original expression can be re-written as:

$$x^3 - 20x + 16 \qquad\qquad \text{By subsituting for } p = -20 \text{ in the original equation}$$

As a check we can put $x = 4$ in this expression:

$$(4)^3 - 20(4) + 16 = 0 \text{ Which confirms } (x-4) \text{ is a factor hence the value of } p \text{ is correct.}$$

1.10 Find a factor of $3x^3 - 12x^2 + 16x - 8$ and hence express $3x^3 - 12x^2 + 16x - 8$ in the   form $(ax + b)(px^2 + qx + r)$.   (1)

First we find a factor of $3x^3 - 12x^2 + 16x - 8$ using the factor theorem.

The factors of 8 are 1, 2 and 4 hence:

$(x \pm 1), (x \pm 2)$ and $(x \pm 4)$ are all possible factors of this expression. Testing these we get:

$(x + 1)$ is not a factor
$(x - 1)$ is not a factor
$(x + 2)$ is not a factor
$(x - 2)$ is a factor   We could stop here as we have found a factor
$(x + 4)$ is not a factor
$(x - 4)$ is not a factor

Hence we can say:
$$3x^3 - 12x^2 + 16x - 8 = (x - 2)(px^2 + qx + r)$$
Expanding the RHS we get

$$(x - 2)(px^2 + qx + r) = px^3 + qx^2 + rx - 2px^2 - 2qx - 2r$$

Hence $3x^3 - 12x^2 + 16x - 8 = px^3 + qx^2 + rx - 2px^2 - 2qx - 2r$

Equating coefficients for like powers of x on both sides, we get:

$$-2 \times r = -8 \Rightarrow r = 4 \qquad \text{Equating terms not involving } x$$
$$r - 2q = 16 \qquad \text{Equating terms involving } x$$
$$\Rightarrow 4 - 2q = 16 \qquad \text{Substituting for } r = 4$$
$$\Rightarrow q = -6$$
And $p = 3$ \qquad Equating terms involving x^3

Hence the required form is:

$$(x-2)(3x^2 - 6x + 4)$$

Subsituting for $p = 3, q = -6$ and $r = 4$ in (1)

1.11 Simplify $\dfrac{6a^2b - 3b^2a}{2a - 6b}$

$$\dfrac{6a^2b - 3b^2a}{2a - 6b}$$

$$= \dfrac{3ab(a - 3b)}{2(a - 3b)}$$

Factorise numerator and denominator first

$$= \dfrac{3ab}{2}$$

Dividing top and bottom by $(a - 3b)$

1.12 Simplify $\dfrac{4s + 16t}{s^2 + 6st + 8t^2}$

$$\dfrac{4s + 16t}{s^2 + 6st + 8t^2}$$

$$= \dfrac{4(s + 4t)}{(s + 4t)(s + 2t)}$$

Factorise numerator and denominator first

$$= \dfrac{4}{(s + 2t)}$$

Divide top and bottom by $(s + 4t)$

1.13 Simplify $\dfrac{p^2 + 2p - 1}{p^2 + 5p + 6}$

$$\dfrac{p^2 + 2p - 1}{p^2 + 5p + 6}$$

$$= \dfrac{p^2 + 2p - 1}{(p + 2)(p + 3)}$$

Denominator does factorise, numerator does not

1.14 Simplify $\dfrac{pq^2}{p^2} \times \dfrac{pr}{q}$

$$\dfrac{pq^2}{p^2} \times \dfrac{pr}{q}$$

$$= \dfrac{p^2q^2r}{p^2q}$$

Multiply terms together

$$= qr$$

Dividing top and bottom by p^2r

1.15 Simplify $\dfrac{2p^3r}{3} \times \left(\dfrac{2}{p}\right)^2$

$$\dfrac{2p^3r}{3} \times \left(\dfrac{2}{p}\right)^2$$

$$= \dfrac{2p^3r}{3} \times \dfrac{4}{p^2}$$

Expand bracket first

$$= \frac{8p^3r}{3p^2}$$ Multiply terms together

$$= \frac{8pr}{3}$$ Divide top and bottom by p^2

$$(3-x) \div \frac{(3-x)}{3x}$$

$$= (3-x) \times \frac{3x}{(3-x)}$$ Invert RH term and change operator to multiplication

$$= \frac{3x(3-x)}{(3-x)}$$ Multiply both terms together

$$= 3x$$ Divide top and bottom by $(3-x)$

$$(p^2 + 2p + 1) \div (p+1) \times \frac{4}{2x^2}$$

$$= (p^2 + 2p + 1) \times \frac{1}{(p+1)} \times \frac{4}{2x^2}$$ Invert second term & change operator to multiplication

$$= \frac{4(p+1)(p+1)}{2x^2(p+1)}$$ Factorise first bracket & multiply terms together

$$= \frac{4(p+1)}{2x^2}$$ Divide top and bottom by $(p+1)$

$$= \frac{2(p+1)}{x^2}$$ Divide top and botom by 2

$$x - \frac{1}{2x}$$

$$= \frac{x}{1} - \frac{1}{2x}$$ Change LH term to fraction

$$= \frac{2x^2}{2x} - \frac{1}{2x}$$ Obtain common denominator by multiplying
top & bottom of LH fraction by 2x

$$= \frac{2x^2 - 1}{2x}$$ Combining terms as a single fraction

$$2(x+1)+3(x-2)-\frac{(x+3)}{(x-1)}$$

$$= 2x+2+3x-6-\frac{(x+3)}{(x-1)} \qquad \text{Expanding two LH brackets}$$

$$= 5x-4-\frac{(x+3)}{(x-1)} \qquad \text{Simplifying}$$

$$= \frac{5x(x-1)-4(x-1)-(x+3)}{x-1} \qquad \text{Rearrange with common denominator}$$

$$= \frac{5x^2-5x-4x+4-x-3}{x-1} \qquad \text{Expanding brackets}$$

$$= \frac{5x^2-10x-1}{x-1} \qquad \text{Simplifying}$$

1.20 Simplify $\left[x(2x^2-3)+\dfrac{3}{4(2x+3)}-(x+1)^2 \right] \times \dfrac{1}{n^2} \div \dfrac{x^2}{n}$

$$\left[x(2x^2-3)+\frac{3}{4(2x+3)}-(x+1)^2 \right] \times \frac{1}{n^2} \div \frac{x^2}{n}$$

$$= \left[\frac{4x(2x+3)(2x^2-3)+3-4(2x+3)(x+1)^2}{4(2x+3)} \right] \times \frac{1}{n^2} \times \frac{n}{x^2} \qquad \text{Using common denominator for bracketed term}$$

$$= \left[\frac{4x(4x^2-9)+3-4(2x^3+5x^2+3x+2x^2+5x+3)}{4(2x+3)} \right] \times \frac{1}{nx^2} \qquad \text{Expanding brackets}$$

$$= \left[\frac{16x^3-36x+3-8x^3-28x^2-32x-12}{4(2x+3)} \right] \times \frac{1}{nx^2} \qquad \text{Simplifying}$$

$$= \left[\frac{8x^3-28x^2-68x-12}{4(2x+3)} \right] \times \frac{1}{nx^2} \qquad \text{Simplifying}$$

$$= \frac{4(2x^3-7x^2-17x-3)}{4(2x+3)} \times \frac{1}{nx^2} \qquad \text{Simplifying}$$

$$= \frac{4(2x+3)(x^2-5x-1)}{4(2x+3)} \times \frac{1}{nx^2} \qquad \text{Factorising numerator}$$

$$= \frac{(x^2-5x-1)}{nx^2} \qquad \text{Dividing top \& Bottom by } 4(2x+3)$$

$$\dfrac{2x^2-4}{(x-2)^2}+\dfrac{2x+6}{x(x-2)}$$

$$=\dfrac{x(2x^2-4)+(2x+6)(x-2)}{x(x-2)} \qquad \text{Multipling both sides by } x(x-2)$$

$$=\dfrac{2x^3-4x+2x^2-4x+6x-12}{x(x-2)} \qquad \text{Expanding brackets}$$

$$=\dfrac{2x^3+2x^2-2x-12}{x(x-2)} \qquad \text{Simplifying}$$

$$=\dfrac{2\left(x^3+x^2-x-6\right)}{x(x-2)} \qquad \begin{array}{l}\text{Taking out common factor of 2}\\ \text{and rearranging}\end{array}$$

This question does not say that $(x+3)$ is a factor of $2x^3-x+5$ therefore we do not know whether it is or not. Testing to see if $(x+3)$ is a factor we get:

$$2(-3)^3-(-3)+5=-54+6+5=-43$$

This result shows that $(x+3)$ is not a factor of $2x^3-x+5$

Hence we must use long division instead. To do this, first rearrange the expression to include a term in x^2 as shown:

$$2x^3+0x^2-x+5$$

Now we can proceed with the long division as follows:

$\,2x^2-6x+17$	Quotient
$x+3)\overline{2x^3+0x^2-x+5}$	x divides into $2x^3$, $2x^2$ times
$\quad\underline{2x^3+6x^2}$	Multiply $x+3$ by $2x^2$ and subtract
$\qquad -6x^2-x$	Bring down the next term $(-x)$
$\qquad\underline{-6x^2-18x}$	Multiply $x+3$ by $-6x$ and subtract
$\qquad\qquad 17x+5$	Bring down the next term (5)
$\qquad\qquad\underline{17x+51}$	Multiply $x+3$ by 17 and subtract
$\qquad\qquad\quad -46$	Remainder

Clearly, the long division process stops when there are no more x terms remaining.

Hence the quotient is $2x^2-6x+17$ and the remainder -46

Or $2x^2-6x+17-\dfrac{46}{x+3}$ (quotient plus a proper fraction)

23. Divide $x^3 + 3x^2 + 2x + 8$ by $x - 2$ and show both the quotient and remainder

Since the question tells us there is a remainder we know that $x - 2$ is not a factor of $2x^2 - 6x + 17$ hence we proceed straight to long division.

$$
\begin{array}{r}
x^2 + 5x + 12 \\
x - 2 \overline{\smash{)}\,x^3 + 3x^2 + 2x + 8} \\
\underline{x^3 - 2x^2} \\
5x^2 + 2x \\
\underline{5x^2 - 10x} \\
12x + 8 \\
\underline{12x - 24} \\
32
\end{array}
$$

Quotient
x divides into x^3, x^2 times
Multiply $x - 2$ by x^2 and subtract
Bring down the next term $(2x)$
Multiply $x - 2$ by $5x$ and subtract
Bring down the next term $(+8)$
Multiply $x - 2$ by 12 and subtract
Remainder

Hence the quotient is $x^2 + 5x + 12$ and the remainder 32

Again we could have expressed this result as a quotient plus a proper fraction as before:

$$x^2 + 5x + 12 + \frac{32}{x - 2}$$

MODULE A2

ALGEBRA 2 – SURDS, INDICES AND LOGARITHMS

Revision Notes

SURDS

Square Roots

If a number is expressed as the product of two equal factors, that factor is called the **square root** of that number.

e.g. $9 = 3 \times 3$ hence 3 is the square root of 9 i.e. $\sqrt{9} = 3$

Additionally -3 is also a square root of 9 but we never write $\sqrt{9} = -3$

The symbol $\sqrt{}$ is used only for the **positive square root**

Thus, although $\sqrt{9} = \pm 3$, the only value assigned to $\sqrt{9}$ is 3

The negative square root of 9 is written as $-\sqrt{9}$

Wherever both square roots are required we would write $\pm\sqrt{9}$

Cube Roots

If a number is expressed as the product of three equal factors, that factor is called the **cube root** of that number.

e.g. $27 = 3 \times 3 \times 3$, so 3 is the cube root of 27 which is written as $\sqrt[3]{27}$

Other Roots

The notation described above is extended for fourth roots and fifth roots etc.

e.g.
$$16 = 2 \times 2 \times 2 \times 2 \Rightarrow \sqrt[4]{16} = 2$$
$$243 = 3 \times 3 \times 3 \times \times 3 \times 3 \Rightarrow \sqrt[5]{243}$$

In general, if a number, n, can be expressed as the product of r equal factors then each of these factors is called the pth root of n and is written as $\sqrt[r]{n}$

Rational Numbers

A **rational number** is either an integer or a fraction whose numerator and denominator are both integers.

e.g. $2, 3, \dots, \dfrac{1}{2}, \dfrac{2}{3}, \dfrac{125}{63}$ etc. are all rational numbers

The square roots of certain numbers are also rational:

e.g. $\sqrt{4} = 2$, $\sqrt{9} = 3$, $\sqrt{\dfrac{9}{49}} = \dfrac{3}{7}$, $\sqrt{\dfrac{16}{81}} = \dfrac{4}{9} = \dfrac{2}{3}$

Irrational Numbers

However, this is not the case with all square roots

e.g. $\sqrt{2} = 1.4142...$, $\sqrt{3} = 1.7320...$, $\sqrt{7} = 2.6457...$

None of these can be expressed as an exact decimal – they cannot be expressed in the form $\dfrac{a}{b}$ where a and b are integers. Such numbers are called **irrational numbers**

In mathematics we always express answers exactly, hence we would express irrational square roots as $\sqrt{2}$, $\sqrt{3}$ etc and not as a decimal which is not exact.

The only exception to this is where we are asked for an approximate answer, say, to 3 significant figures. Otherwise use the exact value $\sqrt{2}$ etc

Simplifying Surds

Simplifying and manipulating surds is an essential skill that the student should master.

If we consider the surd $\sqrt{12}$, we can express this in a more simplified form as follows:

$$\sqrt{12} = \sqrt{4 \times 3} = \sqrt{4} \times \sqrt{3}$$
$$= 2 \times \sqrt{3}$$

Which is usually expressed in the form $2\sqrt{3}$

Similarly $\sqrt{27} = \sqrt{9 \times 3} = \sqrt{9} \times \sqrt{3} = 3\sqrt{3}$

Note that $3\sqrt{3}$ is not the same as $\sqrt[3]{3}$.

The former means 3 times the square root of 3 and the latter means the cube root of 3

Multiplying Surds

We have already seen that $\sqrt{27} = \sqrt{3} \times \sqrt{9}$ and in general $\sqrt{x} \times \sqrt{y} = \sqrt{xy}$

Now consider $\left(\sqrt{3} - 4\right)\left(\sqrt{5} - 2\right)$

These brackets are multiplied out in exactly the same way as multiplying any two linear brackets:

$$\left(\sqrt{3} - 4\right)\left(\sqrt{5} + 2\right) = \sqrt{3}\sqrt{5} + 2\sqrt{3} - 4\sqrt{5} - 8$$
$$= \sqrt{15} - + 2\sqrt{3} - 4\sqrt{5} - 8$$

When the same surd occurs in each bracket we can collect like terms together:

$$\left(\sqrt{5}+6\right)\left(\sqrt{5}-3\right)=\sqrt{5}\sqrt{5}-3\sqrt{5}+6\sqrt{5}-18$$
$$=\sqrt{25}+3\sqrt{5}-18$$
$$=5+3\sqrt{5}-18 \qquad\qquad \text{Since }\sqrt{25}=5$$
$$=3\sqrt{5}-13$$

If we now consider $\left(3-\sqrt{5}\right)\left(3+\sqrt{5}\right)=9+3\sqrt{5}-3\sqrt{5}-\sqrt{5}\sqrt{5}$

$$=9-5 \qquad\qquad \text{Since }\sqrt{5}\sqrt{5}=5$$
$$=4$$

This is a special case because it results in a rational number. This occurs because the brackets are of the form:

$$\left(x-a\right)\left(x+a\right)=x^2+a^2 \text{ hence}$$
$$\left(p-\sqrt{q}\right)\left(p+\sqrt{q}\right)=p^2-q, \text{ which is always rational}$$

This is an important result and often used in manipulating and simplifying surds and therefore should be committed to memory.

Rationalising a Denominator

Any fraction whose denominator contains a surd is more difficult to work with than where surds occur only in the numerator.

Consider $\left(\dfrac{3}{\sqrt{3}}\right)$. If we multiply numerator and denominator by $\sqrt{3}$ we get:

$$\frac{3\sqrt{3}}{\sqrt{3}\sqrt{3}}=\frac{3\sqrt{3}}{3}=\sqrt{3} \text{ which is simpler to work with.}$$

Note: to maintain the integrity of any fraction we must multiply (or divide) the numerator and denominator by the same amount.

Now consider the expression $\dfrac{4\sqrt{3}}{4-\sqrt{3}}$. We saw earlier that if we multiply the denominator by its **conjugate** $\left(4+\sqrt{3}\right)$ we make the denominator a rational number.

$$\frac{4\sqrt{3}}{4-\sqrt{3}}=\frac{4\sqrt{3}\left(4+\sqrt{3}\right)}{\left(4-\sqrt{3}\right)\left(4+\sqrt{3}\right)}$$

$$= \frac{16\sqrt{3} + 4\sqrt{3}\sqrt{3}}{\left(16 - \sqrt{3}\sqrt{3}\right)} = \frac{16\sqrt{3} + 4 \times 3}{16 - 3} \qquad \text{Simplifying}$$

$$= \frac{16\sqrt{3} + 12}{13} \qquad \text{Simplifying}$$

INDICES

The index Notation

If a is any number and n, a positive integer, then a^n means $a \times a \times a$...to n factors. Additionally, n is termed the power of a.

The Laws of Indices

Multiplication

$$a^m \times a^n = a^{m+n}$$

By definition

$$a^m = a \times a \times a\text{...to } m \text{ factors}$$

$$a^n = a \times a \times a\text{...to } n \text{ factors}$$

Hence $a^m \times a^n = (a \times a \times a\text{...to } m \text{ factors}) \times (a \times a \times a\text{...to } n \text{ factors})$

Thus there are $(m + n)$ factors, each of which is a, on the right hand side

$$\therefore a^m \times a^n = a \times a \times a\text{...to } (m + n) \text{ factors}$$

Hence $a^m \times a^n = a^{m-n}$
e.g. $x^3 \times x^2 = x^{3+2} = x^5$

Division

$$a^m \div a^n = a^{m-n}$$

By definition if $m > n$ we have

$$a^m = a \times a \times a\text{...to } m \text{ factors}$$

$$a^n = a \times a \times a\text{...to } n \text{ factors}$$

$$\therefore a^m \div a^n = \frac{a \times a \times a\text{...to } m \text{ factors}}{a \times a \times a\text{...to } n \text{ factors}}$$

If we now cancel out n of the m factors we are left with

$$\therefore a^m \div a^n = a \times a \times a\text{...to } m - n \text{ factors}$$

Hence

$$a^m \div a^n = a^{m-n}$$

e.g. $x^5 \div x^3 = x^{5-3} = x^2$

Fractional and Negative Indices

To cover the possibility of the occurrence of a fractional or negative index, we have to extend or add to the rules of Indices to include these as the previous definitions would preclude these.

If the index is negative or fractional and they are to be regarded as Indices they must obey the rules of Indices.

Hence
$$a^{\frac{1}{n}} \times a^{\frac{1}{n}} \times a^{\frac{1}{n}} \ldots \text{ to } n \text{ factors}$$

$$= a^{\frac{1}{n}+\frac{1}{n}+\frac{1}{n}\ldots \text{ to } n \text{ terms}}$$

$$= a^1$$

$$= a$$

Hence
$$a^{\frac{1}{n}} = \sqrt[n]{a}$$

Similarly

$$a^{\frac{m}{n}} + a^{\frac{m}{n}} + a^{\frac{m}{n}} \ldots \text{to } n \text{ Factors}$$

$$= a^{\frac{m}{n}+\frac{m}{n}+\frac{m}{n}\ldots \text{to } n \text{ terms}}$$

$$= a^{\frac{m}{n} \times n}$$

$$= a^m$$

Hence $a^{\frac{m}{n}} = \sqrt[n]{a^m}$

e.g $a^{\frac{2}{3}} \times a^{\frac{2}{3}} \times a^{\frac{2}{3}} = a^{\frac{2}{3}+\frac{2}{3}+\frac{2}{3}} = a^2$

Hence $a^{\frac{2}{3}}$ must be the cube root of a^2

or $\sqrt[3]{a^2}$

Similarly

$a^0 = 1$ Since

$$a^0 \times a^n = a^{0+n}$$

$$= a^n$$

Hence $a^0 = a^n \div a^n$

$$= 1$$

And

$$a^n \times a^{-n} = a^{n-n} = a^0 = 1$$

Hence $a^{-n} = 1 \div a^n = \dfrac{1}{a^n}$

$e.g \ 2^{-3} = \dfrac{1}{2^3}$

LOGARITHMS

The Laws of Logarithms

We are already familiar with statements like

$$10^2 = 100$$

and

$$10^3 = 1,000$$

Taking the first statement, we usually express this as 10 to the power of 2 = 100 or 10 squared = 100.

However, we could also express this statement as:

2 is the power to which the base 10 is raised to equal 100

Expressed in this way, the power is called a logarithm or log

i.e. 2 is the logarithm to the base 10 of 100

e.g.
$$2^2 = 8 \Rightarrow 2 = \log_2 8$$
$$3^3 = 27 \Rightarrow 3 = \log_3 27$$

Similarly

$$\log_2 16 = 2 \Rightarrow 2^4 = 16$$
$$\log_5 125 = 3 \Rightarrow 5^3 = 25$$

The base of a logarithm can be any positive number or an unspecified number represented by a symbol – for example the letter a.

In general we can express this relationship as

$$b = a^c \Leftrightarrow \log_a b = c$$

The symbol $\Leftrightarrow$ means each of these statements implies the other.

Note that by definition you cannot have the logarithm of a negative number

Sinc3 $\log_a b = c \Rightarrow b = a^c \Rightarrow b$ must be positive as a^c cannot be negative.

Natural Logarithms

We have already seen that some numbers are irrational; $\sqrt{2}$ and π for example – they cannot be expressed in the form $\frac{p}{q}$

Another number which is very important in mathematics is the number "e" that is equal to 2.71828…which is a non - repeating, never ending decimal. (i.e. it is also irrational). It is obtained from the sum:

$1 + \frac{1}{1} + \frac{1}{1.2} + \frac{1}{1.2.3} + ...$ which approaches e as successive terms are added

When **e** is used as the base for logarithms, the convention used to denote this is:

$\ln x$, which is a shortened form of $\log_e x$. This is intended to avoid confusion with any other bases.

Hence $\ln a$ **means** $\log_e a$ **and** $\ln_e a = b \Leftrightarrow e^b = a$

Evaluating Logarithms

Converting to Index Form

It is usually easier to evaluate logarithms by converting them to index form in simple cases:

e.g. evaluate $\log_3 27$. To covert this to index form we proceed as follows:

Let $x = \log_3 27$ then $3^x = 27 \Rightarrow x = 3$ Since $3^3 = 27$

Also for any base a if

$x = \log_a 1 \Rightarrow a^x = 1 \Rightarrow x = 0$. Hence

The logarithm of 1 to any base is zero (0)

Consider also $x = \log_a a \Rightarrow a^x = a$

The only solution is $x = 1$. Hence for any base a

$$\log_a a = 1$$

Using a Calculator

For less straightforward cases, a scientific calculator may be used to evaluate logarithms to the base 10 or the base e.

The calculator key marked "log" will evaluate logarithms to the base 10 and the key marked "ln" will evaluate logarithms to the base e

Additionally the "log" key may be used to evaluate 10^x and the "ln" key will evaluate e^x
These functions are usually accessed by pressing the "mode" key.

The Laws of Logarithms

Since logarithm is simply another word for index or power, logarithms also obey certain rules just as indices do.

Multiplication

Consider the statements:

$x = \log_a b$ and $y = \log_a c$ then:

$a^x = b$ and $a^y = c$

Hence $bc = a^x a^y$

$\Rightarrow \quad bc = a^{x+y}$	By the rules of indices
$\Rightarrow \log_a bc = (x + y)\log_a a$	Taking logs of both sides to base a
$\Rightarrow \log_a bc = x + y$	Since $\log_a a = 1$
$\Rightarrow \log_a bc = \log_a b + \log_a c$	

Remember that *a* represents *any* base. Note that *all* logarithms in the equation *must* have the same base whatever that may be.

Fractions

If $x = \log_a b$ and $y = \log_a c$

Then $\quad \dfrac{b}{c} = \dfrac{a^x}{a^y} \Rightarrow \dfrac{b}{c} = a^{x-y}$	By the laws of indices
$\Rightarrow \log_a\left(\dfrac{b}{c}\right) = (x - y)\log_a a$	Taking logs of both side to base a
$\Rightarrow \log_a\left(\dfrac{b}{c}\right) = x - y$	Since $\log_a a = 1$
$\Rightarrow \log_a\left(\dfrac{b}{c}\right) = \log_a b - \log_a c$	

Other Expressions

We frequently need to deal with expressions such as $\log_a b^n$

Let $x = \log_a b^n \Rightarrow a^x = b^n$

$\Rightarrow a^{\frac{x}{n}} = b$	Since $b = \left(a^x\right)^{\frac{1}{n}} = a^{\frac{x}{n}}$ by laws of indices
$\Rightarrow \dfrac{x}{n}\log_a a = \log_a b$	Taking logs of both side to base a
$\Rightarrow \dfrac{x}{n} = \log_a b$	Since $\log_a a = 1$
$\Rightarrow x = n\log_a b$	

$\Rightarrow \log_a b^n = n\log_a b$

ALGEBRA 2 SOLUTIONS

SURDS

2.1 Simplify $\sqrt{32}$

We can re-write this as $\sqrt{2 \times 16}$

But we know that $\sqrt{16} = 4$

Hence we can say that $\sqrt{32} = 4\sqrt{2}$

2.2 Simplify $\sqrt[3]{192}$

This can be re-written as $\sqrt[3]{3 \times 64} = \sqrt[3]{3} \times \sqrt[3]{64}$

But $64 = 4 \times 4 \times 4$ hence $\sqrt[3]{64} = 4$

So $\sqrt[3]{192} = 4 \times \sqrt[3]{3}$ or $4\sqrt[3]{3}$

2.3 Expand and simplify $(3 - \sqrt{3})(4 + \sqrt{5})$

$$(3 - \sqrt{3})(4 - \sqrt{5}) = 12 + 3\sqrt{5} - 4\sqrt{3} + \left(\sqrt{3}\right)\left(\sqrt{5}\right) \qquad \text{Expanding}$$

$$= 12 + 3\sqrt{5} - 4\sqrt{3} + \sqrt{15} \qquad \text{Simplifying}$$

2.4 Expand and simplify $(2 + \sqrt{5})(3 - \sqrt{5})$

$$(2 + \sqrt{5})(3 - \sqrt{5}) = 2 \times 3 - 2 \times \sqrt{5} + 3 \times \sqrt{5} - \sqrt{5} \times \sqrt{5} \qquad \text{Expanding}$$

$$= 6 - 2\sqrt{5} + 3\sqrt{5} - \sqrt{25} \qquad \text{Simplifying}$$

$$= 6 - 2\sqrt{5} + 3\sqrt{5} - 5$$

$$= 1 + \sqrt{5}$$

2.5 Simplify $\sqrt{2}\left(\sqrt{50} - \sqrt{10}\right)$

$$\sqrt{2}\left(\sqrt{50} - \sqrt{10}\right) = \left(\sqrt{2}\right)\left(\sqrt{50}\right) - \left(\sqrt{2}\right)\left(\sqrt{10}\right) \qquad \text{Expanding}$$

$$= \sqrt{100} - \sqrt{20} \qquad \text{Multiplying surds}$$

$$= 10 - \sqrt{5 \times 4} \qquad \text{Rearranging}$$

$$= 10 - 2\sqrt{5} \qquad \text{Simplifying}$$

2.6 Expand and simplify $\left(3+\sqrt{5}\right)^2$

$$\left(3+\sqrt{5}\right)^2 = \left(3+\sqrt{5}\right)\left(3+\sqrt{5}\right)$$

$$= 9 + 3\left(\sqrt{5}\right) + 3\left(\sqrt{5}\right) + \left(\sqrt{5}\right)\left(\sqrt{5}\right) \qquad \text{Expanding brackets}$$

$$= 9 + 6\sqrt{5} + \sqrt{25} \qquad \text{Simplifying}$$

$$= 9 + 6\sqrt{5} + 5 \qquad \text{Simplifying}$$

$$= 14 + 6\sqrt{5}$$

2.7 Rationalise the denominator of $\dfrac{5}{\sqrt{7}}$

$$\frac{5}{\sqrt{7}} = \frac{5\sqrt{7}}{\left(\sqrt{7}\right)\left(\sqrt{7}\right)} \qquad \text{Multiplying top \& bottom by } \sqrt{7}$$

$$= \frac{5\sqrt{7}}{7}$$

2.8 Rationalise the denominator of $\dfrac{\sqrt{3}+3}{\sqrt{3}-3}$ and simplify

$$\frac{\sqrt{3}+3}{\sqrt{3}-3} = \frac{\left(\sqrt{3}+3\right)\left(\sqrt{3}+3\right)}{\left(\sqrt{3}-3\right)\left(\sqrt{3}+3\right)} \qquad \text{Multiplying top \& bottom by } \left(\sqrt{3}+3\right)$$

$$= \frac{\left(\sqrt{3}\right)\left(\sqrt{3}\right) + 3\left(\sqrt{3}\right) + 3\left(\sqrt{3}\right) + 9}{\left(\sqrt{3}\right)\left(\sqrt{3}\right) + 3\left(\sqrt{3}\right) - 3\left(\sqrt{3}\right) - 9} \qquad \text{Expanding brackets}$$

$$= \frac{3 + 6\left(\sqrt{3}\right) + 9}{3 - 9} \qquad \text{Simplifying}$$

$$= \frac{12 + 6\left(\sqrt{3}\right)}{-6} \qquad \text{Simplifying}$$

$$= -\left(2 + \sqrt{3}\right) \qquad \text{Dividing top \& bottom by } -6$$

2.9 Rationalise and simplify $\dfrac{\sqrt{2}}{\left(\sqrt{3}\right)\left(\sqrt{5}-\sqrt{3}\right)}$

$$\frac{\sqrt{2}}{\left(\sqrt{3}\right)\left(\sqrt{5}-\sqrt{3}\right)} = \frac{\sqrt{2}}{\left(\sqrt{3}\right)\left(\sqrt{5}\right) - \left(\sqrt{3}\right)\left(\sqrt{3}\right)} \qquad \text{Expanding denominator}$$

$$= \frac{\sqrt{2}}{\sqrt{15}-3} \qquad \text{Simplifying}$$

Note that we could continue to rationalise the denominator as follows

$$= \frac{\left(\sqrt{15}+3\right)\left(\sqrt{2}\right)}{\left(\sqrt{15}+3\right)\left(\sqrt{15}-3\right)}$$ Multiplying top & bottom by $\left(\sqrt{15}+3\right)$

$$= \frac{\left(\sqrt{15}\right)\left(\sqrt{2}\right)+3\left(\sqrt{2}\right)}{\left(\sqrt{15}\right)\left(\sqrt{15}\right)-3\sqrt{15}+3\sqrt{15}-9}$$ Expanding

$$= \frac{\sqrt{30}+3\sqrt{2}}{15-9}$$ Simplifying

$$\frac{\sqrt{30}+3\sqrt{2}}{6}$$ Simplifying

2.10 Find the square root of $14+6\sqrt{5}$ in the form $\sqrt{x}+\sqrt{y}$ where x and y are both rational.

Let $\sqrt{14+6\sqrt{5}} = \pm\left(\sqrt{x}+\sqrt{y}\right)$ $\qquad$ (1)

then

$$\left(\sqrt{14+6\sqrt{5}}\right)^2 = \left(\sqrt{x}+\sqrt{y}\right)^2$$ Squaring both sides

$$14+6\sqrt{5} = x+2\sqrt{x}\sqrt{y}+y$$ Expanding

So $x+y=14$ $\qquad$ (2) Equating numbers

and $2\sqrt{xy}=6\sqrt{5}$ $\qquad$ (3) Equating terms containing surds

$\Rightarrow 4xy=36\times5=180$ Squaring both sides of (3)

$\Rightarrow xy=45$ Therefore $y=\dfrac{45}{x}$

$\Rightarrow x+\dfrac{45}{x}=14$ Substituting for $y=\dfrac{45}{x}$ in (2)

$\Rightarrow x^2+45=14x$ Multiplying throughout by x

$\Rightarrow x^2-14x+45=0$ Rearranging

$\Rightarrow (x-9)(x-5)=0$ Factorising LHS

Hence $x=9$ or 5

$\Rightarrow 9+y=14$ Substituting for $x=9$ in (2)

$\Rightarrow y=5$

Therefore $\sqrt{14+6\sqrt{5}} = \pm(\sqrt{9}+\sqrt{5})$ Substituting for $x=9, y=5$ in (1)

Note that we could have used $x=5, y=9$ instead of $x=9, y=5$ in which case we would have the result:

$\sqrt{14+6\sqrt{5}} = \pm\left(\sqrt{5}+\sqrt{9}\right)$. Which is clearly, exactly the same result.

2.11 Express the square root of $18 - 12\sqrt{2}$ in the form $\sqrt{x} - \sqrt{y}$ where x and y are both rational.

Let $\sqrt{18 - 12\sqrt{2}} = \pm\left(\sqrt{x} - \sqrt{y}\right)$ (1)

Then

$\left(\sqrt{18 - 12\sqrt{2}}\right)^2 = \left(\sqrt{x} - \sqrt{y}\right)^2$ Squaring both sides

Hence

$18 - 12\sqrt{2} = x - 2\sqrt{x}\sqrt{y} + y$ Expanding

So $x + y = 18$ (2)   Equating number terms

and $2\sqrt{x}\sqrt{y} = 12\sqrt{2}$ (3)   Equating terms containing surds

$\Rightarrow 4xy = 144 \times 2 = 288$

$\Rightarrow xy = 72$ Therefore $y = \dfrac{72}{x}$

$\Rightarrow x + \dfrac{72}{x} = 18$ Substituting for $y = \dfrac{72}{x}$ in (2)

$\Rightarrow x^2 + 72 = 18x$ Multiplying throughout by x

Or $x^2 - 18x + 72 = 0$ Rearranging

$\Rightarrow (x - 12)(x - 6) = 0$ Factorising LHS

$\Rightarrow x = 12$ or $x = 6$

Hence $12 + y = 18$ Substituting for $x = 12$ in (2)

$\Rightarrow y = 6$

Therefore $\sqrt{18 - 12\sqrt{2}} = \pm\left(\sqrt{12} - \sqrt{6}\right)$ Substituting for $x = 12, y = 6$ in (1)

Note that the method in this question is exactly the same as that in question 11 except that we had to let $\sqrt{18 - 12\sqrt{2}} = \pm\left(\sqrt{x} - \sqrt{y}\right)$ instead of $\sqrt{18 - 12\sqrt{2}} = \pm\left(\sqrt{x} + \sqrt{y}\right)$ because the question asked for the answer in this form.

INDICES

2.12 Simplify $\dfrac{2^4 \times 2^5}{4^3}$

$\dfrac{2^4 \times 2^5}{4^3} = \dfrac{2^4 \times 2^5}{\left((2)^2\right)^3}$ Changing denominator to base 2

$= \dfrac{2^9}{2^6}$ Simplifying denominator

$= 2^9 \times 2^{-6}$ Rearranging

$= 2^{9-6}$

$= 2^3 = 8$ Simplifying

2.13   Simplify $\left(x^3\right)^5 \times x^{-4}$

$$\left(x^3\right)^5 \times x^{-4} = x^{3\times5} \times x^{-4}$$

$$= x^{15} \times x^{-4} \qquad \text{Simplifying bracket}$$

$$= x^{15-4} \qquad \text{Rearranging}$$

$$= x^{11} \qquad \text{Simplifying}$$

2.14   Simplify $\sqrt[3]{a^4 b^5 c^2} \times \dfrac{a^{-\frac{1}{3}}}{a^2 bc}$

$$\sqrt[3]{a^4 b^5 c^2} \times \frac{a^{-\frac{1}{3}}}{a^2 bc} = a^{\frac{4}{3}} \times b^{\frac{5}{3}} \times c^{\frac{2}{3}} \times a^{-\frac{1}{3}} \times a^{-2} \times b^{-1} \times c^{-1} \qquad \text{Rearranging}$$

$$= a^{\frac{4}{3}} \times a^{-\frac{1}{3}} \times a^{-2} \times b^{\frac{5}{3}} \times b^{-1} \times c^{\frac{2}{3}} \times c^{-1} \qquad \text{Grouping like terms together}$$

$$= a^{\frac{4}{3}-\frac{1}{3}-2} \times b^{\frac{5}{3}-1} \times c^{\frac{2}{3}-1} \qquad \text{Rearranging}$$

$$= a^{-1} \times b^{\frac{2}{3}} \times c^{-\frac{1}{3}} \qquad \text{Simplifying}$$

$$Or \quad \frac{b^{\frac{2}{3}} c^{-\frac{1}{3}}}{a}$$

2.15   Simplify $\left(\dfrac{216}{125}\right)^{\frac{1}{3}}$

$$\left(\frac{216}{125}\right)^{\frac{1}{3}} = \frac{216^{\frac{1}{3}}}{125^{\frac{1}{3}}} = \frac{6}{5} \qquad \text{Since } \left(\frac{x}{y}\right)^z = \frac{x^z}{y^z}$$

2.16 Simplify $\left(\dfrac{125}{216}\right)^{-\frac{1}{3}}$

$$\left(\frac{216}{125}\right)^{-\frac{1}{3}} = \left(\frac{125}{216}\right)^{\frac{1}{3}} = \frac{5}{6} \qquad \text{Since } \left(\frac{x}{y}\right)^{-z} = \left(\frac{y}{x}\right)^z = \frac{y^z}{x^z}$$

2.17   Simplify $125^{\frac{1}{3}} \times 25^2 \times 25^{\frac{2}{3}}$

$$125^{\frac{1}{3}} \times 25^2 \times 25^{\frac{2}{3}} = 5 \times 25^{2+\frac{2}{3}} \qquad \text{Since } 125^{\frac{1}{3}} = 5$$

$$= 5 \times 25^{\frac{8}{3}}$$

$$= 5 \times \left(5^2\right)^{\frac{8}{3}} \qquad \text{Changing } 25^{\frac{8}{3}} \text{ to base 5}$$

$$= 5 \times 5^{\left(\frac{8}{3} \cdot 2\right)} \qquad \text{Simplifying}$$

$$= 5 \times 5^{\frac{16}{3}} \qquad \text{Simplifying}$$

$$= 5^{1+\frac{16}{3}} \qquad \text{Since } 5 = 5^1$$

$$= 5^{\frac{19}{3}} \qquad \text{Simplifying}$$

2.18   Simplify $\dfrac{27^{\frac{1}{3}} \times 81^{-\frac{1}{4}}}{3^{-\frac{1}{6}} \times 9^{\frac{2}{3}}}$

We can see that each base (27, 81, 3, 9) is reducible to base 3 as follows:

$$\frac{27^{\frac{1}{3}} \times 81^{-\frac{1}{4}}}{3^{-\frac{1}{6}} \times 9^{\frac{2}{3}}} = 3 \times \left(3^4\right)^{-\frac{1}{4}} \times 3^{\frac{1}{6}} \times \left(3^2\right)^{-\frac{2}{3}} \qquad \text{changing to base 3 and rearranging}$$

$$= 3^1 \times 3^{-1} \times 3^{\frac{1}{6}} \times 3^{-\frac{4}{3}} \qquad \text{Simplifying}$$

$$= 3^{1-1+\frac{1}{6}-\frac{4}{3}} \qquad \text{rearranging}$$

$$= 3^{\frac{1-8}{6}} \qquad \text{changing to common denominator}$$

$$= 3^{-\frac{7}{6}} \qquad \text{Simplifying}$$

Alternatively

$$\frac{27^{\frac{1}{3}} \times 81^{-\frac{1}{4}}}{3^{-\frac{1}{6}} \times 9^{\frac{2}{3}}} = \frac{27^{\frac{1}{3}}}{81^{\frac{1}{4}}} \times 3^{\frac{1}{6}} \times \left(3^2\right)^{-\frac{2}{3}} \qquad \text{Rearranging and changing RH term to base 3}$$

$$= 1 \times 3^{\frac{1}{6}-\frac{4}{3}} \qquad \frac{27^{\frac{1}{3}}}{81^{\frac{1}{4}}} = \frac{3}{3} = 1$$

$$= 3^{\frac{1-8}{6}} \qquad \text{Rearranging with common denominator}$$

$$= 3^{-\frac{7}{6}} \qquad \text{Simplifying}$$

There is often more than one way of simplifying expressions like this but if in doubt, reduce everything to the same base first where possible.

2.19 Simplify $\dfrac{1}{16^{-\frac{1}{4}}} \times \left(\dfrac{27}{8}\right)^{\frac{2}{3}}$

$$\dfrac{1}{16^{-\frac{1}{4}}} \times \left(\dfrac{27}{8}\right)^{\frac{2}{3}} = 16^{\frac{1}{4}} \times \left(\dfrac{3}{2}\right)^{2} \qquad \text{Rearranging}$$

$$= 2 \times \dfrac{9}{4} \qquad \text{Simplifying}$$

$$= \dfrac{9}{2}$$

LOGARITHMS

2.20　Convert the following to logarithmic form

(i)　$2^3 = 8$ 　　$Log_2 8 = 3$

(ii)　$8^{\frac{1}{3}} = 2$ 　　$\log_8 2 = \dfrac{1}{3}$

(iii)　$e^x = 6$ 　　$\ln 6 = x$ 　　　　i.e $\log_e 6 = x$

(iv)　$x^y = 2$ 　　$\log_x 2 = y$

(v)　$p = q^3$ 　　$\log_q p = 3$

(vi)　$10^{-3} = 0.001$ 　　$\log_{0.001} 0.001 = 1$

(vii)　$6^0 = 1$ 　　$\log_6 1 = 0$

(viii)　$e^x = y$ 　　$\ln y = x$ 　　　　i.e $\log_e = x$

(ix)　$a^{2x} = b$ 　　$\log_a b = 2x$

(x)　$c^{2x^2} = p$ 　　$\log_c p = 2x^2$

2.21　Convert the following to index form

(i)　$\log_{10} 1,000 = 3$ 　　$10^3 = 1,000$

(ii)　$\log_5 125 = 3$ 　　$5^3 = 125$

(iii) $\ln x = 6$ $e^6 = x$ $i.e \log_e x = 6$

(iv) $\log_3 27 = 3$ $3^3 = 27$

(v) $\ln x = y$ $e^y = x$ $i.e \log_e x = y$

(vi) $\log_4 64 = 3$ $4^3 = 64$

(vii) $\log_{27} 9 = \dfrac{2}{3}$ $27^{\frac{2}{3}} = 9$

(viii) $2 = \log_3 9$ $3^2 = 9$

(ix) $\log_{64} 16 = \dfrac{2}{3}$ $64^{\frac{2}{3}} = 16$

(x) $\log_p q = r$ $p^r = q$

2.22   Evaluate.

(i) $\log_2 4$

    Let $x = \log_2 4$
    Then $2^x = 4 \Rightarrow x = 2$ Since $2^2 = 4$

(ii) $\log_3 27$

    Let $x = \log_3 27$
    Then $3^x = 27 \Rightarrow x = 3$ Since $3^3 = 27$

(iii) $\log_{10} 0.01$

Let $x = \log_{10} 0.01$

Then $10^x = 0.01 \Rightarrow x = -2$ Since $10^{-2} = \dfrac{1}{100} = 0.01$

(iv) $\log_a a^3$

    Let $x = \log_a a^3$
    Then $a^x = a^3 \Rightarrow x = 3$

(v) $\ln e^{2.5}$

    Let $x = \ln e^{2.5}$
    Then $e^x = e^{2.5} \Rightarrow x = 2.5$ $\ln e^{2.5} = \log_e e^{2.5}$

2.23 Using a calculator Evaluate correct to 3 decimal places the following.

(i) e^3 (20.086)

(ii) e^{-2} (0.135)

(iii) $e^{0.6}$ (1.822)

(iv) $\ln 3$ (1.099)

(v) $\ln 18.2$ (2.901)

(vi) $\ln 0.05$ (−.2996)

(vii) $\log_{10} 4.7$ (0.672)

(viii) $\log_{10} 55.6$ (1.745)

(ix) $\log_{10} 255$ (2.407)

(x) $\ln e$ (1) If $x = \log_e e$ then $e^x = e \Rightarrow x = 1$ *Since* $e^1 = e$

2.24 Express the following in terms of $\log x, \log y$ and $\log z$ where possible :

(i) $\log(x^2 y \sqrt{z})$

$$\log(x^2 y \sqrt{z}) = \log x^2 + \log y + \log \sqrt{z}$$

$$= \log x^2 + \log y + \log z^{\frac{1}{2}}$$

$$= 2\log x + \log y + \frac{1}{2}\log z \qquad \text{Since } \log b^n = n\log b$$

(ii) $\log xy$ $(\log x + \log y)$

(iii) $\log xyz$ $(\log x + \log y + \log z)$

(iv) $\log \dfrac{y}{z}$ $(\log y - \log z)$

(v) $\log \dfrac{xy}{z}$ $(\log x + \log y - \log z)$

(vi) $\log \dfrac{x}{yz}$ $(\log x - \log y - \log z)$

(vii) $\log x^2 y$ $(2\log x + \log y)$

(viii) $\log \dfrac{x^2 y^2}{z^3}$ $(2\log x + 2\log y - 3\log z)$

(i) $x \ln 4x$ $\hspace{4cm}$ $(x \ln x + 2 \ln 2x)$

(ii) $\ln 4x^3$ $\hspace{4cm}$ $(\ln 4 + 3 \ln x)$

(iii) $\ln \dfrac{3x}{x+2}$

$$\ln \frac{3x}{x+2} = \ln 3x - \ln(x+2)$$

(iv) $\ln e^2 x^2 (2x - 3e)$

$$\ln e^2 x^2 (2x - 3e) = \ln e^2 + \ln x^2 + \ln(2x - 3e)$$
$$= 2 \ln e + 2 \ln x + \ln(2x - 3e)$$

(v) $\ln \dfrac{x^3}{x+1}$

$$\ln \frac{x^3}{x+1} = \ln x^3 - \ln(x+1)$$
$$= 3 \ln x - \ln(x+1)$$

2.27 Prove that (i) $\log_a\left(\dfrac{1}{x}\right) = -\log_a x$

$\hspace{3cm}$ (ii) $\log_a x = \log_b x \div \log_b a$

(i)   Let $r = \log_a \dfrac{1}{x}$

$\hspace{1cm} \Rightarrow a^r = \dfrac{1}{x}$

$\Rightarrow r \log_a a = \log_a 1 - \log_a x$

$\hspace{1.5cm} \Rightarrow r = -\log_a x$ $\hspace{2cm}$ Since $\log_a 1 = 0$

$\hspace{0.5cm} \therefore \log_a \dfrac{1}{x} = -\log_a x$ $\hspace{2cm}$ Reversing the substitution for $r = \log_a \dfrac{1}{x}$

(ii)   Let $r = \log_a x$

$\hspace{0.5cm}$ Then $a^r = x$

$\Rightarrow r \log_b a = \log_b x$ $\hspace{2cm}$ Note the change to base b here

$\hspace{1cm} \Rightarrow r = \dfrac{\log_b x}{\log_b a}$

Hence

$\hspace{1cm} \log_a x = \dfrac{\log_b x}{\log_b a}$ $\hspace{2cm}$ Reversing the substitution for $r = \log_a x$

We introduced base b because the question involves logs to the base b. Don't forget we can use any base we wish when taking logs.

(i) $\log_7 343$

(ii) $(\log_a 81)(\log_3 a)$

(i) Let $x = \log_7 343$

Then

$\qquad 7^x = 343$

$\qquad \Rightarrow x = 3$

Hence

$\log_7 343 = 3$ $\qquad\qquad\qquad\qquad$ Since $7^3 = 343$

$\qquad$ (ii)

$\qquad$ Let $x = \log_a 81$ $\qquad\qquad$ (1)

$\qquad$ and $y = \log_3 a$ $\qquad\qquad$ (2)

$\qquad$ Then

$\qquad a^x = 81$ $\qquad\qquad\qquad$ (3)

$\qquad 3^y = a$ $\qquad\qquad\qquad$ (4)

$\qquad$ Hence

$\qquad \left(3^y\right)^x = 81$ $\qquad\qquad$ Substituting for $a = 3^y$ from (4) in (3)

$\qquad$ or $3^{xy} = 81$

$\qquad$ and $xy = 4$ $\qquad\qquad$ Since $3^4 = 81 \Rightarrow xy = 4$

$\qquad$ Hence $(\log_a 81)(\log_3 a) = 4$ $\quad$ Since $x = \log_a 81$ & $y = \log_3 a$ from (1) and (2)

2.27   Solve the equation $3^{2x+1} - 11 \times 3^x - 4 = 0$

The key to solving this equation is in recognising that the powers of x suggest that we can ultimately reduce this to a quadratic equation as follows:

$3^{2x+1} - 11 \times 3^x - 4 = 0$ $\qquad\qquad$ (1)

Can be re-written as

$3 \times 3^{2x} - 11 \times 3^x - 4 = 0$ $\qquad$ (2) $\qquad$ Since $3^{2x+1} = 3^{2x} \times 3^1 = 3 \times 3^{2x}$

Now let $3^x = b$

Then (2) becomes

$3b^2 - 11b - 4 = 0$ $\qquad\qquad\qquad\qquad$ Hence we have a quadratic in b

$\Rightarrow (3b+1)(b-4) = 0$ $\qquad\qquad\qquad$ Factorising

$\Rightarrow b = -\dfrac{1}{3}$ or $b = 4$

If $b = -\dfrac{1}{3}$ then $3^x = -\dfrac{1}{3}$ $\qquad\qquad$ Since $b = 3^x$

This result has no solution since 3^x cannot be negative

Note that x may be negative but **not** 3^x as 3^x is positive for all values of x

The second solution gives

$3^x = 4$

$\Rightarrow x \log 3 = \log 4$

$\Rightarrow x = \dfrac{\log 4}{\log 3} = \dfrac{0.6021}{0.4471} = 1.2619$

Hence the solution to (1) is $x = 1.2619$

2.28 (i) Prove that $\log_a x = \log_b x \log_a b$

(ii) Given that $\log_2 3 = 1.585$, evaluate $\log_2 32$ and $\log_8 9$

(i) Let $\log_b x = r$ (1)

    and $\log_a b = s$

     Then $b^r = x$ (2)

     and $a^s = b$ (3)

        $\Rightarrow a^{rs} = x$ Substituting for $b = a^s$ in (2)

$\Rightarrow \; rs \log_a a = \log_a x$

     $\Rightarrow \log_a x = rs$ Since $\log_a a = 1$ (see below)

           $= \log_b x \log_a b$ (4)   Substituting for r and s *in* (1) and (2)

Note: If $z = \log_a a \Rightarrow a^z = a \Rightarrow z = 1 \Rightarrow \log_a a = 1$

(ii) $\log_{27} 32 = \log_2 32 \times \log_{27} 2$ From part (i) Eq (4) with $b = 2$, $a = 27$ & $x = 32$

$$= \log_{27} 2 \qquad \text{Since } \log_2 32 = 5 \text{ because } 2^5 = 32 \quad (1)$$

$$\text{Let } z = \log_{27} 2$$

$$\text{Then } 27^z = 2$$

$$\left(3^3\right)^z = 2$$

$$3^{3z} = 2$$

Hence $3z \log_2 3 = \log_2 2$ Taking logs to base 2

$$\text{Giving } z = \frac{1}{3\log_2 3} \qquad \text{Since } \log_2 2 = 1$$

$$\log_{27} 2 = \frac{\log_2 32}{3\log_2 3} \qquad \text{Reversing the substitution for } z$$

$$= \frac{\log_2 (2)^5}{3\log_2 3}$$

$$= \frac{5}{3\log_2 3} \qquad \text{If } x = \log_2 (2)^5 \text{ then } 2^x = 2^5 \Rightarrow x = 5$$

$$= \frac{5}{3 \times 1.585} \qquad \text{Since we are given } \log_2 3 = 1.585$$

$$= 1.0515 = \log_{27} 32 \qquad \text{Since } \log_{27} 2 = \log_{27} 32 \text{ from (1)}$$

For the second part of the question we follow the same procedure as the first part.

$$\log_{\frac{1}{8}} \frac{1}{9} = \frac{\log_2 \frac{1}{9}}{\log_2 \frac{1}{8}}$$

$$= \frac{\log_2 (3)^{-2}}{\log_2 (2)^{-3}} \qquad \text{Using the same method as before}$$

$$= \frac{-2\log_2 3}{-3\log_2 2}$$

$$= \frac{2\log_2 3}{3\log_2 2} \qquad \text{Simplifying}$$

$$= \frac{2 \times 1.585}{3} \qquad \text{Since } \log_2 2 = 1$$

$$= 1.057$$

MODULE A3

ALGEBRA 3 - QUADRATIC EQUATIONS

Revision Notes

Solution by Factorising

If a, b and c are any real numbers then any quadratic equation can be written in the form $ax^2 + bx + c$.

Consider the equation $x^2 + 3x - 4 = 0$. We can **factorise** the LHS since:

$(x + 4)(x - 1) \equiv x^2 + 3x - 4$.

$(x + 4)(x - 1) = x(x - 1) + 4(x - 1) = x^2 - x + 4x - 4 = x^2 + 3x - 4$

Therefore we can say $(x + 4)(x - 1) = 0$

If the product of two quantities is zero, then one or both of these must zero

Hence we can say in this case that:

$(x + 4) = 0 \; or \; (x - 1) = 0$
$\Rightarrow x = -4 \; or \; x = 1$

If we replace x by 1 and then by -4 in the given equation, we can see that:

$(1)^2 + 3(1) - 4 = 0 \; and \; (-4)^2 + 3(-4) - 4 = 0$

Hence these two values of x are the solution to the equation and are called the **roots** of the equation.

All solutions must be checked in this way in the original equation to ensure they are valid solutions.

Solution by Completing the Square

If we consider the equation $x^2 + 3x + 1 = 0$, we cannot this factorise as before since we cannot find two bracketed terms by inspection which when multiplied together will yield the original equation.

However we can solve this equation by make the LHS of this equation a perfect square as follows:

1. Rearrange the equation with the terms containing x on the LHS and terms not containing x on the RHS.
   $x^2 + 3x = -1$

2. Add (half the coefficient of x)^2 to each side of the equation. In this case we add $\left(\dfrac{3}{2}\right)^2$ to each side. Hence we have:

$$x^2 + 3x + \left(\frac{3}{2}\right)^2 = -1 + \left(\frac{3}{2}\right)^2$$

The LHS is now a **perfect square** and factorises into $\left(x + \frac{3}{2}\right)^2$

Hence we can say that:

$$\left(x + \frac{3}{2}\right)^2 = -1 + \left(\frac{3}{2}\right)^2$$

Now taking the square root of both sides we get

$$\left(x + \frac{3}{2}\right) = \pm\sqrt{\frac{5}{2}}$$

$$\Rightarrow x = -\frac{3}{2} - \sqrt{\frac{5}{2}} \ \text{or} \ x = -\frac{3}{2} + \sqrt{\frac{5}{2}}$$

If we have an equation where the coefficient of x^2 is not 1, divide the complete equation through by this coefficient first.

e.g $5x^2 - 7x - 6 = 0$ becomes

$$x^2 - \frac{7}{5}x - \frac{6}{5} = 0$$

$$\Rightarrow x^2 - \frac{7}{5}x + \left(\frac{7}{10}\right)^2 = \frac{6}{5} + \left(\frac{7}{10}\right)^2 \qquad \text{Completing the square}$$

$$\Rightarrow \left(x - \frac{7}{10}\right)^2 = \frac{6}{5} + \frac{49}{100} \qquad \text{Forming LHS into perfect square}$$

$$\Rightarrow \left(x - \frac{7}{10}\right)^2 = \frac{169}{100}$$

$$\Rightarrow \left(x - \frac{7}{10}\right) = \pm\frac{13}{10} \qquad \text{Taking square root of both sides}$$

$$\Rightarrow x = 2 \ \text{or} \ x = -\frac{3}{5}$$

Solving with use of Formula

Looking at the general case:

$$ax^2 + bx + c = 0$$

We can also solve this by completing the square just as before

$$x^2 + \frac{x}{a} + \frac{c}{a} = 0 \qquad \text{Dividing throughout by } a$$

$$\Rightarrow ax^2 + bx = -c \qquad \text{Rearranging}$$

$$\Rightarrow x^2 + \frac{b}{a}x + \left(\frac{b}{2a}\right)^2 = \left(\frac{b}{2a}\right)^2 - \frac{c}{a} \qquad \text{Completing the square}$$

$$\Rightarrow \left(x + \frac{b}{2a}\right)^2 = \frac{b^2}{4a^2} - \frac{c}{a} \qquad \text{Forming LHS into perfect square}$$

$$\Rightarrow \left(x + \frac{b}{2a}\right) = \pm\sqrt{-\frac{c}{a} + \frac{b^2}{4a^2}} \qquad \text{Taking square roots of both sides}$$

$$\Rightarrow x = -\frac{b}{2a} \pm \sqrt{\frac{b^2 - 4ac}{4a^2}} \qquad \text{Rearranging}$$

Or

$$x = -\frac{b}{2a} \pm \frac{\sqrt{b^2 - 4ac}}{2a} \qquad \text{Rearranging}$$

Sketching Quadratic Graphs

The first thing to remember about quadratic graphs is that they are always either $\cup$ shaped or $\cap$ shaped.

To establish which way up the graphs will be, all you have to do is look at the coefficient of x^2. If it is positive the graph will be $\cup$ shaped. If it is negative the graph will be $\cap$ shaped.

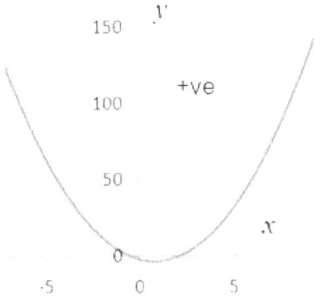

Fig 3.1 Graph of $y = 2x^2 - 3x - 2$ Fig 3.2 $y = -x^2 + 5x - 6$

To sketch the graph in Fig 3.1 we need to put $x = 0$ and $y = 0$ in $y = 2x^2 - 3x - 2$ to determine where the graph intersects the x and y axes. If we put $x = 0$ and $y = 0$ we get:

(i) $\quad y = 2 \times (0)^2 - 3 \times (0) - 2 = -2$

(ii) $\quad 0 = 2x^2 - 3x - 2 \Rightarrow (2x+1)(x-2) = 0$

$\quad\quad \Rightarrow x = -\dfrac{1}{2}$ and $x = 2$

Marking these points on the graph we can sketch a smooth curve that passes through them.

Finding the Maximum and Minimum Values for the Function

The Maximum or Minimum values of a function can be found by completing the square as described earlier.

$y = 2x^2 - 3x - 2 = x^2 - \dfrac{3}{2}x - 1$ $\quad\quad$ Divide throughout by 2 first

$\quad = \left(x - \dfrac{3}{4}\right)^2 = 1 + \dfrac{3}{4}$ $\quad\quad$ Completing the square

$\quad = \left(x - \dfrac{3}{4}\right)^2 = \dfrac{7}{4}$ $\quad\quad$ Simplifying

In this case we want to find the minimum value which will occur when $\left(x - \dfrac{3}{4}\right)^2$ is zero. This can only occur when $x = \dfrac{3}{4}$. Note that a maximum or minimum value is also a **turning point** and often called the **Vertex** of the function

To find the corresponding value of y we simply substitute for $x = \dfrac{3}{4}$ in the original equation:

$y = 2\left(\dfrac{3}{4}\right)^2 - 3\left(\dfrac{3}{4}\right) - 2$

$\quad = 2 \times \dfrac{9}{16} - \dfrac{9}{4} - 2 = -\dfrac{25}{8}$

Hence the coordinates of the Vertex in this case are $\left(\dfrac{3}{4}, -\dfrac{25}{8}\right)$ which is a minimum value.

The Roots of Quadratic Equations

A root of an equation occurs when it crosses the x axis. We have already defined the meaning of a, b and c when discussing the use of the formula for solving quadratic equations. However, there are properties of the roots of these equations which are extremely important:

When $b^2 - 4ac > 0$ there will be two distinct roots

When $b^2 - 4ac = 0$ There is only one root

When $b^2 < 4ac$ There are no real roots

$b^2 - 4ac > 0$	$b^2 - 4ac = 0$	$b^2 - 4ac < 0$
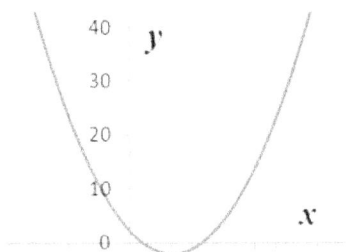	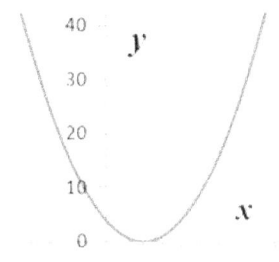	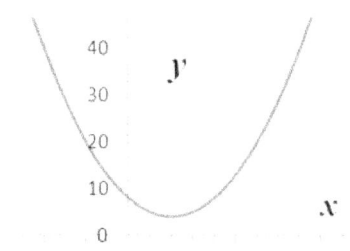
Fig 3.3 $y = x^2 - 4x + 2$ $= (x-2)^2 - 2$	Fig 3.4 $y = x^2 - 4x + 4$ $= (x-2)^2$	Fig 3.5 $y = x^2 - 4x + 8$ $= (x-2)^2 + 4$
Two Roots	One Root	No Roots
The graph crosses the x axis twice so there are two roots	The graph just touches the x axis so there is one root	The graph doesn't touch the x axis at all so there are no roots

Roots and Range

Find the range of values of k for which

(i) $f(x) = 0$ has two distinct roots

(ii) $f(x) = 0$ has one solution only

(iii) $f(x) = 0$ has no solutions

Where $f(x) = 4x^2 + 3x + k = 0$

Case (i) For $f(x)$ to have two distinct roots, we require that $b^2 > 4ac$

Hence we require that $9 > 4 \times 4 \times k$

Or $\dfrac{9}{16} > k \Rightarrow k < \dfrac{9}{16}$

Case (ii)

Here we require that $b^2 = 4ac$

So we need $9 = 16k \Rightarrow k = \dfrac{9}{16}$

Case(iii) For $f(x)$ to have no real roots, we require that $b^2 < 4ac$

So we need $9 < 16k \Rightarrow k > \dfrac{9}{16}$

WORKED SOLUTIONS

$(i) \ x^2 + 5x = 6$

$(ii) \ x(x - 3) = x^2 - 6$

$(iii) x^2 - 2ax + a^2 = 0$

$(iv) \ \dfrac{1}{x^2} - 1 = \dfrac{1}{x} - 1$

$(v) \ \dfrac{1}{x + 1} + \dfrac{2}{x + 2} - 1 = 0$

(i) First of all rearrange the equation

$x^2 + 5x = 6$

$\Rightarrow x^2 + 5x - 6 = 0$

And we can see by inspection that it factorises

$x^2 + 5x - 6 = 0$

$\Rightarrow (x + 6)(x - 1) = 0 \qquad$ Factorising

$\Rightarrow x = -6 \ or \ x = 1$

$(ii) x(x - 3) = x^2 - 6$

$\Rightarrow x^2 - 3x = x^2 - 6 \qquad$ Expanding bracket

$\Rightarrow x^2 - 3x - x^2 + 6 = 0 \qquad$ Rearranging

$\Rightarrow -3x + 6 = 0 \qquad$ Simplifying

$\Rightarrow 3x = 6 \qquad$ Rearranging

$\Rightarrow x = 2$

Note that technically that there is another solution to this equation
$x = \infty$ where ∞ means infinitely large

This is because we cancelled out the x^2 terms on the LHS and RHS of the equation. In most maths problems this has no meaning and is discounted. However, if $x = \tan \alpha$ then $x = \infty \Rightarrow \alpha = 90°$ this solution would be valid.

$(iii) \ x^2 - 2ax + a^2 \equiv (x - a)^2 \qquad$ Factorising

$i.e \ (x - a)(x - a) = 0$

$\Rightarrow x = a, a \qquad$ a is a repreated solution

$(iv) \ \dfrac{1}{x^2} - 1 = \dfrac{1}{x} - 1$

$\Rightarrow 1 - x^2 = x - x^2 \qquad$ Multiplying throughout by x^2 to simplify

$i.e - x^2 + x^2 - x + 1 = 0$

$\Rightarrow x = 1, \infty \qquad$ Because we have eliminated x^2 from both sides

$(v) \dfrac{1}{x+1} + \dfrac{2}{x+2} - 1 = 0$

$\Rightarrow (x+2) + 2(x+1) - (x+2)(x+1) = 0$ Multiplying throughout by $(x+2)(x+1)$

$\Rightarrow x + 2 + 2x + 2 - x^2 - 3x - 2 = 0$ Expanding brackets

$\Rightarrow -x^2 + 2 = 0$ Simplifying

$\Rightarrow x^2 - 2 = 0$ Multiplying throughout by -1

i.e $x = \sqrt{2}, -\sqrt{2}$

3.2 Solve $x^2 - 5x - 6 = 0$

Using the formula associated with the standard equation $ax^2 + bx + c$, the solution is given by:

$$x = \dfrac{-b}{2a} \pm \dfrac{\sqrt{b^2 - 4ac}}{2a}$$

In this case we have:

$a = 1$

$b = -5$

$c = -6$

Hence the solution is given by

$$x = \dfrac{5}{2} \pm \dfrac{\sqrt{5^2 - 4 \times 1 \times (-6)}}{2}$$ Substituting these values into the formula

$$= \dfrac{5}{2} \pm \dfrac{\sqrt{25 + 24}}{2}$$ Simplifying

$$= \dfrac{5}{2} \pm \dfrac{7}{2}$$

i.e. $x = 6, -1$

3.2 Prove that for any real values of a and x the value of the expression $(a^2 + 1)x^2 - 2a^2x + a^2 - 1$ can never be less than -1

Let the value of this expression be z, then we have

$(a^2 + 1)x^2 - 2a^2x + a^2 - 1 = z$

$\Rightarrow (a^2 + 1)x^2 - 2a^2x + a^2 - 1 - z = 0$ Rearranging

For x to be real we require

$b^2 \geq 4ac$ We cannot discount the roots being equal hence we use $b^2 \geq 4ac$ and not $b^2 > 4ac$
Where

$a = a^2 + 1$

$b = -2a^2$

$c = a^2 - 1 - b$

Hence we require $(-2a^2)^2 \geq 4(a^2 + 1)(a^2 - 1 - b)$

$\Rightarrow 4a^4 \geq 4(a^2 + 1)(a^2 - 1 - b)$	Simplifying
$\Rightarrow a^4 \geq a^4 - a^2 - a^2b + a^2 - 1 - b$	Dividing through by 4 and multiplying out brackets
$or\ a^2b + b \geq -1$	Simplifying and rearranging
$\Rightarrow b(a^2 + 1) \geq -1$	Factorising

$i.e\ b \geq \dfrac{-1}{a^2 + 1}$

The minimum value of a occurs when $a = 0$ Because we want b to be a maximum
& any other value would give a smaller
result for b

Hence $b \geq -1$

i.e $(a^2 + 1)x^2 - 2a^2x + a^2 - 1 \geq -1$

3.4. If the roots of the equation $ax^2 + bx + c = 0$ are p and q and are unequal, find in terms of a, b, c the condition that:

$p + q^2 = q + p^2$

The method here is to try and rearrange this expression to obtain sums and products of the roots. i.e. $p + q$ and pq and then substituting as shown below.

$p + q^3 = q + p^3$	
$\Rightarrow p - q = p^3 - q^3$	Rearranging
$\Rightarrow p - q = (p - q)(p + q)^2 - p^2q + pq^2$	Expressing p and q as products and sums of p and q
	Note the manipulation here
	$p^3 - q^3 = (p - q)(p + q)^2 - p^2q + pq^2$
$\Rightarrow p - q = (p - q)(p + q)^2 - pq(p - q)$	Factorising $p^2q + pq^2$

Hence

$(p + q)^2 - pq = 1$	(1)	Dividing both sides by $p - q$
$p + q = \dfrac{-b}{a}$		Sum of roots
$pq = \dfrac{c}{a}$		Product of roots

Substituting for $p+q$ and pq in (1) we get

$$\left(\frac{-b}{a}\right)^2 - \frac{c}{a} = 1$$

$$\Rightarrow \frac{b^2}{a^2} - \frac{c}{a} = 1$$

$$\Rightarrow b^2 - ca = a^2$$

$or\ a^2 = b^2 - ca$ This is the required condition

3.5 If α and β are the roots of the equation $x^2 + 3x - 2 = 0$, find the values of $\alpha^3 + \beta^3$ and $\alpha^3 \beta^3$. Write down the equation whose roots are α^3 and β^3.

By inspection we can see that $\alpha + \beta = -3$ $\dfrac{-3}{1}$

and $\alpha\beta = -2$ $\dfrac{-2}{1}$

to express $\alpha^3 + \beta^3$ we can use the expansion

$(\alpha+\beta)^3 \equiv \alpha^3 + 3\alpha^2\beta + 3\alpha\beta^2 + \beta^3$ $(\alpha+\beta)(\alpha+\beta)(\alpha+\beta)$

$\equiv \alpha^3 + \beta^3 + 3\alpha\beta(\alpha+\beta)$ Rearranging

$\therefore \alpha^3 + \beta^3 \equiv (\alpha+\beta)^3 - 3\alpha\beta(\alpha+\beta)$

$= (-3)^3 - 3(-2)(-3)$ Substituting for $\alpha+\beta$ and $\alpha\beta$

$= -45$

$\alpha^3\beta^3 \equiv (\alpha\beta)^3$

$= (-2)^3 = -8$

The required equation has roots α^3 and β^3 with the sum of its roots

$\alpha^3 + \beta^3 = -45$

and the product of its roots $\alpha^3\beta^3 = -8$

Hence the required equation is $x^2 - (-45)x + (-8)$

$= x^2 + 45x - 8$

3.6 If α and β are the roots of the equation $ax^2 - bx + c = 0$, find the equation whose roots are:

$\alpha + \dfrac{1}{\alpha}$ and $\beta + \dfrac{1}{\beta}$

The roots of this equation are: α and β

$\therefore \alpha + \beta = \dfrac{b}{a}$ and $\alpha\beta = \dfrac{c}{a}$

The method of attack here and in (most cases) is to express the roots of the required equation in terms of (α +β) and $\alpha\beta$

To form the required equation we first need to find the sum of the roots

$$\left(\alpha + \frac{1}{\alpha}\right) + \left(\beta + \frac{1}{\beta}\right)$$

$$= \alpha + \beta + \left(\frac{1}{\alpha} + \frac{1}{\beta}\right) \qquad \text{Rearranging}$$

$$= \alpha + \beta + \frac{\alpha + \beta}{\alpha\beta} \qquad \text{Changing bracket to a single fraction}$$

$$= \frac{b}{a} + \left(\frac{b}{a} \div \frac{c}{a}\right) \qquad \text{Substituting for } \alpha + b; \alpha\beta$$

$$= \frac{b}{a} + \frac{b}{c}$$

We now need to find the product of the roots as follows:

$$\left(\alpha + \frac{1}{\alpha}\right)\left(\beta + \frac{1}{\beta}\right) \equiv \alpha\beta + \frac{1}{\alpha\beta} + \frac{\alpha}{\beta} + \frac{\beta}{\alpha} \qquad \text{Multiplying out brackets}$$

$$\equiv \alpha\beta + \frac{1}{\alpha\beta} + \frac{\alpha^2 + \beta^2}{\alpha\beta}\alpha \qquad \text{Combining last two terms as single fraction}$$

$$\equiv \alpha\beta + \frac{1}{\alpha\beta} + \frac{(\alpha + \beta)^2 - 2\alpha\beta}{\alpha\beta} \qquad \text{Since } (\alpha + \beta)^2 = \alpha^2 + 2\alpha\beta + \beta^2$$

$$\text{so we have to subtract } 2\alpha\beta$$

$$= \frac{c}{a} + \frac{a}{c} + \frac{\left(\frac{b}{a}\right)^2 - 2\left(\frac{c}{a}\right)}{\frac{c}{a}} \qquad \text{Substitute for } \alpha + \beta; \alpha\beta$$

$$= \frac{c}{a} + \frac{a}{c} + \frac{b^2}{ac} - 2 \qquad \text{Simplifying}$$

We now have all the information we need to construct the required equation which we can write down as:

$$x^2 - b\left(\frac{1}{a} + \frac{1}{c}\right)x + \frac{c}{a} + \frac{a}{c} + \frac{b^2}{ac} - 2 = 0$$

or

$$acx^2 - b(a + c)x + a^2 + b^2 + c^2 - 2ac = 0 \qquad \text{Multiplying throughout by } ac$$

3.7 If α and β are the roots of the equation $x^2 - px + q = 0$, form the equation whose roots are $\dfrac{\alpha}{\beta^2}$ and $\dfrac{\beta}{\alpha^2}$.

By inspection we know that:

$\alpha + \beta = p$

$\alpha\beta = q$

The sum of the roots of the required equation is:

$$\frac{\alpha}{\beta^2} + \frac{\beta}{\alpha^2} = \frac{\alpha^3 + \beta^3}{\alpha^2\beta^2}$$

to express $\alpha^3 + \beta^3$ we can use the same expansion as before

$(\alpha+\beta)^3 \equiv \alpha^3 + 3\alpha^2\beta + 3\alpha\beta^2 + \beta^3$ $(\alpha+\beta)(\alpha+\beta)(\alpha+\beta)$

$\equiv \alpha^3 + \beta^3 + 3\alpha\beta(\alpha + \beta)$ Rearranging

$\therefore \alpha^3 + \beta^3 \equiv (\alpha+\beta)^3 - 3\alpha\beta(\alpha + \beta)$

$= p^3 - 3q(p)$

$= p(p^2 - 3q)$

The product of the roots is:

$$\frac{\alpha}{\beta^2} \cdot \frac{\beta}{\alpha^2} = \frac{1}{\alpha\beta}$$

$$= \frac{1}{q}$$

Hence the required equation is:

$$x^2 - \frac{(p)(p^2 - 3q)}{q^2}x + \frac{1}{q} = 0 \qquad \text{Substituting for } \alpha + \beta; \alpha\beta$$

Or

$$q^2x^2 - p(p^2 - 3q) + q = 0$$

3.8 (i) Show that the roots of the equation $2qx^2 - 2(p - q)x - 3p = 2q$ are real when p and q are real.

   (ii) If one root of this equation is double the other, prove that either $p = 2q$ or $4p = 11q$

8 (i) It is important to approach all questions with a systematic plan and not be intimidated by them. This question is no exception.

First of all we should recall that for the roots of a quadratic equation to be real we need the following condition:

If the equation was of the standard form: $ax^2 + bx + c = 0$

We would need $b^2 \geq 4ac$ for the roots to be real. Now let's go back to the question we have to solve.

First of all rearrange the equation to put it in the standard form $ax^2 + bx + c$

$$2qx^2 + 2(p+q)x + 3p = 2q$$
$$\Rightarrow 2qx^2 + 2(p+q)x + 3p - 2q = 0$$

Now we can identify the following:

$a = 2q$
$b = -2(p+q)$
$c = 3p - 2q$

Hence for real roots we require:

$4(p+q)^2 \geq 4(2q)(3p-2q)$ This is simply $b^2 \geq 4ac$ from the standard form

$\Rightarrow (p+q)^2 \geq 6pq - 4q^2$ Simplifying

$\Rightarrow p^2 + 2pq + q^2 - 6pq + 4q^2 \geq 0$ Expanding brackets and rearranging

$\Rightarrow 2p^2 - 4pq + 5q^2 \geq 0$

So far so good but now we have to show that this condition is satisfied when p and q are real. How do we do this? Well we know that the square of any real number cannot be negative so we need to rearrange our last result to achieve this. The problem is that the last expression is not a perfect square. This is one way we can overcome this problem.

$p^2 - 4pq + 5p^2 = (p^2 - 4pq + 4q^2) + q^2$ Rearranging

$= (p - 2q)^2 + q^2$ Rearranging as the sum of two squares

$\Rightarrow (p - 2q)^2 + q^2 \geq 0$

Providing p and q are both real

8(ii) The second part of the question is relatively straightforward.

Let the two roots be α and 2α

Then the sum of the roots is 3α and the product of the roots is $2\alpha^2$

Hence $3\alpha = -\dfrac{2(p+q)}{2q}$ Sum of roots ($\alpha + 2\alpha$)

$\Rightarrow \alpha = -\dfrac{p+q}{3q}$

$\Rightarrow \alpha^2 = \dfrac{(p+q)^2}{9q^2}$ (1)

And $2\alpha^2 = \dfrac{3p - 2q}{2q}$ Product of roots $\alpha \times 2\alpha$

$\Rightarrow \alpha^2 = \dfrac{3p - 2q}{4q}$ (2)

We can now equate (1) and (2)

The objective is to rearrange the following expression to arrive at a quadratic equation that can be expressed as a product of two factors to find p and q.

$$\frac{(p+q)^2}{9q^2} = \frac{3p-2q}{4q}$$

$\Rightarrow 4(p+q)^2 = 9q(3p-2q)$ Multiplying both sides by $36q^2$

$\Rightarrow 4p^2 + 8pq + 4q^2 = 27pq - 18q^2$ Expanding brackets

$\Rightarrow 4p^2 - 19pq + 22q^2 = 0$ Rearranging

$\Rightarrow (p - 2q)(4p - 11q) = 0$ Factorising

Hence either $p = 2q$ or $4p = 11q$

3.9 (i) Show that if λ is positive but not greater than 3, the roots of the equation $(\lambda - 2)x^2 - (8 - 2\lambda)x - (8 - 3\lambda) = 0$ are real.

(ii) Find the range of values of λ for which one root is real and positive and the other root is real and negative.

9(i) Using the standard form of the quadratic equation $ax^2 + bx + c = 0$, we require that $b^2 \geq 4ac$. In this case we have

$a = (\lambda - 2)$

$b = -(8 - 2\lambda)$

$c = -(8 - 3\lambda)$

Hence in this question for the roots to be real we require

$(8 - 2\lambda)^2 \geq -4(\lambda - 2)(8 - 3\lambda)$

$\Rightarrow 64 - 32\lambda + 4\lambda^2 \geq -32\lambda + 12\lambda^2 + 64 - 24\lambda$ Expanding brackets

$\Rightarrow 24\lambda - 8\lambda^2 \geq 0$ Simplifying

$\Rightarrow 3\lambda - \lambda^2 \geq 0$ Simplifying

$\Rightarrow \lambda(3 - \lambda) \geq 0$ Factorising

The only way in which the condition can be met is if $\lambda \leq 3$. Hence we make our closing statement in the proof.

For this condition to be satisfied we require $\lambda \leq 3$

9(ii) To find the range of values of λ for which one root is real and positive and the other root is real and negative we need to look at our standard quadratic equation again to plan our method of attack.

$ax^2 + bx + c = 0$

For both roots to be real we would require

$b^2 > 4ac$

But this case we need:

$$\sqrt{b^2 - 4ac} > b$$ This step is the key to solving this question

Why? Well let's examine the standard form a little more

We have the general solution $x = \dfrac{-b}{2a} \pm \dfrac{\sqrt{b^2 - 4ac}}{2a}$

i.e $x_1 = \dfrac{-b}{2a} + \dfrac{\sqrt{b^2 - 4ac}}{2a} > 0 \ when \ \sqrt{b^2 - 4ac} > b$

and $x_2 = \dfrac{-b}{2a} - \dfrac{\sqrt{b^2 - 4ac}}{2a} < 0 \ when \ \sqrt{b^2 - 4ac} > b$

In this case we have

$a = (\lambda - 2)$
$b = -(8 - 2\lambda)$
$c = -(8 - 3\lambda)$

Hence we proceed with the solution

$\sqrt{(8 - 2\lambda)^2 + 4(\lambda - 2)(8 - 3\lambda)} > 8 - 2\lambda$ $\sqrt{b^2 - 4ac} > b$

$\Rightarrow (8 - 2\lambda)^2 + 4(\lambda - 2)(8 - 3\lambda) > (8 - 2\lambda)^2$ Squaring both sides

$\Rightarrow 4(\lambda - 2)(8 - 3\lambda) > 0$ Dividing both sides by $(8 - 2\lambda)^2$

$\Rightarrow (\lambda - 2)(8 - 3\lambda) > 0$

$\Rightarrow (\lambda - 2) > 0$

i.e $\lambda > 2$

or $(8 - 3\lambda) > 0$

i.e $\lambda < \dfrac{8}{3}$

Hence $2 < \lambda < \dfrac{8}{3}$ is the required condition

3.10. Find the equation whose roots are the squares of the roots of the equation
$$x^2 - px + q = 0 \qquad \qquad (1)$$

Let the roots of the equation be α and β and let the equation be

$(x - \alpha)(x - \beta) = 0 \qquad \qquad (2)$

$\Rightarrow x^2 - \alpha x - \beta x + \alpha\beta = 0$

or

$x^2 - (\alpha + \beta)x + \alpha\beta = 0 \qquad \qquad (3)$

Any numerical multiple of this equation will have the same roots so that if the equation: $x^2 - px + q = 0$ is known to have the **same** roots α and β as (2)
We can then say

$$x^2 - px + q = 0 \equiv \lambda(x - \alpha)(x - \beta) \text{ for some number } \lambda$$

Since the LHS and RHS are identical we can equate the coefficients and immediately see by comparing coefficients of x^2 that λ =1. Hence:

$$x^2 - px + q \equiv x^2 - (\alpha + \beta)x + \alpha\beta \qquad \text{Using eq's (1) and (3)}$$

Similarly, comparing the other coefficients we obtain

$$\alpha + \beta = p$$
$$\alpha\beta = q$$

The squares of the roots will then be α^2 and β^2

Hence the sum of the squares of the roots is

$$\alpha^2 + \beta^2 = (\alpha + \beta)^2 - 2\alpha\beta$$
$$= p^2 - 2q$$

And the product of the squares of the roots is

$$\alpha\beta = q^2$$

Hence the equation whose roots are the squares of $x^2 - px - q$ is:

$$x^2 - (2q - p^2)x + p^2 = 0$$

3.11 Find the value of k so that the equation $4x^2 - 8x + k = 0$ shall have equal roots.

For equal roots we require:

$$b^2 = 4ac \text{ Where } a = 4, b = -8, c = k$$

Hence

$$(-8)^2 = 4(4)(k)$$
$$64 = 16k$$
$$\Rightarrow k = 4$$

3.12 If x is real, show that the expression $y = \dfrac{x^2 + x + 1}{x + 1}$ can have no real value between - 3 and 1.

Rearranging this as a quadratic in x,

$$(x+1)y = x^2 + x + 1 \qquad \text{Cross multiplying both side by } (x+1)$$

$$x^2 + x + 1 - xy - y = 0 \qquad \text{Expanding}$$

$$x^2 + (1-y)x + 1 - y = 0 \qquad \text{Rearranging}$$

For x to be real

$$(1-y)^2 \geq 4(1)(1-y) \qquad b^2 > 4ac$$

$$1 - 2y + y^2 \geq 4 - 4y \qquad \text{Expanding brackets}$$

or

$$y^2 + 2y - 3 \geq 0 \qquad \text{Rearranging}$$

Hence

$$(y+3)(y-1) \geq 0 \qquad (1) \qquad \text{Factorising}$$

If y lies between -3 and 1, $y + 3 > 0$, and $y - 1 < 0$ giving $(y-1)(y+3) < 0$ and inequality (1) is not satisfied.

Hence the given expression cannot have no real value between -3 and 1.

3.13 Prove that the roots of the equation $(p-q-r)x^2 + px + q + r = 0$ are real if p, q and r are real.

To solve this we must use the condition for real roots and then try to manipulate the LHS side into an expression squared as show below. Any expression that is squared cannot be negative.

This is a frequently used technique for proving an expression cannot be negative and should be remembered.

The condition for real roots is
$$b^2 \geq 4ac$$

Hence we require

$$p^2 \geq 4(p - q - r)(q + r)$$

$$\text{i.e } p^2 \geq 4(pq - q^2 - qr + pr - qr - r^2) \qquad \text{Expanding brackets}$$

$$\Rightarrow p^2 - 4p(q+r) + 4(q+r)^2 \geq 0 \qquad \text{Simplifying}$$

$$\text{or } \left(p - 2(q+r)\right)^2 \geq 0 \qquad \text{Factorising}$$

The LHS is the square of a real quantity and therefore cannot be negative

3.14   Solve the equation $\sqrt{3-x} - \sqrt{7+x} = \sqrt{16+2x}$

The technique here is to try to eliminate the square roots by squaring both sides of the equation. Repeated squaring may be needed to eliminate any remaining square roots.

Squaring both side of the equation we obtain

$$\left(\sqrt{3-x}-\sqrt{7+x}\right)^2 = \left(\sqrt{16+2x}\right)^2$$

$3-x+7+x-2\sqrt{(3-x)(7+x)} = 16+2x$ Expanding brackets

$-2x-6 = -2\sqrt{(3-x)(7+x)}$ Rearranging

$x+3 = \sqrt{(3-x)(7+x)}$ Dividing both side by -2

Hence

$(x+3)^2 = (3-x)(7+x)$ Squaring both sides

$x^2+6x+9 = 21-4x-x^2$ Expanding brackets

$\Rightarrow x^2+10x-12 = 0$

$\Rightarrow (x+6)(x-1) = 0$ Expanding brackets

$\Rightarrow x = -6 \text{ or } x = 1$

We have to be careful here. While $x = -6$ does satisfy the original equation, $x = 1$ does not! Check it for yourself.

It does satisfy $\sqrt{3-x}+\sqrt{7+x} = \sqrt{16+2x}$ and using the same procedure on this equation will lead to the same result.

This demonstrates that you should always check the results you obtain in the original equation.

Hence we conclude:

The correct answer is $x = -6$ as $x = 1$ does not satisfy the given equation

3.15 Solve the equation $x^2+3x-2 = \dfrac{8}{x^2+3x}$

To ease the algebra note that we can use a substitution here

Let $y = x^2+3x$ and we can now write the equation as:

$$y-2 = \frac{8}{y}$$

or

$y^2-2y-8 = 0$

Hence

$(y-4)(y+2) = 0$ Factorising

$\Rightarrow y = 4, -2$

Case1

$y = 4$

$\Rightarrow x^2 + 3x = 4$

$\Rightarrow x^2 + 3x - 4 = 0$

$\Rightarrow (x+4)(x-1) = 0$ Factorising

Giving

$x = -4, x = 1$

Case2

$y = -2$

$\Rightarrow x^2 + 3x = -2$

$\Rightarrow x^2 + 3x + 2 = 0$

$\Rightarrow (x+2)(x+1) = 0$

Giving

$x = -2, -1$

Hence the roots of the original equation are:

$-1, -2, -4$ and 1

3.16 Solve $x^{10} - 31x^5 - 31 = 0$

This is a "disguised" quadratic quadratic equation. We can solve this by substitution in a similar manner to the previous question

We want our new equation to be in a quadratic form hence we make the substitution.

Let $y = x^5$ Remember this method

Then we can write the original equation as:

$y^2 - 31y - 32 = 0$

$\Rightarrow (y - 32)(y + 1) = 0$ Factorising

$\Rightarrow y = 32, -1$

Hence

$x^5 = 32 \Rightarrow x = \sqrt[5]{32} \Rightarrow x = 2$

Or

$x^5 = -1 \Rightarrow x = -1$ Since $(-1)^5 = -1$

3.17   Solve the equation $e^{2x} + 7e^x - 8 = 0$

This is another "disguised" quadratic equation and we proceed in a similar manner to the previous question.

Let $y = e^x$

$$y^2 + 7y - 8 = 0$$
$$\Rightarrow (y+8)(y-1) = 0 \qquad \text{Factorising}$$
$$\Rightarrow y = 1 \; or - 8$$

Hence

$$e^x = 1$$

$i.e \; x \ln e = \ln 1$ \qquad Taking logs of both sides

$$\Rightarrow x = 0$$

Or

$$e^x = -8$$

$i.e \; x \ln e = \ln(-8)$ \qquad Taking logs of both sides

Clearly $\ln(-8)$ does not exist

Hence the only solution is $x = 0$

3.18 Find a, b and c such that $2x^2 - 9x + 14 \equiv a(x-1)(x-2) + b(x-1)$

Note: The symbol $\equiv$ is used to denote identity between two expressions. Hence whenever this symbol is used to separate two expressions, we can equate the coefficients of like powers of the variable. So we can proceed as follows:

$$2x^2 + 9x + 14 \equiv a(x^2 - 3x + 2) + b(x-1) + c \qquad \text{Expanding brackets}$$
$$\equiv ax^2 - (3a - b)x + 2a - b + c \qquad \text{Rearranging}$$

Equating the coefficients of x^2 and x and the term that is independent of x

We get

$$a = 2 \qquad\qquad (1)$$
$$3a - b = 9 \qquad\quad (2)$$
$$2a - b + c = 14 \qquad (3)$$

Substituting for $a = 2$ in (2)

$$3(2) - b = 9$$
$$\Rightarrow b = -3$$

Substituting for $a = 2$ and $b = -3$ in (3)

$$2(2) - (-3) + c = 14$$
$$\Rightarrow c = 7$$

3.19 Solve the equation $(3x^2 + 2x)^2 + 8 = 9(3x^2 + 2x)$

Let $y = 3x^2 + 2x$ (1)

Then

$y^2 + 8 = 9y$ Substituting for $y = 3x^2 + 2x$ in original equation

$\Rightarrow y^2 - 9y + 8 = 0$ Rearranging

$\Rightarrow (y - 8)(y + 1) = 0$ Factorising

$\Rightarrow y = 8, 1$

$Case1$

$3x^2 + 2x = 8$ Substituting for $y = 8$ in (1)

$\Rightarrow (3x - 4)(x + 2) = 0$ Factorising

$\Rightarrow x = \dfrac{4}{3}$ or $x = -2$

$Case2$

$3x^2 + 2x = 1$ Substituting for $y = 1$ in (1)

$\Rightarrow (3x - 1)(x + 1) = 0$ Factorising

$\Rightarrow x = \dfrac{1}{3}$ or $x = -1$

Hence the complete solution set is $\dfrac{4}{3}, -2, \dfrac{1}{3}, -1$

3.19.1 If the equation $ax^2 + 6abx + +ac + 8b^2 = 0$ has equal roots, prove that the   roots of the equation $ac(x + 1)^2 = 4b^2x$ are also equal.

For the roots of a quadratic equation to be equal we require:

$b^2 - 4ac = 0$ (in the standard form $ax^2 + bx + c = 0$)

We are told that the first equation has equal roots therefore:

$(6ab)^2 = 4a^2(ac + 8b^2)$

$\Rightarrow 36a^2b^2 = 4a^2(ac + 8b^2)$ Squaring LH bracket

$\Rightarrow 9b^2 = ac + 8b^2$ Dividing both sides by $4a^2$

$\Rightarrow b^2 = ac$

Rearranging the second equation we get

$$ac(x^2 + 2x + 1) = 4b^2x$$

$$\Rightarrow acx^2 + 2acx + ac - 4b^2x = 0 \qquad \text{Rearranging}$$

$$acx^2 + (2ac - 4b^2)x + ac = 0 \qquad \text{Rearranging in standard form}$$

For the roots to be real we require:

$$(2ac - 4b^2)^2 - 4(ac)(ac) = 0$$

$$(2ac - 4b^2)^2 - 4a^2c^2 = 0 \qquad \text{Simplifying}$$

$$4a^2c^2 - 16ab^2c + 16b^4 - 4a^2c^2 = 0 \qquad \text{Expanding LH bracket}$$

$$16b^4 = 16ab^2c \qquad \text{Simplifying and rearranging}$$

$$b^2 = ac \qquad \text{Dividing both sides by } b^2$$

Hence the second equation will also have equal roots

3.20 Find t from the equation $t - 1.324\sqrt{t} - 2.896 = 0$

$$t - 1.324\sqrt{t} - 2.896 = 0 \qquad (1)$$

$$\Rightarrow t - 2.896 = 1.324\sqrt{t} \qquad \text{Rearranging}$$

$$\Rightarrow (t - 2.896)^2 = \left(1.324\sqrt{t}\right)^2 \qquad \text{Squaring both sides}$$

$$\Rightarrow t^2 - 5.792t + 8.387 = 1.753t \qquad \text{Expanding brackets}$$

$$\Rightarrow t^2 - 5.792t - 1.753t + 8.387 = 0 \qquad \text{Rearranging}$$

$$\Rightarrow t^2 - 7.545t + 8.387 = 0 \qquad (2)$$

This is a quadratic equation which we need to solve by means of the formula:

$$x = \frac{-b \pm \sqrt{b^2 - 4ac}}{2a} \quad \text{where } ax^2 + bx + c = 0$$

In this case $a = 1$, $b = -7.545$ and $c = 8.387$

$$t = \frac{7.545 \pm \sqrt{(7.545)^2 - 4 \times 8.387}}{2}$$

$$= \frac{7.545 \pm \sqrt{56.927 - 33.548}}{2} \qquad \text{Expanding bracket}$$

$$= \frac{7.545 \pm \sqrt{23.379}}{2} \qquad \text{Simplifying}$$

$$= \frac{7.545 \pm 4.835}{2}$$

$$= \frac{12.38}{2} \ or \ \frac{2.71}{2}$$

$i.e \ t = 6.19 \ or \ 1.355$

Note that 6.19 is the only solution to (1) although both solution satisfy (2)
Another approach to solving this problem is as follows:

Let $z = \sqrt{t}$

Then (1) becomes $t^2 - 1.324t - 2.895 = 0$

Again we can solve this equation by using the formula as before to yield the same results.

The reason that 6.19 is the only solution can be seen from the graph below:

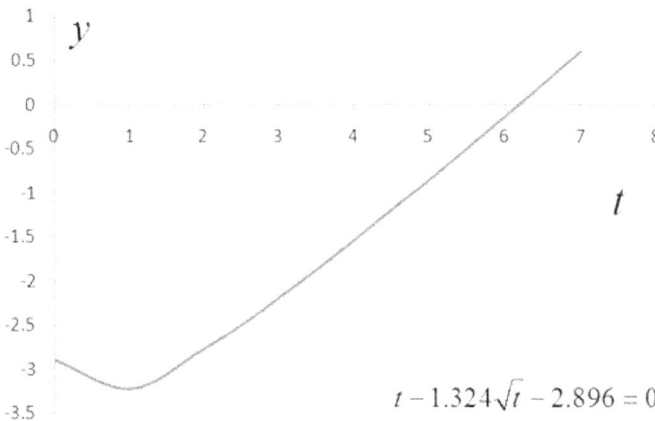

$$t - 1.324\sqrt{t} - 2.896 = 0$$

Fig 3.6

MODULE A4

ALGEBRA – INEQUALITIES

Revision Notes

Linear Inequalities

Inequalities is a topic that many people fear and find confusing but these difficulties can be overcome by adhering to a few simple rules and applying a logical approach.

First of all a recap on what the inequality signs mean:

< Less than

> Greater than

≤ Less than or equal to

≥ Greater than or equal to

There are two important rules you must know. The first one is that multiplying or dividing both sides of an inequality by -1 reverses the inequality sign. For example we know that $4 > 3$ but is we multiply or divide both sides of this by -1 we would get $-4 > -3$ which is clearly wrong so we change the inequality sign to < I.e. $-4 < -3$. So fix this in your mind now.

1. **Multiplying or dividing an inequality by -1 means you have to change the direction of the operator.**

2. **Never divide both sides by a variable** (explained later)

If you remember the first rule you can then treat the inequality in the same way as an equation. This is best illustrated with an example.

Ex 1 Solve $5x - 4 < 3x - 2$

$$5x - 4 < 3x - 2$$
$$\Rightarrow 5x - 3x < -2 + 4$$
$$\Rightarrow 2x = 2 \Rightarrow x < 1$$

Ex 2 Solve $x - 4 < 3x - 2$

$$x - 4 < 3x - 2$$
$$\Rightarrow x - 3x < -2 + 4$$
$$\Rightarrow -2x < 2 \Rightarrow -x = 1$$

Now replace the = sign with the inequality operator

$$-x < 1$$

However to make this a bit tidier we need to multiply both sides of this inequality by -1

To get $x > -1$

Noting that we have changed the inequality sign using rule No.1

Linear Inequalities are as simple as that to solve!

Now let's take a look at a graphical solution to Q1:

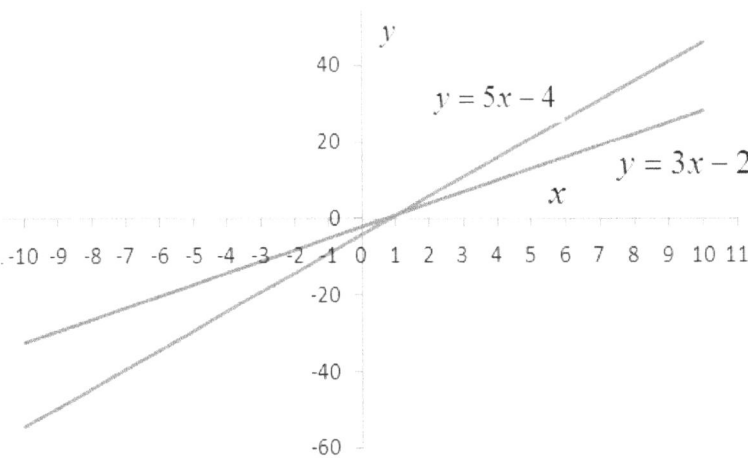

Fig 4.1

We can clearly see here that the two lines cross at $x = 1$. In the lower left quadrant and you can see that the blue line $(y = 5x - 4)$ is physically lower than the red line $(y = 3x - 2)$ when $x < 1$ indicating that $5x - 4 < 3x - 2$ for all values of x less than this.

Failure to observe rule 2 can have dire consequences as shown below

Consider the inequality

$20x < 5x^2$

First we can divide both sides by 5 to give:

$4x < x^2$

The great temptation now would be to divide both sides by x which gives

$4 < x$ but this overlooks the possibility that x could be negative or zero.

It is far safer to take $4x$ from both sides:

$$4x - 4x < x^2 - 4x$$

$$\Rightarrow 0 < x^2 - 4x$$

$$\Rightarrow 4x - x^2 > 0 \qquad \text{By rule 1 (multiplying both sides by } -1 \text{ \& reversing the inequality sign)}$$

Quadratic Inequalities

With quadratic inequalities it is advisable to solve them graphically as shown below:

Ex 3. Find the range of values of x which satisfy $2x^2 + 5x - 12 > 0$

To solve this graphically we let $y = 2x^2 - 5x - 12$ and then draw the graph:

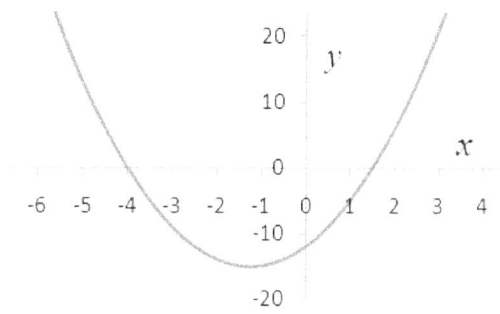

Fig 4.2 Graph of $y = 2x^2 + 5x - 12$

The graph crosses the x axis where $2x^2 + 5x - 12 = 0$

Or when $(2x - 3)(x + 4) = 0$ \qquad Factorising

$$\Rightarrow x = \frac{3}{2} \text{ or } x = -4$$

We now need to determine when this function is positive (above the x axis) and there are two

positions where this happens- at $x = -4$ and at $x = \frac{3}{2}$

Hence to satisfy the inequality $2x^2 + 5x - 12 > 0$ we require $x < -4 \text{ or } x > \frac{3}{2}$

WORKED SOLUTIONS

$3x + 4 < 2x$

$\Rightarrow 3x - 2x + 4 < 2x - 2x$

$\Rightarrow x < -4$

Hence the answer is $x < -4$

This question is similar to question 1 above and we can tackle it the same way

$7x + 13 \geq 4x - 3$

$\Rightarrow 7x - 4x \geq -3 - 13$

$\Rightarrow 3x \geq -16$

$\Rightarrow x \geq -\dfrac{16}{3}$ Note in this case the question asks for $\geq$ and not just $>$

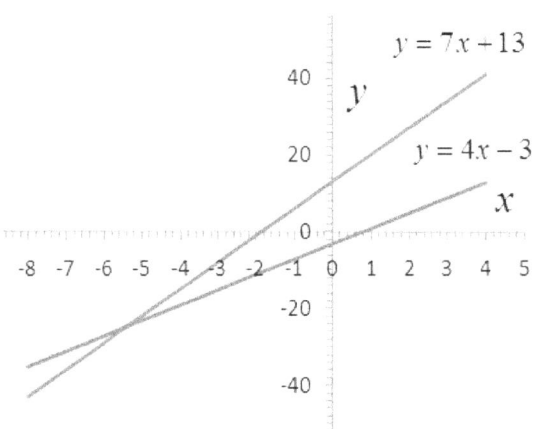

Fig 4.3 Graphs of $y = 7x + 13$ and $y = 4x - 3$

$x^2 - 4x - 5$ Can be factorised into $(x - 5)(x + 1)$

In other words both brackets must be positive or both brackets must be negative.

In this case, this to be positive x must be greater than 5 or less than 1

$x > 5 \ or \ x < 1$

If you find this difficult to see, try putting in values between these. I.e. less than 5 or greater than 1 to verify this result. However, a graphical solution always gives a picture of what is going on as shown below.

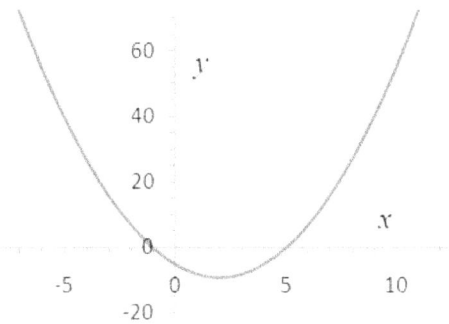

Fig 4.4 Graph of $y = x^2 - 4x - 5$

4.4 Determine for which values of x, $-x^2 + 2x + 3 \geq 0$

$-x^2 - 2x - 3 \geq 0$

$\Rightarrow x^2 + 2x + 3 \geq 0$ Multiplying throughout by -1

$\Rightarrow (x+1)(x-3) \geq 0$ Factorising

$\Rightarrow x \geq -1 \; or \; x \leq 3$

For this to be positive or zero we require $x \geq 1 \; or \; x \leq 3$
I.e When $-1 \leq x \leq 3$

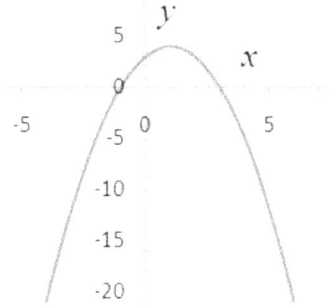

Fig 4.5 Graph of $y = -x^2 - 2x - 3$

4.5 Find the range of values that satisfy $-x^2 - 5x + 2 \geq -2$

$-x^2 - 5x + 2 \geq -2 \Rightarrow -x^2 - 5x + 4 \geq 0$ Rearranging

$$\Rightarrow -x^2 - 5x + 4 \geq 0$$
$$\Rightarrow (-x - 4)(x - 1) \geq 0$$
$$\Rightarrow x \geq -4 \ or \ x \leq 1$$

Hence $-x^2 - 5x + 2 \geq -2$ is satisfied when x is less than or equal to 1 or greater than or equal to 4
I.e when $1 \leq x \geq 4$

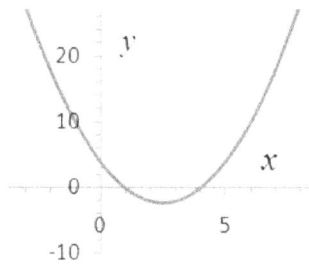

Fig 4.6 Graph of $y = -x^2 - 5x + 4$

4.6 Determine for which values of x, $\dfrac{2x^2 - 4x + 5}{x^2 + 3} \geq 2$

$$\dfrac{2x^2 - 4x + 5}{x^2 + 3} \geq 2$$

$$\Rightarrow \dfrac{2x^2 - 4x + 5}{x^2 + 3} - 2 \geq 0 \qquad \text{Subtracting 2 from each side}$$

$$\Rightarrow \dfrac{2x^2 - 4x + 5 - 2(x^2 + 3)}{x^2 + 3} \geq 0 \qquad \text{Changing expession to a common denominator}$$

$$\Rightarrow \dfrac{2x^2 - 4x + 5 - 2 - 2x^2 - 6}{x^2 + 3} \geq 0 \qquad \text{Rearranging}$$

$$\Rightarrow \dfrac{-4x - 3}{x^2 + 3} \geq 0 \qquad \text{Simplifying}$$

Since the denominator will be positive for all values of x we only require the numerator to be positive
I.e $-4x - 3 \geq 0$
Which occurs when $4x + 3 \leq 0$ \qquad Mulipying both sides by -1(Rule 1)
I.e when $4x \leq -3$
or $x \leq -\dfrac{3}{4}$

Shown below in Fig 4.7 is the graph of $y = \dfrac{-4x - 3}{x^2 + 3}$ and in Fig 4.8 the graph of $y = -4x - 3$

Both of these confirm the result $x \leq -\dfrac{3}{4}$ for the required condition to be met

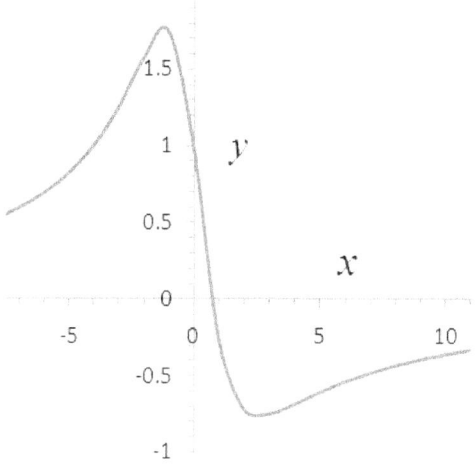

Fig 4.7 graph of $y = \dfrac{-4x - 3}{x^2 + 3}$

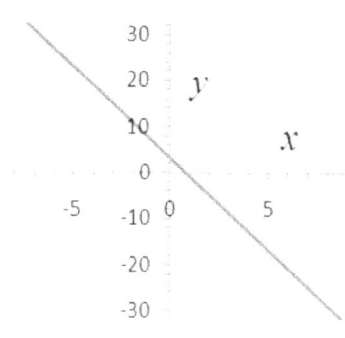

Fig 4.8 Graph of $y = -4x - 3$

4.7 Determine for which values of x, $\dfrac{x^2 - 4x + 3}{x - 2} > 0$

$$\frac{x^2 - 4x + 3}{x - 2} \Rightarrow \frac{(x - 3)(x - 1)}{x - 2} \qquad \text{Factorising}$$

If $x > 3$ then the numerator and the denominator will be positive

But if $x < 2$ the denominator will be negative if $1 < x < 3$

The numerator will be negative as well if $1 < x < 3$

Both of these can only be true if $1 < x < 2$

$\therefore$ either $1 < x < 2$ or $x > 3$

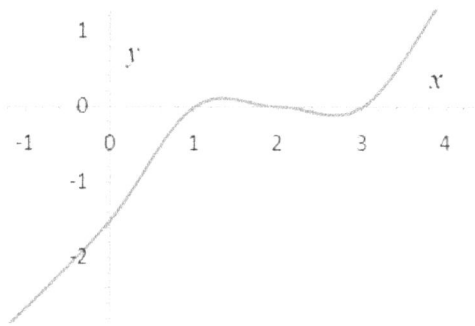

Fig 4.9 Graph of $y = \dfrac{x^2 - 4x + 3}{x - 2}$

This expression does not factorise therefore we will have to tackle it by completing the square.

First we divide through by 3 to obtain:

$$x^2 + 2x + \frac{2}{3} > 0$$

$$\Rightarrow x^2 + 2x > -\frac{2}{3} \qquad\qquad \text{Rearranging}$$

$$\Rightarrow x^2 + 2x + (1)^2 > -\frac{2}{3} + (1)^2 \qquad\qquad \text{Adding (half the coefficient of } x)^2 \text{ to both sides}$$

$$\Rightarrow (x+1)^2 > \frac{1}{3} \qquad\qquad (1) \qquad\qquad \text{Completing the square}$$

$$\Rightarrow x + 1 > \pm\sqrt{\frac{1}{3}} \qquad\qquad \text{Taking the square root of both sides}$$

$$\Rightarrow x + 1 > \pm\frac{1}{\sqrt{3}} \qquad\qquad \text{Simplifying}$$

$$\Rightarrow x > -1 \pm \frac{1}{\sqrt{3}} \qquad\qquad \text{Rearranging}$$

$$\Rightarrow x > \frac{1}{\sqrt{3}} - 1 \ or \ x < -\left(1 + \frac{1}{\sqrt{3}}\right)$$

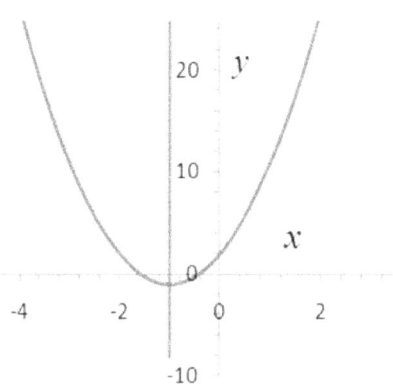

Fig 4.10 Graph of $y = 3x^2 + 6x + 2$

The minimum value of the expression can be established in one of two ways

(i) All quadratic graphs are symmetrical about a line passing through the vertex as shown in red in Fig 4.10. This clearly lies mid way between the two roots:

It follows that we can add the two roots together and divide by two to find the position of the vertex.

$$\frac{1}{2}\left[\frac{1}{\sqrt{3}}-1+\left[-\left(1+\frac{1}{\sqrt{3}}\right)\right]\right]$$

$$=\frac{1}{2}\left(\frac{1}{\sqrt{3}}-1-1-\frac{1}{\sqrt{3}}\right)$$

$$=\frac{1}{2}\times(-2)$$

$$-1$$

Hence by putting $x=-1$ in the original expression we get:

$$3(-1)^2+6(-1)+2$$

$$=3-6+2=-1$$

(ii) We can also obtain the minimum value from (1)

$$(x+1)^2>\frac{1}{3}$$

$$\Rightarrow(x+1)^2+\frac{1}{3} \qquad\qquad (2)$$

The minimum value will occur when $=-1$ \qquad\qquad When $(x+1)^2=0$

Any other value of x will give a greater value for $(x+1)^2$

Finally, we need to substitute this value in the original expression to yield the same result as before.

4.9 For what value of p does the equation $(p+1)x^2+4x+(p-2)=0$ have real roots?

For this equation to have real roots, we require that $b^2\geq 4ac$

$$\Rightarrow 4^2\geq 4(p+1)\times(p-2)$$

$$\Rightarrow 16\geq 4(p^2-p-2)\qquad\qquad\text{Mutltiplying bracketed terms together}$$

$$\Rightarrow 16\geq 4p^2-4p-8\qquad\qquad\text{Expanding the bracket and simplifying}$$

$$\Rightarrow 24-4p^2+4p\geq 0\qquad\qquad\text{Rearranging}$$

$$\Rightarrow 4p^2-4p-24\leq 0\qquad\qquad\text{Multiplying throughout by }-1\text{ and changing }\geq\text{ sign to }\leq\text{ (by rule 1)}$$

$$\Rightarrow p^2-p-6\leq 0\qquad\qquad\text{Dividing throughout by 4}$$

$$\Rightarrow (p+2)(p-3)\leq 0\qquad\qquad\text{Factorising}$$

Hence $p+2\leq 0$ *or* $p-3\leq 0$

$$\Rightarrow p\leq -2 \text{ } or \text{ } p\leq 3$$

$$Or -2\leq p\leq 3$$

$(x-1)^2 > 4x^2$

$\Rightarrow x^2 - 2x + 1 > 4x^2$ Expanding bracketed term

$\Rightarrow x^2 - 2x + 1 - 4x^2 > 0$ Rearranging

$\Rightarrow -3x^2 - 2x + 1 > 0$ Simplifying   (1)

$\Rightarrow 3x^2 + 2x - 1 < 0$ Multiplying throughout by -1 and changing $>$ sign to $<$ (by rule 1)

$\Rightarrow (3x-1)(x+1) < 0$ Factorising

$\Rightarrow x < \dfrac{1}{3}$ or $x > -1$

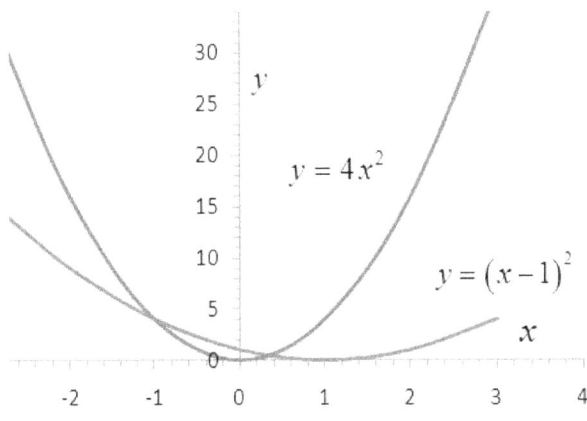

Fig 4.11 Graph of $y = (x-1)^2$ and $y = 4x^2$

Note that we could also solve the problem graphically by plotting a graph of (1) shown below:

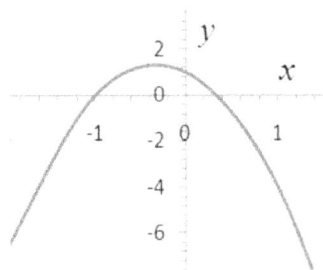

Fig 4.12 graph of $y = -3x^2 - 2x + 1 > 0$

Which yields the same result.

$(x-1)(x-3) < x(3-x)$

$\Rightarrow x^2 - 4x + 3 < 3x - x^2$ Expanding brackets

$\Rightarrow 2x^2 - 7x + 3 < 0$ Rearranging

$\Rightarrow (2x-1)(x-3) < 0$ Factorising

$\Rightarrow (2x-1) < 0 \Rightarrow x > \dfrac{1}{2}$

$or\ (x-3) < 0 \Rightarrow x < 3$

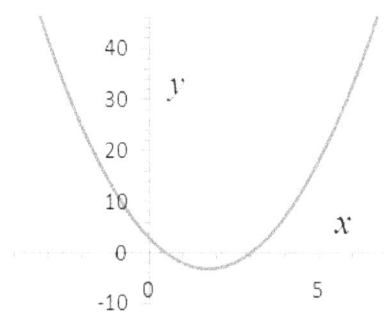

Fig 4.13 Graph of $2x^2 - 7x + 3$

There are two sets of inequalities this time so we will deal each of them separately

(i) $2x - 1 < x^2 - 4$

$\Rightarrow -x^2 + 2x + 3 < 0$ Rearranging

$\Rightarrow x^2 - 2x - 3 > 0$ Multiplying throughout by -1 and changing $<$ sign to $>$ (by rule 1)

$\Rightarrow (x-4)(x+1) > 0$ Factorising

Now let $f(x) = (x-3)(x+1)$ (1)

(ii) $x^2 - 4 < 12$

$\Rightarrow x^2 - 16 < 0$ Rearranging

$\Rightarrow (x+4)(x-4) < 0$ Factorising

Now let $g(x) = (x+4)(x-4)$ (2)

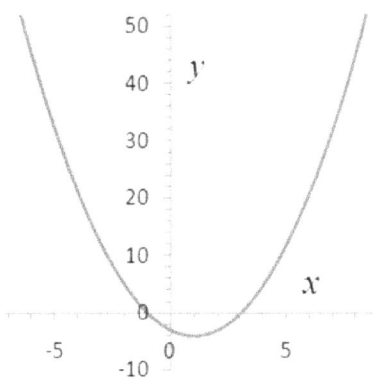

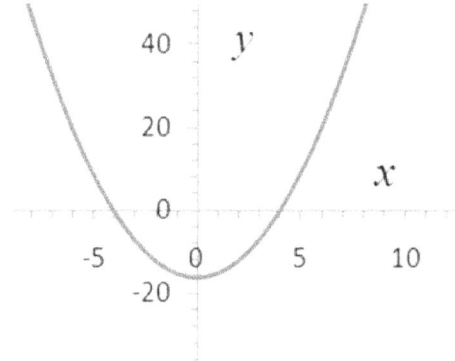

Fig 4.14 Graph for $f(x) > 0$ Fig 4.15 Graph for $g(x) < 0$

$f(x)$ shows that $x < -1$ and $x > 3$ $g(x)$ shows that $x > -4$ and $x < 4$

Putting these results on a number line gives a clearer picture of what is happening:

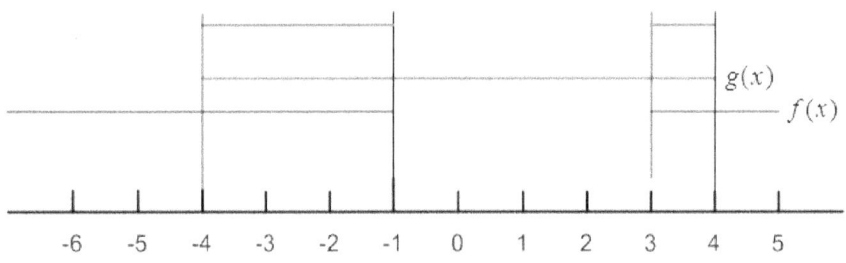

Fig 4.16 Solution sets of $f(x)$ and $g(x)$

The blue line represents the solution set of $f(x)$ and the green line represents the solution set of $g(x)$. Where these solution sets overlap are the values of x that satisfy (i) and (ii) and are shown in red.

This occurs where $-4 < x < -1$ and $3 < x < 4$

4.13 Find the range of values of p for which the equation $x^2 - px + (p+3) = 0$ has real roots

For this equation to have real roots we require:

$p^2 \geq 4 \times 1 \times (p+3)$ Using the condition that $b^2 \geq 4ac$

$\Rightarrow p^2 \geq 4p + 12$ Rearranging

$\Rightarrow p^2 - 4p - 12 \geq 0$ Rearranging

$\Rightarrow (p-6)(p+2) \geq 0$ Factorising

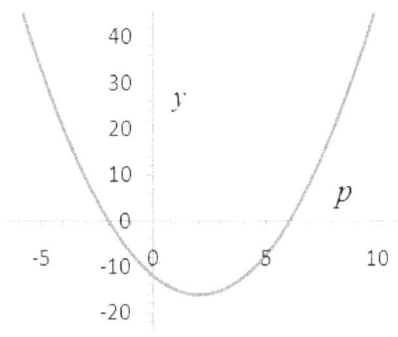

Fig 4.17 Graph of $y = p^2 - 4p - 12$

From which we can see the required condition is $x \leq -2$ and $x \geq 6$

4.14 (i) Find the set of values of p for which $f(x) \equiv 3x^2 - 4x - p$ is greater than 1 for all real values of x

(ii) Show that for all p, the minimum value of $f(x)$ occurs when $x = \frac{2}{3}$

(i) We require that $3x^2 - 4x - p > 1$

Or $3x^2 - 4x - p - 1 > 0$ (1)

Hence we require that:

$16 > 4 \times 3 \times (-p - 1)$ Using the condition that $b^2 > 4ac$

$\Rightarrow 16 > -12p - 12$ Expanding bracket and simplifying

$\Rightarrow 12p + 28 > 0$ Simplifying

$\Rightarrow p > -\frac{7}{3}$

(ii) Rearranging Eq (1) above we obtain:

$x^2 - \frac{4}{3}x - p$ Dividing throughout by 3 to make the cofficient of x^2, 1

$\Rightarrow \left(x - \frac{2}{3}\right)^2 - \left(\frac{2}{3}\right)^2 - p$ Rearranging by completing the square

Since $\left(x - \frac{2}{3}\right)^2$ cannot be negative the minimum value occurs when $x = \frac{2}{3}$. I.e. when $\left(x - \frac{2}{3}\right)^2 = 0$

4.15 If $a > 0$, prove that the quadratic expression $ax^2 + bx + c$ is positive for all real values of x providing $b^2 < 4ac$. Hence find the range of values of p for which the function $f(x) = 4x^2 + 4px - (3p^2 + 4p - 3)$ is positive for all values of x.

Dividing $ax^2 + bx + c$ throughout by a we get $x^2 + \frac{b}{a}x + \frac{c}{a}$ (1)

Now substituting for $a = 1$, $b = \frac{b}{a}$ and $c = \frac{c}{a}$ and completing the square we get:

$$\left(x+\frac{b}{2a}\right)^2-\left(\frac{b}{2a}\right)^2+\frac{c}{a}$$

Note $\left(x+\frac{b}{2a}\right)^2=x^2+\frac{b}{a}+\left(\frac{b}{2a}\right)^2$ so we have to subtract $\left(\frac{b}{2a}\right)^2$

to maintain the integrity of Eq 1

$$\left(x+\frac{b}{2a}\right)^2-\frac{b^2}{4a^2}+\frac{c}{a}$$ Rearranging

$$=\left(x+\frac{b}{2a}\right)^2-\frac{\left(b^2-4ac\right)}{4a^2}$$ Rearranging

Now if $b^2<4ac$ then $\dfrac{\left(b^2-4ac\right)}{4a^2}$ must be negative since the denominator is positive.

Hence: $-\dfrac{\left(b^2-4ac\right)}{4a^2}$ must be positive and therefore $\left(x+\dfrac{b}{2a}\right)^2-\dfrac{\left(b^2-4ac\right)}{4a^2}$ must also be positive

for all values of x.

For part (ii) of the question we use the given condition that $b^2<4ac$.

In this case $a=4$, $b=4p$ and $c=-(3p^2+4p-3)$

$$16p^2<4\times4\times-(3p^2+4p-3)$$

$$\Rightarrow16p^2<-16(3p^2+4p-3)$$ Simplifying

$$\Rightarrow16p^2<-48p^2-64p+48$$ Expanding bracket

$$\Rightarrow16p^2+48p^2+64p-49<0$$ Rearranging

$$\Rightarrow64p^2+64p-48<0$$ Simplifying

$$\Rightarrow4p^2+4p-3<0$$ Dividing throughout by 16

$$\Rightarrow(2p+3)(2p-1)<0$$ Factorising

Now let $y=(2p+3)(2p-1)$

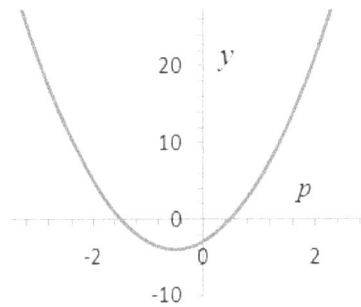

Fig 4.18 Graph of $(2p+3)(2p-1)$

The graph illustrates that the required range for $y<0$ is $-\dfrac{3}{2}<x<\dfrac{1}{2}$

$(x-1)(x+2) < x(4-x)$

$\Rightarrow x^2 + x - 2 < 4x - x^2$ Expanding brackets

$\Rightarrow 2x^2 - 3x - 2 < 0$ Rearranging & simplifying

$(2x+1)(x-2) < 0$

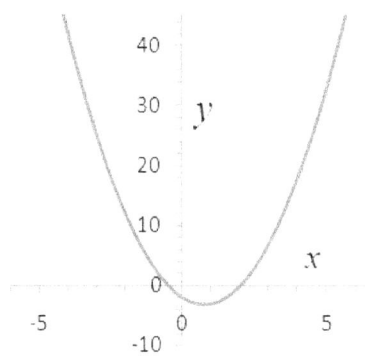

Fig 4.19 Graph of $y = 2x^2 - 3x - 2$

The graph show the solution is $x > -\dfrac{1}{2}$ and $x < 2$ or $-\dfrac{1}{2} < x < 2$

For $x^2 - kx + (k+3) = 0$ to have real roots we require that $b^2 \geq 4ac$

$k^2 \geq 4 \times 1 \times (k+3)$

I.e. that $\Rightarrow k^2 \geq 4k + 12$ Simplifying

    $\Rightarrow k^2 - 4k - 12 \geq 0$ Rearranging

    $\Rightarrow (k+2)(k-6) \geq 0$ Factorising

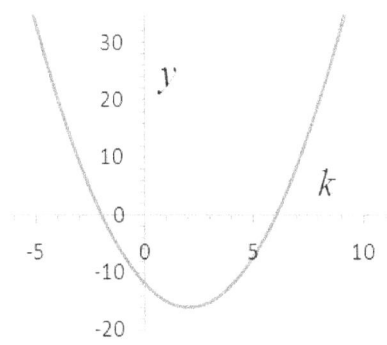

Fig 4.20 Graph of $y = k^2 - 4k - 12$

This shows the solution to be $k \le -2$ and $k \ge 6$

4.18 If x is real and $x^2 + (2-k)x + 1 - 2k = 0$ show that k cannot lie between certain limits and find these limits.

Here we require that $b^2 = 4ac$

$$(2-k)^2 = 4 \times 1 \times (1 - 2k)$$

$\Rightarrow 4 - 4k + k^2 = 4 - 8k$ \qquad Expanding bracket

$\Rightarrow k^2 + 4k = 0$ \qquad Simplifying

$\Rightarrow k(k-4) = 0$ \qquad Factorising

Hence either $k = 0$ or $k = 4$

So for x to be real, k cannot lie between 0 and -4

I.e. the limits are $-4 < k < 0$ as shown in Fig 4.21 below.

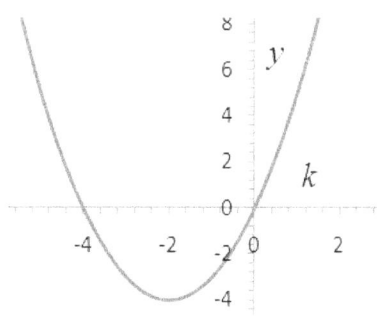

Fig 4.21 Graph of $y = k^2 + 4k$

4.19 Find the values of k so that the expression $2x^2 + 6x + 1 + k(x^2 + 2)$ will be positive for all values of x.

First we need to rearrange this expression in the form $ax^2 + bx + c$

$$2x^2 + 6x + 1 + k(x^2 + 2) = 2x^2 + 6x + 1 + kx^2 + 2k$$ \qquad Expanding bracket

$$= 2x^2 + kx^2 + 6x + 2k + 1$$ \qquad Rearranging

$$= (2+k)x^2 + 6x + 2k + 1$$ \qquad Factorsising x^2 terms

Where: $a = 2 + k$, $b = 6$ and $c = 2k + 1$

In order that this expression is positive we require $2x^2 + 6x + 1 + k(x^2 + 2) > 0$

First we will divide this expression throughout by $a(2+k)$

$$x^2 + \frac{6}{2+k} + \frac{2k+1}{2+k}$$

Since the coefficient of x is positive the graph must be $\bigcup$ shaped and therefore must have a least value at the bottom of the $\bigcup$

If we now complete the square we obtain:

$$x^2 + \left(\frac{3}{2+k}\right)^2 + \frac{2k+1}{2+k} - \left(\frac{3}{2+k}\right)^2 \qquad \text{Remember we have added}\left(\frac{3}{2+k}\right)^2 \text{ so we have to subtract}$$

it again to maintain the integrity of the expression.

Now we can continue to complete the square

$$\left(x + \frac{3}{2+k}\right)^2 + \frac{2k+1}{2+k} - \left(\frac{3}{2+k}\right)^2$$

Now we know that $\left(x + \frac{3}{2+k}\right)^2$ cannot be less than zero so the least value is

$$\frac{2k+1}{2+k} - \left(\frac{3}{2+k}\right)^2 \text{ and this occurs when it is greater than zero. I.e. when}$$

$$\frac{2k+1}{2+k} - \left(\frac{3}{2+k}\right)^2 > 0$$

$$\Rightarrow \frac{2k+1}{2+k} - \frac{9}{(2+k)^2} > 0 \qquad\qquad \text{Expanding bracket}$$

$$\Rightarrow \frac{(2k+1)(2+k) - 9}{(2+k)^2} > 0 \qquad\qquad \text{Rearranging with common denominator of } (2+k)^2$$

$$\Rightarrow \frac{2k^2 + 5k - 7}{(2+k)^2} > 0 \qquad\qquad \text{Expanding brackets}$$

$$\Rightarrow \frac{(2k+7)(k-1)}{(2+k)^2} > 0 \qquad\qquad \text{Factorising numerator}$$

Since the denominator cannot be negative we only require:

$$(2k+7)(k-1) > 0$$

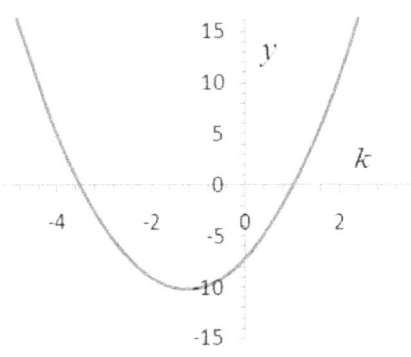

Fig 4.22 Graph of $y = (2k + 7)(k - 1)$

This shows the required condition to be $k > 1$ and $k < -\dfrac{7}{2}$

4.20 Find the range of values of the real number k that in order that the equation $x^2 + 2kx - 2k + 3$ will have real roots. Sketch the resulting graph.

In order that this equation will have real root we require that $b^2 \geq 4ac$

Where $a = 1$, $b = 2k$ and $c = -2k + 3$

$4k^2 \geq 4 \times 1 \times (-2k + 3)$

$\Rightarrow 4k^2 \geq -8k + 12$ Expanding bracket

$\Rightarrow 4k^2 + 8k - 12 \geq 0$ Rearranging

$\Rightarrow k^2 + 2k - 3 \geq 0$ Dividing throughout by 4

$\Rightarrow (k + 3)(k - 1) \geq 0$ Factorising

$\Rightarrow x \geq -3$ and $x \geq 1$

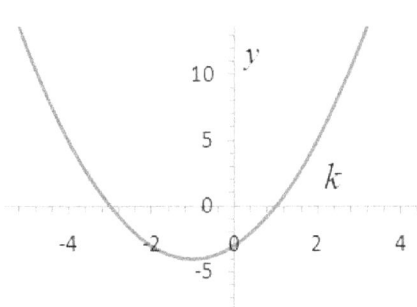

Fig 4.23 Graph of $y = k^2 + 2k - 3$

MODULE SR1

SERIES 1

INTRODUCTION

A set of numbers, each of which can be obtained from a definite rule is called a **sequence** or a **progression.** The sets:

$$(i) \ 1, 3, 5, 7, ...,$$

$$(ii) \ 1, 2, 4, 8, ...,$$

$$(iii) \ 1^2, 2^2, 3^2, 4^2, ...,$$

All are **sequences**. In the first set each number is obtained by adding 2 to the preceding one. In the second set each term is twice the preceding term. Each term in the third set is the square of successive integers.

We can deduce a formula to find the general or n^{th} term for each of the above series:

In (i) the formula would be $2n - 1$ .e.g. $2 \times 3 - 1 = 5$ which gives the third term

In (ii) it would be 2^{n-1} e.g. $2^{4-1} = 2^3 = 8$ which gives the fourth term

In (iii) it would be n^2 e.g. the fourth term would be $4^2 = 16$

If the formula for the nth term is known it is possible to write down further terms by assigning successive integral values to n hence the series whose n^{th} term is $3n - 1$ is one whose terms are:

$2, 5, 8, 11, ...,$ This is achieved by putting $n = 1, 2, 3$ and 4 in this formula.

Series play a very important part in Mathematical Analysis and in Number Theory but here we will restrict our discussions to some of the simpler ones.

Arithmetic Progressions

A **sequence** where each term is obtained by adding or subtracting a constant quantity from a preceding term is called an **Arithmetic Progression (AP).** Hence the series:

$1, 3, 5, 7, ...,$ Can be written as $a, a + d, a + 2d, a + 3d, ...,$ where a is the first term and d is the **common difference** between successive terms.

When three terms are in arithmetic progression the middle term is called the **arithmetic mean** of the other two and hence a is the arithmetic mean between $a - d$ and $a + d$.

The coefficient of d is always one (1) less than the number of the term in the series. Thus:

$a + 2d$ Is the third term, $a + 3d$ is the fourth term and $a + 5d$ is the sixth term etc.

If the sequence consists of n terms and we define the nth or last term by l then

$$l = a + (n - 1)d \qquad (1.1)$$

We also frequently need to find the sum S_n of a **series** and we consider the sequence:
$a, a+d, a+2d, a+3d,...,$ its sum is given by:

$$S_n = a + (a+d) + (a+2d) + (a+3d) + ... + (l-2d) + (l-d) + l \qquad (1.2)$$

Since if l is the last term, the previous term to it will be $l-d$ and the term before that will be $l-2d$ and so on. Summing the terms like this is called a **series.**

If we now write this series in reverse order and add the two together, we get:

$$S_n = a + (a+d) + (a+2d) + (a+3d) + ... + (l-2d) + (l-d) + l \qquad (1.3)$$

$$S_n = l + (l-d) + (l-2d) + ... + (a+3d) + (a+2d) + (a+d) + a \qquad (1.4)$$

We can see that the corresponding terms reduce to $(a+l)$ in each case and there are n of these, hence we are left with:

$$2S_n = n(a+l) \qquad (1.5)$$

This usually written as:

$$S_n = \frac{n}{2}(a+l) \qquad (1.6)$$

Or using (1.1) above:

$$S_n = \frac{n}{2}\left(2a + (n-1)d\right) \qquad (1.7)$$

(1.1), (1.5), (1.6) and (1.7) are important results and should be committed to memory as should the proof shown above. They are listed again here for convenience:

$$\text{The } n^{th} \text{ or last term } l = a + (n-1)d \qquad (1.8)$$

$$S_n = \frac{n}{2}(a+l) \qquad (1,9)$$

$$S_n = \frac{n}{2}\left(2a + (n-1)d\right) \qquad (1.10)$$

You should be aware that there is another notation for summing a series given by the Greek letter $\sum$ (Sigma) which you are certain to come across at some time so that: $\sum_{n=1}^{10}(4n-1)$ means sum the series $4n-1$ starting with $n=1$ up to and including the 10th term $(n=10)$

WORKED SOLUTIONS

1.1. Find the twelfth term of the series whose n^{th} terms is given by $(3n-2)$

This is the series 1, 4, 7,

To find the 12th term we simply put $n = 12$ in $(3n-2)$

To get $3 \times 12 - 2 = 34$

1.2. Find the 10th term of the series $2, 5, 8, 11, \ldots$, and find the formula for the nth term

In this case we have $a = 2$, $n = 10$ and $d = 3$

Using (1.1) above we have $a + (n-1)d$ so:

The tenth term is $2 + (10-1) \times 3 = 2 + 9 \times 3 = 29$ the last term l in this case is 10, the 10th term

The formula for the nth term is also obtained from (1.1)

$2 + (n-1) \times 3 = 2 + 3n - 3 = 3n - 1$ This time we are looking for the nth term so

1.3. Find the last term of the sequence that starts with 5, has a common difference of 2 and has 15 terms

This is the sequence 5, 7, 9,

We are given $a = 5$, $d = 2$ and $n = 15$

The nth term is given by $a + (n-1)d$

So the fifteenth term is given by $5 + (15-1) \times 2 = 5 + 14 \times 2 = 33$

1.4. Find the nth term of the sequence 15, 18, 21.

We are given $a = 15$

The nth term is given by (1)

$15 + (n-1) \times 3$
$= 15 + 3n - 3$
$= 3n + 12$

1.5. A series has the 8th term 16 and 16th term 40. Find the sum of the first 6 terms and the nth term.

Using (1.1) we have

$$a+(8-1)d = 16 \qquad\qquad \text{Value of the } 8^{th}\text{ term}$$
$$a+7d = 16 \qquad\qquad (1.12)$$
$$\text{and } a+(16-1)d = 40 \qquad\qquad \text{Value of the } 16^{th}\text{ term}$$
$$a+15d = 40 \qquad\qquad (1.13)$$
$$\Rightarrow 8d = 24 \qquad\qquad \text{Subtracting } (1.12) \text{ from } (1.13)$$
$$\Rightarrow d = 3$$
$$2a+22d = 56 \qquad\qquad \text{Adding } (1.12) \text{ and } (1.13)$$
$$2a+22\times 3 = 56 \qquad\qquad \text{Substituting for } d = 3$$
$$\Rightarrow 2a = \frac{-10}{2}$$
$$\Rightarrow a = -5$$
$$\text{Then } S_6 = \frac{6}{2}\times + \{2\times(-5)+(6-1)\times 3\}$$
$$= 3(-10+15)$$
$$= 15$$

The nth term is given by $-5+(n-1)\times 3 = 3n-8$ \qquad Using (1.8)

1.6. Find the common difference and the nth term in a sequence that starts with -3, ends with 43 and has 24 terms.

Using (5.1) we get:

$$-3+(24-1)d = 43 \qquad\qquad \text{Using (1.8) with } a = -3$$
$$\Rightarrow -3+23d = 46$$
$$\Rightarrow d = 2$$

The nth term is given by $-3+(n-1)\times 2 = 2n-5$ \qquad Using (1.8) with $a = -3$ and $d = 2$

1.7. Find $\displaystyle\sum_{n=1}^{15} 3n - 2$

As explained previously $\displaystyle\sum_{n=1}^{15} 3n - 2$ means sum the series whose n^{th} term is given by $3n-2$. So we are being asked to sum the series from $n-1$ to $n = 15$.

Now the first term is obtained by putting $n = 1$ in $3n-2$. I.e. $3\times 1 - 2 = 1$

The second term is obtained in the same way by putting $n = 2$ in $3n-2$ I.e. $3\times 2 - 2 = 4$

Therefore the common difference d is the difference between these two terms I.e. 3

Now we know $a = 1$ and $d = 2$ we can use Eq. (5.6) $S_n = \dfrac{n}{2}\left(2a + (n-1)d\right)$ to find the required sum.

$$S_{15} = \dfrac{15}{2}\left\{(2 \times 1) - (15 - 1) \times 3\right\}$$
$$= \dfrac{15}{2}(2 + 14 \times 3) = 44 \times \dfrac{15}{2} = 330 \qquad \text{Simplifying}$$

Alternatively we could also find the 15th (last) term by putting $n = 15$ in $3n - 2$ to obtain 43 and then use Eq. (5.5) $S_n = \dfrac{n}{2}(a + l)$

Substituting for $a = 1$, $d = 3$ and $n = 15$

We get $S_{15} = \dfrac{15}{2}(1 + 43) = \dfrac{15}{2} \times 44 = 330$ as before

Arithmetic Mean

When three terms are in arithmetic progression the middle one is called the ***arithmetic mean*** of the other two.

Hence a is the arithmetic mean between $a - d$ and $a + d$

1.8 . Find three numbers in arithmetic progression such that their sum is 27 and their product is 504

Let the three numbers be $a - d$, a and $a + d$.

Then the sum of the 3 numbers is $3a$ which is equal to $27 \Rightarrow a = 9$

The product of the three terms is $(a - d)a(a + d) = a\left(a^2 - d^2\right) = 504$ Remember this trick

$$9\left(9^2 - d^2\right) = 504$$
$$\Rightarrow 81 - d^2 = \dfrac{504}{9} = 56$$

Now since we have $a = 9$ we can say that $\Rightarrow d^2 = 81 - 56$
$$= 25$$
$$\Rightarrow d = \pm\sqrt{25}$$
$$= \pm 5$$

The three numbers must therefore be 9 and ± 5. if this progression is $a - d$, a and $a + d$

Then the numbers are 4, 9 and 14

1.9 Insert 3 arithmetic means between 8 and 18

Let the 3 arithmetic means be P, Q and R

Then $8 + P + Q + R + 18$ forms an arithmetic progression

Then we have the first term $a = 8$ (1.14)

and the fifth term $a + 4d = 18$ (1.15)

Subtracting (1.14) from (1.15) we obtain $4d = 10$

$\Rightarrow d = 2.5$

Hence the arithmetic means are 10.5,13,15.5

1.10 Insert 4 arithmetic means between 2 and 12

Let the four arithmetic means be P, Q, R and S

The we have the first term $a = 2$ (1.16)

and the sixth term $a + 5d = 12$ (1.17)

Subtracting (1.16) from (1.17) we obtain $5d = 10$

$\Rightarrow d = 2$

Hence the arithmetic means are: 4,6,8,10

Geometric Progressions

An example of a geometric progression (GP) is the series $+ 4 + 8 + 16 + 32 + ...$

Each term can be obtained by multiplying the previous term by a constant factor of 2. This factor is called the **common ratio** and denoted by **r** and is found by dividing any term by the preceding one.

$e.g\ 8 \div 4 = 2;\ 32 \div 16 = 2; etc$

A GP therefore has the form : $a + ar + ar^2 + ar^3...$

Where a is the first term and r the common ratio.

The index or the power of r is one less than the number of the terms in the series.

Thus ar^2 is the third term and ar^5 is the sixth term. Hence the last or n^{th} term of the series is given by $l = ar^{n-1}$.   (1.18)

To obtain the sum (S_n) of a GP to n terms we have:

$$S_n = a + ar + ar^2 + ar^3 + \dots + ar^{n-2} + ar^{n-1} \qquad (1.19)$$

Multiplying (1.18) throughout by r gives

$$rS_n = ar + ar^2 + ar^3 + \dots + ar^{n-1} + ar^n \qquad (1.20)$$

Subtracting (1.20) from (1.19), all the terms on the right hand side of (1.19) apart from a and ar^n, cancel out in pairs yielding the result:

$$S_n - rS_n = a - ar^n$$
$$\Rightarrow \quad S_n(1-r) = a(1-r^n)$$
$$\Rightarrow S_n = \frac{a(1-r^n)}{1-r} \qquad (1.21)$$

WORKED SOLUTIONS

1.11 Find the find the number of terms in the series whose first term is 2, has a common ratio of 3 and whose last term is 486, and find the sum of the series.

Using $l = ar^{n-1}$ and substituting $a = 2$ and $l = 486$ we get:

$$486 = 2 \times 3^{n-1}$$
$$\Rightarrow 243 = 3^{n-1}$$
$$\Rightarrow (n-1)\ln 3 = \ln 243 \qquad \text{Taking logs of both sides}$$
$$\Rightarrow n-1 = \frac{\ln 243}{\ln 3} = 5 \qquad \text{Rearranging}$$
$$\Rightarrow n = 6$$

The sum of the series is given by $S_n = \dfrac{a(1-r^n)}{1-r}$

Hence $S_n = \dfrac{2(1-3^6)}{1-3}$

$$= \frac{2(1-729)}{-2}$$
$$= \frac{2 \times (-728)}{-2}$$
$$= 728$$

1.12 Find the sum of the first ten terms of the geometrical progression
3, -6, 12....

In this case, the first term a is 3 and the common ratio r is -2

Hence $S_{10} = \dfrac{3\left(1-(-2)^{10}\right)}{1-(-2)}$

$= \dfrac{3\left(1-1,024\right)}{3}$

$= -1,023$

1.13 The third term of a GP is 18 and the sixth is 486. Find the first term and establish a formula for the sum to n terms

The first six terms of the series are : $a + ar + ar^2 + ar^3 + ar^4 + ar^5$

Then we have $ar^2 = 18$ (1.22)

And $ar^5 = 486$

Then $\dfrac{ar^5}{ar^2} = \dfrac{486}{18}$

Giving $r^3 = 27$

$\Rightarrow r = 3$

$a \times 3^2 = 18$ Substituting for $r = 3$ in (1.22)

$\Rightarrow a = \dfrac{18}{9} = 2$

$S_n = a\left(\dfrac{1-r^n}{1-r}\right) = 2 \times \left(\dfrac{1-3^n}{1-3}\right) = 2 \times \left(\dfrac{1-3^n}{-2}\right) = 3^n - 1$ Substituting for $r = 3$ and $a = 2$ in (1.21)

1.14. Find three numbers in an ascending geometric progression such that their sum is 39 and their product is 729

Let the three numbers be $\dfrac{a}{r}$, a, and ar (1.23)   Remember this method

Then their product is $\dfrac{a}{r} \times a \times ar = a^3$ (1.24)

Hence $a^3 = 729$

Giving $a = 9$ Taking the cube root of 729

Since the sum of these numbers is 39 we have:

$\dfrac{9}{r} + 9 + 9 \times r = 39$ Substituting for $a = 9$ in (1.23) & adding

$\Rightarrow 9r^2 + 9r - 39r + 9 = 0$ Rearranging

$\Rightarrow 9r^2 - 30r + 9 = 0$ Rearranging

$\Rightarrow 3r^2 - 10r + 3 = 0$ Dividing throuout by 3

$\Rightarrow (3r - 1)(r - 3) = 0$ Factorising

$\Rightarrow r = 3 \ or \ \dfrac{1}{3}$

The three numbers required are therefore $\dfrac{9}{3}$, 9 and 9×3 *or* 3, 9 and 27

Note that the answer of $\dfrac{1}{3}$ whilst valid has been discounted as the question specifies an ascending series.

1.15. The second and third terms of a GP are 24 and $12(b+1)$ respectively. Determine the value of b if the sum of the first three terms of the progression is 76.

Since the second and third terms are 24 and $12(b+1)$,

we can determine the value of r as being $\dfrac{12(b+1)}{24} = \dfrac{b+1}{2}$ Dividing 3rd term by 2nd term

Hence the first term a will be $\dfrac{24}{r} = \dfrac{24}{\dfrac{b+1}{2}} = \dfrac{48}{b+1}$ Simplifying

It would be easy to jump in here straight away and use the result:

$S_n = \dfrac{a(1 - r^n)}{1 - r}$ and substituting for a and r we would get:

$S_3 = \dfrac{48}{b+1}\left(\dfrac{1 - \left(\dfrac{b+1}{2}\right)^3}{1 - \dfrac{b+1}{2}} \right) = 76$ which is starting to look rather heavy to solve.

However, if we think about it for a moment, we have already found the first three terms which we can add togther directly as follows:

We are given that the sum of the first three terms is 76

Hence $S_3 = \dfrac{48}{b+1} + 24 + 12(b+1) = 76$

$\Rightarrow \dfrac{48}{b+1} + 24 + 12(b+1) - 76 = 0$ Rearranging

$\Rightarrow \dfrac{48 + 24b + 24 + 12(b^2 + 2b + 1) - 76(b+1)}{b+1} = 0$ Using common denominator

$\Rightarrow \dfrac{48 + 24b + 24 + 12b^2 + 24b + 12 - 76b - 76}{b+1} = 0$ Expanding brackets

$\Rightarrow \dfrac{12b^2 - 28b + 8}{b+1} = 0$ Simplifying

$\Rightarrow \dfrac{4\left(3b^2 - 7b + 2\right)}{b+1} = 0$ Rearranging

$\Rightarrow \dfrac{4(3b-1)(b-2)}{b+1} = 0$ Factorising

$\Rightarrow b = \dfrac{1}{3}, 2$

1.16. If the first, second and fifth terms of an AP, are three consecutive terms of a GP, find the common ratio.

Let the first five terms of the AP be:

$a, a+d, a+2d, a+3d, a+4d$

Then the common ratio of the GP can be obtained as follows:

$r = \dfrac{a+d}{a}$ Ratio of the second and first terms of the AP

$r = \dfrac{a+4d}{a+d}$ Ratio of the fifth and second terms of the AP

Hence $\dfrac{a+d}{a} = \dfrac{a+4d}{a+d}$

$\Rightarrow (a+d)^2 = a(a+4d)$ Rearranging

$\Rightarrow a^2 + 2ad + d^2 = a^2 + 4ad$ Expanding brackets

$\Rightarrow 2ad + d^2 = 4ad$

$\Rightarrow d^2 = 2ad$

$\Rightarrow d = 2a$

Substituting for $d = 2a$ in the first, second and fifth terms of the AP we get:

$a, a+2a, a+8a$

or $a, 3a, 9a$

We can now see that this result gives a GP with a common ratio of 3

1.17. The sum of the last three terms of a GP having n terms is 1024 times the sum of the first three terms. If the third term is 5 find the last term of the progression.

Let the first term be a and the common ratio r

Then the first three terms will be a, ar, ar^2

Then $S_3 = \dfrac{a(1-r^2)}{1-r} = a + ar + ar^2$

$\Rightarrow \dfrac{1-r^2}{1-r} = 1 + r + r^2$ Dividing throughout by a

Hence $1 - r^2 = (1-r)(1 + r + r^2)$ Multiplying throughout by $1 - r$

$\Rightarrow 1 - r^2 = 1 + r + r^2 - r - r^2 - r^3$ Expanding brackets

$\Rightarrow -2r^2 - r^3 = 0$ Simplifying

$\Rightarrow -r^2(2 - r) = 0$ Factorising

$\Rightarrow r = 2$

Hence $ar^2 = 5$ Since we are given that the third term is 5

$\Rightarrow 4a = 5$ Since $r = 2$

$\Rightarrow a = 1.25$

Hence the series is 1.25, 2.5, 5, ...

And the sum of the first three terms is $1.25 + 2.5 + 5 = 8.75$

We are also told that the sum of the last three terms is $1024 \times$ the sum of the first three terms

Therefore the sum of the last three terms must be $1,024 \times 8.75 = 8,960$

Now let the first term of the the last three terms be $\dfrac{p}{r}$

Then the last three terms will be $\dfrac{p}{r}$, $p + pr$ Remember this technique from (1.23)

And their sum will be $\dfrac{p}{2} + p + 2p$ Since $r = 2$

Then $\dfrac{7}{2}p = 8,960$

Hence $p = 2,560$

Therefore the first of these terms will be $\dfrac{p}{2} = \dfrac{2,560}{2} = 1,280$

And the second term will be $p = 2,560$

And the last term will be $2p = 5,120$

Note that we have treated the last three terms of the progression as a 'new progression' with a new value of $\dfrac{p}{2}$ for the first term. This is similar to the 'trick' mentioned in Q1.13.

Note also that if the sum of the last three terms is $1,024 \times$ the sum of the first three terms, then the last term will be $1,024 \times$ the third term. i.e. $1,024 \times 5 = 5,120$. So this question could have been answered in about 10 seconds!

1.18. A nail, 50 mm long, is driven into a piece of wood by successive blows of a hammer. The first blow drives the nail in 30.5 mm and each successive blow drives the nail in $\frac{2}{5}$ of the previous distance except the last blow which is less. Determine how many blows are needed to completely drive the nail in to the wood.

Here we are clearly dealing with a GP with a common ratio (r) of $\frac{2}{5}$ and first term (a) of 32. So the series is:

$$30.5, \; 30.5 \times \frac{2}{5}, \; 30.5 \times \left(\frac{2}{5}\right)^2, \; .... \; 30.5 \times \left(\frac{2}{5}\right)^{n-1}$$

Hence $S_n = \dfrac{30.5 \times \left(1 - \left(\frac{2}{5}\right)^n\right)}{1 - \left(\frac{2}{5}\right)} = 50$ Using (1.21)

$= \dfrac{5}{3} \times 30.5 \times \left(1 - \left(\frac{2}{5}\right)^n\right) = 50$ Rearranging

$\Rightarrow 50.83 \times \left(1 - \left(\frac{2}{5}\right)^n\right) = 50$

$\Rightarrow 1 - \left(\frac{2}{5}\right)^n = 0.9837$

$\Rightarrow \left(\frac{2}{5}\right)^n = 0.01613$

$n \log 0.4 = \log(0.01613)$ Taking logs of both sides

$\Rightarrow n = \dfrac{\log(0.01613)}{\log 0.4} = 4.504$

Therefore the least number of blows required is 5.

1.19 If the first, third and sixth terms of an AP are in geometrical progression, determine the common ratio of the GP.

The terms from the AP are:

$a, \; a + 2d, \; a + 5d$ (1.25)

If they are to be in Geometrical Progression then:

$\dfrac{a + 2d}{a} = r$ and   (1.26)

$\dfrac{a + 5d}{a + 2d} = r$ (1.27)

If we now subtract (1.27) *from* (1.26) we get:

$$\frac{a+2d}{a} - \frac{a+5d}{a+2d} = 0 \qquad\qquad (1.28)$$

$\Rightarrow (a+2d)^2 - a(a+5d) = 0$ Multiplying (1.28) throughout by $a(a+2d)$

$\Rightarrow a^2 + 4ad + 4d^2 - a^2 - 5ad = 0$ Expanding brackets

$\Rightarrow 4d^2 - ad = 0$ Simplifying

$\Rightarrow d(4d - a) = 0$ Factorising

$\Rightarrow a = 4d$

Substituting for $a = 4d$ *in* (1.25) we get:

$4d$, $6d$ and $9d$ which is a geometrical progression with common ratio of $\dfrac{6d}{4d} = \dfrac{3}{2}$

Geometric Mean

Just as in an AP, it is possible to insert any number between two given quantities such that the resulting series will be in geometrical progression. Such numbers are referred to as **Geometric Means**.

We have already seen an example of this in Q (1.14) above but we ill look at another two examples now as so that you will be familiar with the term **Geometric mean** *to make the meaning clear and in case you may be asked this in an exam paper.*

Note the special case of three terms in a GP. If x, y, z are three terms in a GP then the common

ratio r is equal to $\dfrac{y}{x}$ or to $\dfrac{z}{y}$

Then $\dfrac{y}{x} = \dfrac{z}{y} \Rightarrow y^2 = xz \Rightarrow y = \sqrt{xz}$ \qquad\qquad (1.29)

This is the condition for x, y, z to be three consecutive terms of a GP.

y is said to be the geometric mean of x and z, so the geometric mean of two numbers is the square root of their product.

E.g., the geometric mean of 4 and 9 is $\sqrt{4 \times 9} = 6$

Let the first be $a = 162$

And let the common ratio of the GP be r

We are given the last term, $l = 1,250$

The from (1.18) we have $1,250 = 162r^4$ Since there are five terms in this series

Hence $r = \left(\dfrac{2,250}{162}\right)^{\frac{1}{4}} = \dfrac{5}{3}$ Taking the positive answer

So the second terms would be $162 \times \left(\dfrac{5}{3}\right) = 270$

and the third term would be $162 \times \left(\dfrac{5}{3}\right)^{2} = 450$

and the third term would be $162 \times \left(\dfrac{5}{3}\right)^{3} = 750$

This is an almost identical question to the last one in terms of the solution.

Let the first be $a = 15$

And let the common ratio of the GP be r

We are given the last term, $l = 3,645$

The from (1.18) we have $3.645 = 15r^5$ Since there are six terms in this series

Hence $r = \left(\dfrac{3.645}{15}\right)^{\frac{1}{5}} = 3$ Taking the positive answer

So the second terms would be $3 \times 15 = 45$

and the third term would be $15 = 3^2 \pm 135$

the fourth term would be $15 \times 3^3 = 405$

the fifth term would be $15 \times 3^4 = 1215$

Using the result from (1.29):

$r = \sqrt{18 \times 72}$
$= \sqrt{1.296} = 36$

Convergence of Geometric Series

If we consider the series $1 + \dfrac{1}{2} + \dfrac{1}{4} + \dfrac{1}{8} + \dfrac{1}{16} + \ldots$

This is a geometrical progression with common ratio of $\dfrac{1}{2}$

If we stop at the third term, the sum of the first three terms is $1\dfrac{3}{4}$ and this is less than 2 by the value of the third term - $\dfrac{1}{4}$.

Similarly the sum of the first four terms is $1\dfrac{7}{8}$ which is less than 2 by the fourth term - $\dfrac{1}{8}$

When we stop at the fifth term, the sum of the series is $1\dfrac{15}{16}$ which is less than 2 by the fifth term - $\dfrac{1}{16}$

We can see that the sum of this series never exceeds 2 and is never equal to 2 but can be made as close to 2 as we wish by taking a sufficient number of terms.

In this case, the value 2 is called the **limit of the sum** of this series and similar series where such a limit exists are said to be **convergent.**

If we now consider the geometrical progression:

$a + ar + ar^2 + \ldots$

Then by (1.21), the sum to n terms is given by

$$S_n = \frac{a\left(1 - r^n\right)}{1 - r} = \frac{a}{1 - r} - \frac{ar^n}{1 - r} \qquad \text{Splitting the first fraction into two fractions}$$

Now suppose that r lies between 0 and 1. Then r^n decreases as n increases and since r^n cannot be negative, it must tend towards some positive limit l

Now since $r^{n+1} = r.r^n$ then both r^n and r^{n+1} are ultimately equal to l, we have $l = rl$ and since r cannot be equal to 1, then l must be zero.

In a similar way, we can show that if r lies between -1 and 0 we can show that the limiting value of

r^n is also zero. Thus the value of the term $\dfrac{ar^n}{1 - r}$ becomes nearer and nearer to zero as n increases.

Hence the limit of the sum of the progression is $S = \dfrac{a}{1 - r}$ for $-1 < r < 1$:

This is because we can make S_n as near to $\dfrac{a}{1-r}$ as we please by making n sufficiently large.

Hence we can conclude that: **the geometric series whose first term is a and has common ratio r, converges when $-1 < r < 1$ and the limit of the sum of the series is:**

$$S = \frac{a}{1-r} \qquad\qquad (1.30)$$

Clearly there can be no limit to the sum of a geometrical progression whose common ratio lies outside the range $-1 < r < 1$ as each term in such a series would get larger and larger. Indeed, each term would exceed the sum of all the preceding ones

For example the series $1 + 3 + 9 + 27 + ...$ gets bigger and bigger as the number of terms increase and this series is said to **divergent.**

Note that the limit of the sum of a convergent geometrical progression is sometimes called **the sum to infinity** and a convergent geometrical progression is sometimes referred to as an **infinite** geometrical progression. These are rather loose terms and should be avoided whenever possible.

WORKED EXAMPLES

1.22. Evaluate $0.3^{\bullet}$ as a fraction.

The expression in the question $0.3^{\bullet}$ means 0.3 recurring. i.e. .3333333 .....

Hence $0.3^{\bullet} = \dfrac{3}{10} + \dfrac{3}{100} + \dfrac{3}{1000} + ...$

Which we can rewrite as:

$$\frac{3}{10}\left(1 + \frac{1}{10} + \frac{1}{100} + \frac{1}{1000} + ...\right) \qquad\qquad (1.31)\ \text{Factorising}$$

Hence the series inside the bracket is a convergent geometrical progression with the first term of 1 and common ratio of $\dfrac{1}{10}$

Using (1.30) $S = \dfrac{1}{1 - \dfrac{1}{10}} = \dfrac{10}{9}$

Hence from (1.31) the value of $0.3^{\bullet}$ is $\dfrac{3}{10} \times \dfrac{10}{9} = \dfrac{1}{3}$

1.23. Find the first term and the common ratio of a convergent progression in which (i) the limit of the sum is 4 and (ii) the limit of the sum of the series formed by the cubes of the terms is 192

Let the first term be a and the common ratio be r

From (1.30) we have $S = \dfrac{a}{1-r} = 4$

Since, in this case we are dealing with the cubes of these terms, the first term will be a^3 and the common ratio r will be r^3

Hence in this case from (1.30) $S = \dfrac{a^3}{1-r^3} = 192$ $\qquad\qquad$ (1.32)

And also from (1.30)

$a = 4(1-r)$

$\Rightarrow 4r = 4 - a$

$\Rightarrow r = \dfrac{4-a}{4} \Rightarrow r^3 = \left(\dfrac{4-a}{4}\right)^3$ $\qquad\qquad$ (1.33)

$\Rightarrow \dfrac{a^3}{1-\left(\dfrac{4-a}{4}\right)^3} = 192$ $\qquad\qquad$ Substituting for r^3 in (1.32)

$\Rightarrow a^3 = 192 - 192\left(\dfrac{4-a}{4}\right)^3$

$\Rightarrow a^3 = 192 - \dfrac{192(4-a)^3}{64}$ $\qquad\qquad$ Rearranging

$\Rightarrow a^3 = 192 - 3(4-a)^3$

$\Rightarrow a^3 + 3(4-a)^3 = 192$

$\Rightarrow a^3 + 3(-a^3 + 12a^2 - 48a + 64) = 192$ $\qquad\qquad$ Expanding brackets

$\Rightarrow a(-a+12)(a-6) = 0$ $\qquad\qquad$ Factorising

$\Rightarrow a = 0, 12 \text{ or } 6$

Checking these answers in (1.33) the only solution so that the series will be convergent, is if $a = 6$ otherwise r would be zero or 2 and for convergence we need $-1 < r < 1$

Substituting for $a = 6$ in (1.33) gives

$r^3 = \left(\dfrac{4-6}{4}\right)^3 = \left(\dfrac{-2}{4}\right)^3 = \left(-\dfrac{1}{2}\right)^3 = -\dfrac{1}{8}$

$\Rightarrow r = -\dfrac{1}{2}$

1.24. Show that there are two geometrical progressions in which the second term is $-\dfrac{4}{3}$ and where the sum of the first three terms is $\dfrac{28}{9}$. Show that only one of these progressions is convergent and find its sum.

Here we have the series $a, -\dfrac{4}{3}, ar^2, \ldots$

Where the second term $ar = -\dfrac{4}{3}$ so that

$$r = -\frac{4}{3a} \qquad\qquad (1.34)$$

Adding the first three terms together we get;

$$a - \frac{4}{3} + ar^2 = \frac{28}{9}$$
$$a(1 + r^2) = \frac{28}{9} + \frac{4}{3}$$
$$= \frac{40}{9} \qquad\qquad (1.35)$$

But from (1.34) $r = -\dfrac{4}{3a}$ and substituting for this in (1.35) we get:

$$a\left(1 + \left(-\frac{4}{3a}\right)^2\right) = \frac{40}{9}$$
$$\Rightarrow a\left(1 + \frac{16}{9a^2}\right) = \frac{40}{9}$$
$$\Rightarrow a + \frac{16}{9a} = \frac{40}{9} \qquad\qquad \text{Expanding bracket}$$
$$\Rightarrow a^2 + \frac{16}{9} = \frac{40a}{9} \qquad\qquad \text{Multiplying throughtout by } a$$
$$\text{Hence } 9a^2 + 16 = 40a \qquad\qquad \text{Multiplying throught by 9}$$
$$\Rightarrow 9a^2 - 40a + 16 = 0$$
$$\Rightarrow (9a - 4)(a - 4) = 0$$
$$\text{Hence } a = \frac{9}{4} \text{ or } 4$$

Substituting for these values in (1.34) gives:

When $a = \dfrac{4}{5}$, $r = -\dfrac{4}{3\left(\dfrac{4}{5}\right)} = -\dfrac{5}{3}$ Hence r is not within the range $-1 < r < 1$ and therefore is not

convergent.

When $a = 4$, $r = -\dfrac{4}{3 \times 4} = -\dfrac{1}{3}$ and as this is within the range $-1 < r < 1$ it is therefore convergent.

Using (1.30) we can now substitute for $a = 4$ and $r = -\dfrac{1}{3}$ to find the sum: $S = \dfrac{4}{1 - \left(-\dfrac{1}{3}\right)} = 3$

1.25 The limit of the sum of a convergent geometrical progression is k and the limit of the sum of the squares of its terms is l. Find the first term and the common ratio of the progression.

Let the first term be a and the common ratio r.

Then by (1.30) $k = \dfrac{a}{1 - r}$

Since, in this case we are dealing with the squares of these terms, the first term will be a^2 and the common ratio r will be r^2.

Hence in this case $l = \dfrac{a^2}{1 - r^2}$

Squaring the expression for k and dividing by that for l we obtain:

$$\dfrac{k^2}{l} = \dfrac{a^2}{(1-r)^2} \div \dfrac{a^2}{1-r^2}$$

$$\dfrac{k^2}{l} = \dfrac{1-r^2}{(1-r)^2} = \dfrac{(1+r)(1-r)}{(1-r)^2} \qquad \text{Factorising the numerator}$$

$$= \dfrac{1+r}{1-r} \qquad \text{Simplifying}$$

$$k^2(1-r) = l(1+r) \qquad \text{Multiplying thoughout by } 1-r$$

$$k^2 - k^2r = l + lr \qquad \text{Expanding brackets}$$

$$\Rightarrow lr + k^2r = k^2 - l \qquad \text{Rearranging}$$

$$\Rightarrow r(l + k^2) = k^2 - l \qquad \text{Factorising LHS}$$

$$\Rightarrow r = \dfrac{k^2 - l}{l + k^2}$$

Substituting for this value of r in (1.30) we get:

$$k^2(1-r) = l(1+r)$$
$$\Rightarrow k^2 - k^2 r = l + lr$$
$$\Rightarrow lr + k^2 r = k^2 - l$$
$$\Rightarrow r(l + k^2) = k^2 - l$$
$$\Rightarrow r = \frac{k^2 - l}{l + k^2}$$

Substituting for this value of r in $k = \dfrac{a}{1-r}$ we get

$$a = k\left(1 - \frac{k^2 - l}{l + k^2}\right)$$
$$= k\left(\frac{l + k^2 - k^2 + 1}{l + k^2}\right)$$
$$= \frac{2kl}{l + k^2}$$

MODULE D1

CALCULUS D1 GENERAL DIFFERENTIATION

Revision Notes

DIFFERENTIATION FROM FIRST PRINCIPLES

A convenient notation for a small increase or increment in the value of a variable of the variable x is the symbol δx called "delta x".

Note that δx does not mean δ multiplied by x. It means $\delta x = x_1 - x$ where x_1 differs from x by a small quantity.

When y is a function of x ($y = f(x)$) as shown in fig1, the symbol δy is used to denote the change in the value of y corresponding to a change δx in the value of the independent variable x.

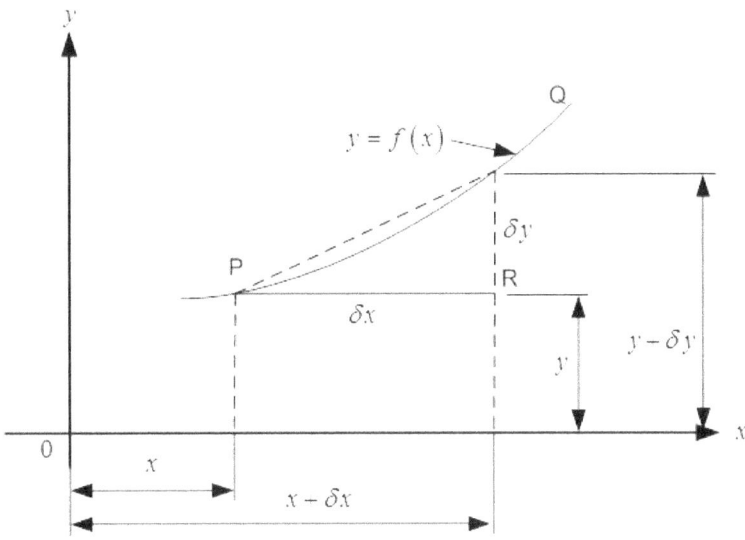

Fig 1.1 Differentiation from first principles

The point **P** on the graph has coordinates of x and y, and **Q** is another point on the graph with coordinates $x + \delta x$ and $y + \delta y$.

The average rate of change of y as x changes to $x + \delta x$ is measured by the angle **QPR** which is clearly equal to $\dfrac{\delta y}{\delta x}$.

The gradient (or slope of the tangent) of the curve at point **P** is the limiting value of the ratio $\dfrac{\delta y}{\delta x}$ as δx approaches zero

To illustrate this we will use example for the function $y = x^2$ at the point **P** which has coordinates (x, y).

So at the point **P** we have $y = x^2$ $\qquad\qquad$ (1.1)

And at the point **Q** with coordinates $(x + \delta x, y + \delta y)$

Therefore we have

$$y + \delta y = (x + \delta x)^2 \qquad\qquad (1.2)$$

Hence $x^2 + \delta y = (x + \delta x)^2$ $\qquad$ Substituting for $y = x^2$ from (1.1)

$$\Rightarrow \delta y = (x + \delta x)^2 - x^2$$

$$= x^2 + 2x\delta x + (\delta x)^2 - x^2$$

$$= \delta x (2x + \delta x)$$

$$\Rightarrow \frac{\delta y}{\delta x} = 2x + \delta x$$

As δx approaches zero, the ratio $\dfrac{\delta y}{\delta x}$ approaches the limiting value $2x$.

It is usual to use the convention that "as $\delta x \to 0, \dfrac{\delta y}{\delta x} \to 2x$"

Hence we can say that the gradient (or slope) of the function $y = x^2$ with coordinates (x, y) is $2x$.

Therefore at the point where $x = 3$, the gradient is 6 or $\dfrac{\delta y}{\delta x} = 6$

The process we have just discussed is known as differentiation from first principles and the limit determined is generally denoted by the symbol $\dfrac{dy}{dx}$. However, alternative notation are

$\dfrac{d}{dx} f(x), f'(x)$ and less frequently $D_x f(x)$

Alternative terms for this quantity are the **derivative** or the **derived function**

As a further example of differentiation form first principles we will differentiate the function $y = 3x^2$

Therefore $y + \delta y = 3(x + \delta x)^2$

$$\Rightarrow 3x^2 + \delta y = 3(x + \delta x)^2 \qquad\qquad \text{Substituting for } y = 3x^2$$

$$\Rightarrow \delta y = 3(x + \delta x)^2 - 3x^2$$

$$= 3\left(x^2 + 2x\delta x + (\delta x)^2 \right) - 3x^2$$

$$= 3(2x + \delta x)\delta x$$

Hence $\qquad \dfrac{\delta y}{\delta x} = 6x + 3\delta x$

Since the limiting value as $\delta x \to 0$ is $6x$ then

$$\frac{dy}{dx} = \lim_{x \to 0}\left(\frac{\delta y}{\delta x} \right) = 6x$$

We could also express this result as

$\dfrac{d}{dx}(3x^2) = 6x$ or if $f(x) = 3x^2$ then $f'(x) = 6x$

The differential coefficient of x^n where n is a positive integer

If $y = x^n$ then

$$y + \delta y = (x + \delta x)^n$$

$$\Rightarrow x^n + \delta y = (x + \delta x)^n \qquad \text{Substituting for } y = x^n$$

$$\Rightarrow \delta y = (x + \delta x)^n - x^n$$

Since n is a positive integer, $(x + \delta x)^n$ can be expanded by use of the Binomial Theorem in a terminating series as follows:

$$\delta y = x^n + nx^{n-1}(\delta x) + \frac{n(n-1)}{2!}x^{n-2}(\delta x)^2 + \ldots + (\delta x)^n - x^n$$

$$= nx^{n-1}(\delta x) + \frac{n(n-1)}{2!}x^{n-2}(\delta x)^2 + \ldots + (\delta x)^n$$

$$\Rightarrow \frac{\delta y}{\delta x} = nx^{n-1} + \frac{n(n-1)}{2!}x^{n-2}(\delta x) + \ldots + (\delta x)^{n-1}$$

Since all of the terms on the right hand side contain a product with δx except the first term, the limit of this expression as $\delta x \to 0$ is nx^{n-1} which yields the result that if:

$y = x^n$, then

$$\frac{dy}{dx} = nx^{n-1}$$

The same result holds true if n is a negative integer or fractional.

Example (i)

Using the above result

Find $\dfrac{dy}{dx}$ when $y = 3x^2 + 4$

$$\frac{dy}{dx} = 6x \qquad \text{The "4" disappears because it is a constant (has no } x \text{ term in it)}$$

Note that we have differentiated this function term by term

Sometimes a question may be posed in a different form e.g.

Find $\dfrac{d}{dx}(3x^2 + 2x + 4)$. They both mean the same thing – differentiate the given function with respect to x.

Example (ii)

Differentiate the function $y = \dfrac{2}{x^2}$

We can re-write this as $y = 2x^{-2}$

Hence $\dfrac{dy}{dx} = -4x^{-3}$ or $-\dfrac{4}{x^3}$

Example (iii)

Find $\dfrac{d}{dx}\left(4\sqrt{x}\right)$

Let $y = 4\sqrt{x} = 4x^{\frac{1}{2}}$

Differentiating this we get:

$$\frac{dy}{dx} = 4 \times \frac{1}{2} x^{-\frac{1}{2}} = 2x^{-\frac{1}{2}} = \frac{2}{\sqrt{x}}$$

The differential coefficients of sin x and cos x sin x

Let $y = \sin x$. Then

$$y + \delta y = \sin(x + \delta x)$$

$$\Rightarrow \sin x + \delta y = \sin(x + \delta x) \qquad \text{Substituting for } y = \sin x$$

$$\Rightarrow \delta y = \sin(x + \delta x) - \sin x$$

$$= 2\cos(x + \tfrac{1}{2}\delta x)\sin(\tfrac{1}{2}\delta x)$$

Since $\sin A - \sin B = 2\cos\frac{1}{2}(A + B)\sin\frac{1}{2}(A - B)$ where $A = (x + \frac{1}{2}\delta x)$ and $B = x$

Hence $\dfrac{\delta y}{\delta x} = \cos(x + \tfrac{1}{2}\delta x)\left(\dfrac{\sin(\tfrac{1}{2}\delta x)}{\tfrac{1}{2}\delta x}\right)$ $\qquad$ Dividing both sides by δx and rearranging

Now as $\delta x \to 0$, $\dfrac{\sin(\tfrac{1}{2}\delta x)}{\tfrac{1}{2}\delta x} \to 1$ and $\cos(x + \tfrac{1}{2}\delta x) \to \cos x$

Hence if $y = \sin x$ then $\dfrac{dy}{dx} = \cos x$

cos x

If $y = \cos x$ then

$$y + \partial y = \cos(x + \partial x)$$

$$\Rightarrow \partial y = \cos(x + \partial x) - y$$

$$= \cos(x + \partial x) - \cos x \qquad \text{Substituting for } y = \cos x$$

This can also be written in the form:

$$\frac{\delta y}{\delta x} = -\sin(x + \tfrac{1}{2}\partial x)\left(\frac{\sin\tfrac{1}{2}\partial x}{\tfrac{1}{2}\partial x}\right)$$

The limit of $\dfrac{dy}{dx}$ as $\partial x \to 0$ is $-\sin x$

Hence if $y = \cos x$ then $\dfrac{dy}{dx} = -\sin x$

Differentiation of a sum

Let $y = u + C$ where u is a given function of x and C is a constant. Now, if x increases to $x + \partial x$, then u increases to $u + \delta u$ and y will increase to $y + dy$. Hence:

$$y + \partial y = u + \delta u + C$$

Therefore $u + C + \partial y = u + \delta u + C \qquad \text{Substituting for } y = u + C$

$$\Rightarrow \partial y = \delta u$$

$$\Rightarrow \frac{\partial y}{\partial x} = \frac{\delta u}{\delta x} \qquad\qquad \text{Dividing both sides by } \partial x$$

In the limit, as $x \rightarrow 0$ we then have:

$$\frac{dy}{dx} = \frac{du}{dx}$$

Thus an additive constant (in this case C) disappears on differentiation.

Now let $u = u + v$ where u and v are given functions of x. If x increases to $x + \partial x$ then u and v increase to $u + \delta u$ and $v + \delta v$ respectively and y increases to $y + \partial y$. Hence:

$$y + \partial y = u + \delta u + v + \delta v$$

Therefore $u + v + \partial y = u + \delta u + v + \delta v$ Substituting for $y = u + v$

$$\Rightarrow \partial y = +\delta u + \delta v \qquad\qquad \text{Dividing both sides by } \partial x$$

In the limit, as $x \rightarrow 0$ we then have:

$$\frac{dy}{dx} = \frac{du}{dx} + \frac{dv}{dx}$$

Thus the differential coefficients of the sum of two functions, is the sum of the differential coefficients of all of the separate functions.

Clearly, the same result holds if the plus sign is replaced by a minus sign. Thus the differential coefficients of the difference of two functions, is the difference of the differential coefficients of all of the separate functions.

Now if $y = u + v + w$ where w is a third function of x, we can re-write this as:

$$y = (u + v) + w$$

By the previous result

$$\frac{dy}{dx} = \frac{d}{dx}(u + v) + \frac{dw}{dx} \qquad\qquad \text{Since } (u + v) \text{ is simply another function of } x$$

$$= \frac{du}{dx} + \frac{du}{dx} + \frac{dv}{dx}$$

This procedure can be extended to the sum of any number of functions so that the differential coefficient of the sum of any finite number of functions is the sum of differential coefficients of the separate functions.

Example (iv)

Differentiate $4x^2 + 2x - 6$

$$\frac{dy}{dx} = 8x + 2$$

Example (v)

Find $\dfrac{d}{dx}\left(x^{\frac{3}{2}} - 6\sqrt{x} + 2 \right)$

Let $y = \left(x^{\frac{3}{2}} - 6x^{\frac{1}{2}} + 2 \right)$ Rearranging

$$\Rightarrow \frac{dy}{dx} = \tfrac{3}{2} x^{\frac{1}{2}} - 6 \times \tfrac{1}{2} x^{-\frac{1}{2}}$$

$$= \tfrac{3}{2}\sqrt{x} - \frac{3}{x^{\frac{1}{2}}} \qquad\qquad \text{Simplifying}$$

$$= \tfrac{3}{2}\sqrt{x} - \frac{3}{\sqrt{x}}$$

$$= \frac{3x - 6}{2\sqrt{x}}$$

Note differentiation of products and quotients will be discussed in later modules

WORKED SOLUTIONS

1.1 Differentiate $x^3 + 2x^2$ from first principles

Let $y = x^3 + 2x^2$ Then

$$y + \partial y = (x + \partial x)^3 + 2(x + \partial x)^2$$

$$\Rightarrow \partial y = (x + \partial x)^3 + 2(x + \partial x)^2 - y$$

$$\Rightarrow \partial y = (x + \partial x)^3 + 2(x + \partial x)^2 - (x^3 + 2x^2) \qquad \text{Substituting for } y = x^3 + 2x^2$$

$$= x^3 + 3x^2\partial x + 3x(\partial x)^2 + (\partial x)^3 + 2x^2 + 4x\partial x + 2(\partial x)^2 - x^3 - 2x^2 \qquad \text{Expanding}$$

$$= 3x^2\partial x + 3x(\partial x)^2 + (\partial x)^3 + 4x\partial x + 2(\partial x)^2$$

$$\Rightarrow \frac{\delta y}{\delta x} = 3x^2 + 3x\partial x + (\partial x)^2 + 4x + 2\partial x \qquad \text{Dividing both sides by } \delta x \text{ and rearranging}$$

The limit of $\dfrac{dy}{dx}$ as $\partial x \to 0$ is $3x^2 + 4x$

Hence if $y = x^3 + 2x^2$ then $\dfrac{dy}{dx} = x^3 + 4x$

1.2 Differentiate $\dfrac{1}{x}$ from first principles

Let $y = \dfrac{1}{x}$ then

$$y + \partial y = \frac{1}{x + \partial x}$$

$$\Rightarrow \partial y = \frac{1}{x + \partial x} - \frac{1}{x} \qquad \text{Substituting for } y = \frac{1}{x}$$

$$= \frac{x - x - \partial x}{x(x + \partial x)} = \frac{-\partial x}{x^2 + x\partial x} \qquad \text{Rearranging}$$

$$\Rightarrow \frac{\delta y}{\delta x} = \frac{-1}{x^2 + x\partial x}$$

The limit of $\dfrac{dy}{dx}$ as $\partial x \to 0$ is $\dfrac{-1}{x^2}$

Hence if $y = \dfrac{1}{x}$ then $\dfrac{dy}{dx} = -\dfrac{1}{x^2}$

1.3 Differentiate $\dfrac{1}{x^2}$ from first principles

Let $y = \dfrac{1}{x^2}$ then

$$y + \delta y = \frac{1}{(x + \delta x)^2}$$

$\Rightarrow \delta y = \dfrac{1}{x^2 + 2x\delta x + (\delta x)^2} - y$ Rearranging and expanding the term $(x + \delta x)^2$

$\Rightarrow \delta y = \dfrac{1}{x^2 + 2x\delta x + (\delta x)^2} - \dfrac{1}{x^2}$ Substituting for $y = \dfrac{1}{x^2}$

$\quad = \dfrac{x^2 - x^2 - 2x\delta x - (\delta x)^2}{x^2 \left(x^2 + 2x\delta x + (\delta x)^2 \right)}$ Rearranging

$\quad = \dfrac{-2x\delta x - (\delta x)^2}{x^4 + 2x^3\delta x + x^2(\delta x)^2}$ Simplifying

$\Rightarrow \dfrac{\delta y}{\delta x} = \dfrac{-2x - \delta x}{x^4 + 2x^3\delta x + x^2(\delta x)^2}$ Dividing both sides by δx and rearranging

The limit of $\dfrac{dy}{dx}$ as $\delta x \to 0$ is $\dfrac{-2x}{x^4} = -\dfrac{2}{x^3}$

Hence if $y = \dfrac{1}{x^2}$ then

$$\frac{dy}{dx} = -\frac{2}{x^3}$$

1.4 Differentiate $2x^2 + \sin 2x$ from first principles

Let $y = 2x^2 + \sin 2x$ then

$y + \delta y = 2(x + \delta x)^2 + \sin 2(x + \delta x)$

$\Rightarrow \delta y = 2(x + \delta x)^2 + \sin 2(x + \delta x) - y$ Rearranging

$\Rightarrow \delta y = 2(x + \delta x)^2 + \sin 2(x + \delta x) - (2x^2 + \sin 2x)$ Substituting for $y = 2x^2 + \sin 2x$

$\quad = 2x^2 + 4x\delta x + 2(\delta x)^2 + \sin 2(x + \delta x) - 2x^2 - \sin 2x$

$\quad = 4x\delta x + 2(\delta x)^2 + \sin 2(x + \delta x) - \sin 2x$

$\quad = 4x\delta x + 2(\delta x)^2 + 2\cos(2x + \delta x)\sin(\delta x)$

Since $\sin A - \sin B = 2\cos\frac{1}{2}(A + B)\sin\frac{1}{2}(A - B)$ where $A = (2x + \delta x)$ and $B = 2x$

Hence $\dfrac{\delta y}{\delta x} = \dfrac{4x\delta x + (\delta x)^2 + 2\cos(2x + \delta x)\sin(\delta x)}{\delta x}$ Dividing both sides by δx and rearranging

$\quad = 4x + \delta x + 2\cos(2x + \delta x)\left(\dfrac{\sin(\delta x)}{\delta x} \right)$

Now as $\delta x \to 0$, $\dfrac{\sin\left(\frac{1}{2}\delta x\right)}{\frac{1}{2}\delta x} \to 1$ and $\cos\left(2x + \delta x\right) \to \cos 2x$

Hence if $y = 2x^2 + \sin 2x$ then $\dfrac{dy}{dx} = 4x + 2\cos 2x$

1.5 Differentiate $8x^4 + 3x^3 - 5x^2 + 2x - 4$ with respect to x

Let $y = 8x^4 + 3x^3 - 5x^2 + 2x - 4$

Then $\dfrac{dy}{dx} = 32x^3 + 9x^2 - 10x + 2$

1.6 Differentiate $\dfrac{1}{2}x^3 + \dfrac{1}{5}x^2 - \dfrac{2}{3}x$ with respect to x

Let $y = \dfrac{1}{2}x^3 + \dfrac{1}{5}x^2 - \dfrac{2}{3}x$

Then $\dfrac{dy}{dx} = \dfrac{3}{2}x^2 + \dfrac{2}{5}x - \dfrac{2}{3}$

1.7 Differentiate $5 - \dfrac{3}{x^2} + \dfrac{1}{4x} - 2x$ with respect to x

Let $y = 5 - \dfrac{3}{x^2} + \dfrac{1}{4x} - 2x$

$\quad = 5 - 3x^{-2} + \dfrac{1}{4}x^{-1} - 2x$ Rearranging

Then $\dfrac{dy}{dx} = 6x^{-3} - \dfrac{1}{4}x^{-2} - 2$

$\quad = \dfrac{6}{x^3} - \dfrac{1}{4x^2} - 2$ Simplifying

1.8 Differentiate $x^2\left(\dfrac{1}{x} + \dfrac{2}{x^3} - 3x\right)$ with respect to x

Let $y = x^2\left(\dfrac{1}{x} + \dfrac{2}{x^3} - 3x\right)$

$\quad = x + 2 - 3x^3$ Simplifying

Then $\dfrac{dy}{dx} = 1 - 9x^2$

1.9 Differentiate $4\sqrt{x} + 2\sqrt{7}$ with respect to x

Let $y = 4\sqrt{x} + 2\sqrt{7}$

$\quad = 4x^{\frac{1}{2}} + 2\sqrt{7}$ Rearranging

Then $\dfrac{dy}{dx} = 2x^{-\frac{1}{2}} = \dfrac{2}{\sqrt{x}}$

1.10 Differentiate $2x^{\frac{3}{2}} - \left(\sqrt{x}\right)^{3}$ with respect to x

Let $y = 2x^{\frac{3}{2}} - \left(\sqrt{x}\right)^{3} = 2x^{\frac{3}{2}} - x^{\frac{3}{2}} = x^{\frac{3}{2}}$ Rearranging

Then $\dfrac{dy}{dx} = \frac{3}{2}x^{\frac{1}{2}} = \frac{3}{2}\sqrt{x}$

1.11 Evaluate $\dfrac{d}{dx}\left(1+\dfrac{1}{x}\right)^{2}$

Let $y = \left(1+\dfrac{1}{x}\right)^{2} = 1 + \dfrac{2}{x} + \dfrac{1}{x^{2}} = 1 + 2x^{-1} + x^{-2}$ Expanding bracket & Rearranging

Then $\dfrac{dy}{dx} = -2x^{-2} - 2x^{-3}$

$\qquad\quad = -\dfrac{2}{x^{2}} - \dfrac{2}{x^{3}}$

1.12 If $y = x^{2n} + 2nx - 4n$ where n is a constant find $\dfrac{dy}{dx}$

$\dfrac{dy}{dx} = 2nx^{2n-1} + 2n$

1.13 Find $\dfrac{ds}{dt}$ for $s = ut + \dfrac{1}{2}ft^{2}$ where u and f are constants

$\dfrac{ds}{dt} = u + ft$

1.14 If $s = 60t - 10t^{2}$ find $\dfrac{ds}{dt}$ and determine the value of t when $\dfrac{ds}{dt} = 24$

$\dfrac{ds}{dt} = 60 - 20t$

Hence $60 - 20t = 24$

And $t = \dfrac{60-24}{20} = \dfrac{36}{20} = \dfrac{18}{10} = \dfrac{9}{5}$

1.15 If $y = 2 - 3x^{2}$ prove that $x\dfrac{dy}{dx} - 2y + 4 = 0$

$\dfrac{dy}{dx} = -6x$

Hence $x\dfrac{dy}{dx} - 2y + 4 = x(-6x) - 2(2 - 3x^{2}) + 4$ Substituting for $y = 2 - 3x^{2}$

$\qquad\qquad\qquad\qquad = -6x^{2} - 4 + 6x^{2} + 4 = 0$

1.16 If $y = \sqrt{x}$ prove that $2x\dfrac{dy}{dx} = y$

$\qquad\qquad y = x^{\frac{1}{2}}$ Rewriting

So $\dfrac{dy}{dx} = \dfrac{1}{2}x^{-\frac{1}{2}}$

Hence $2x\dfrac{dy}{dx} = 2x\left(\dfrac{1}{2}x^{-\frac{1}{2}}\right) = x^{\frac{1}{2}} = \sqrt{x} = y$

1.17 If $y = x^4 + x$ prove that $x^2\dfrac{d^2y}{dx^2} - 2x\left(\dfrac{dy}{dx} - 3\right) = 4y$

$\dfrac{dy}{dx} = 4x^3 + 1$

$\dfrac{d^2y}{d^2x} = 12x^2$

$x^2\dfrac{d^2y}{d^2x} - 2x\left(\dfrac{dy}{dx} - 3\right) = x^2\left(12x^2\right) - 2x\left(4x^3 + 1 - 3\right)$

$\qquad\qquad\qquad = 12x^4 - 8x^4 + 4x$

$\qquad\qquad\qquad = 4x^4 + 4x = 4(x^4 - x) = 4y$

1.18 If $v = \dfrac{4}{3}\pi r^3$ v Find $\dfrac{dv}{dr}$ and $\dfrac{d^2v}{dr^2}$

$\dfrac{dv}{dr} = 4\pi r^2$

$\dfrac{d^2v}{dr^2} = 8\pi r$

1.19 Find $\dfrac{dp}{dv}$ when (i) $pv = k$ and (ii) when $pv^{\sqrt{2}} = k$ where k is a constant value

(i) $p = \dfrac{k}{v} = kv^{-1}$ hence $\dfrac{dp}{dv} = -kv^{-2} = -\dfrac{k}{v^2}$

(ii) $p = \dfrac{k}{v^{\sqrt{2}}} = kv^{-\sqrt{2}}$ hence $\dfrac{dp}{dv} = -\sqrt{2}kv^{-\sqrt{2}-1} = -\sqrt{2}\dfrac{k}{v^{\sqrt{2}+1}}$

1.20 Differentiate $\dfrac{x^2 - 3x + 2}{x}$ with respect to x

Let $y = \dfrac{x^2 - 3x + 2}{x} = x - 3 + 2x^{-1}$ Dividing top & Bottom by x

Then $\dfrac{dy}{dx} = 1 - 2x^{-2} = 1 - \dfrac{2}{x^2}$

1.21 Differentiate $\dfrac{4x^4 + 5x^2 - 6}{x^2}$ with respect to x

Let $y = \dfrac{4x^4 + 5x^2 - 6}{x^2} = 4x^2 + 5 - 6x^{-2}$ Dividing top & Bottom by x^2

Then $\dfrac{dy}{dx} = 8x + 12x^{-3} = 8x + \dfrac{12}{x^3}$ Rearranging

1.22 Differentiate $y = \sqrt{x^2 + 4}$ with respect to x

Let $z = x^2 + 4$

Then $y = \sqrt{z}$

And $\dfrac{dy}{dx} = \dfrac{dy}{dz} \cdot \dfrac{dz}{dx}$ Using the chain rule

Where $\dfrac{dy}{dz} = \dfrac{1}{2} z^{-\frac{1}{2}} = \dfrac{1}{2\sqrt{z}} = \dfrac{1}{2\sqrt{x^2 + 4}}$ Substituting for $z = x^2 + 4$

And $\dfrac{dz}{dx} = 2x$

So that $\dfrac{dy}{dx} = 2x\left(\dfrac{1}{2\sqrt{x^2 + 4}}\right) = \dfrac{x}{\sqrt{x^2 + 4}}$ Expanding bracket

1.23 Find $\dfrac{dy}{dx}$ if $y = (2x + 3)^5$

Let $z = 2x + 3$

Then $y = z^5$

And $\dfrac{dy}{dx} = \dfrac{dy}{dz} \cdot \dfrac{dz}{dx}$ Using the chain rule

Where $\dfrac{dy}{dz} = 5z^4 = 5(2x + 3)^4$ Substituting back for $z = 2x + 3$

And $\dfrac{dz}{dx} = 2$

So that $\dfrac{dy}{dx} = 5(2x + 3)^4 \times 2 = 10(2x + 3)^4$

1.24 Differentiate $y = \sqrt{ax + b}$ with respect to x where a and b are constants

$y = \sqrt{ax + b} = (a + b)^{\frac{1}{2}}$

Let $z = ax + b$

Then $y = z^{\frac{1}{2}}$

And $\dfrac{dy}{dx} = \dfrac{dy}{dz} \cdot \dfrac{dz}{dx}$ Using the chain rule

Where $\dfrac{dy}{dz} = \dfrac{1}{2} z^{-\frac{1}{2}} = \dfrac{1}{2\sqrt{z}} = \dfrac{1}{2\sqrt{(ax + b)}}$ Substituting back for $z = ax + b$

And $\dfrac{dz}{dx} = a$

So that $\dfrac{dy}{dx} = a \times \dfrac{1}{2(ax + b)} = \dfrac{a}{2\sqrt{(ax + b)}}$

1.25 Differentiate $\sqrt{x^2 + 3x + 5}$ with respect to x

Note: In practice, the introduction of the letter z is not necessary, provided that the function of x is treated as a single variable as shown below:

Let $y = \sqrt{x^2 + 3x + 5} = \left(x^2 + 3x + 5\right)^{\frac{1}{2}}$

Then $\dfrac{dy}{dx} = \dfrac{1}{2}\left(x^2 + 3x + 5\right)^{-\frac{1}{2}} \times (2x + 3) = \dfrac{2x + 3}{2\sqrt{x^2 + 3x + 5}}$

Note: Here we have differentiated the bracketed term as a single variable (the z previously) and then differentiated the expression inside the bracket. Finally the two are multiplied together.

1.26 Find $\dfrac{d}{dx}\left(\ln\left(\cos 2x\right)\right)$

Note that $\dfrac{d}{dx}(\ln x) = \dfrac{1}{x}$

$\dfrac{d}{dx}\left(\ln\left(\cos 2x\right)\right) = \dfrac{1}{\cos 2x}(-2\sin 2x)$ Using chain rule $z = \cos 2x$

$= \dfrac{-2\sin 2x}{\cos 2x} = -2\tan 2x$ Simplifying

1.27 Differentiate $y = \cos^3(2x)$ with respect to x

$\dfrac{dy}{dx} = 3\cos^2 2x\left(-2\sin 2x\right)$ Using chain rule $z = \cos 2x$

$= -6\sin 2x \cos^2 2x$

1.28 Differentiate $y = e^{\sin x}$ with respect to x

Note that $\dfrac{d}{dx}\left(e^x\right) = e^x$

$\dfrac{dy}{dx} = e^{\sin x}\cos x$ Using chain rule $z = \sin x$

1.29 Differentiate $y = e^{\sin^2 x}$ with respect to x

$\dfrac{dy}{dx} = e^{\sin^2 x}\left(2\sin x \cos x\right)$ Using chain rule $z = \sin^2 x$

$= 2e^{\sin^2 x}\sin x \cos x = e^{\sin^2 x}\sin 2x$ Since $2\sin x \cos x = \sin 2x$

1.30 Differentiate $y = 3e^{\cos^3 x}$ with respect to x

$\dfrac{dy}{dx} = \left(3e^{\cos^3 x}\right)\left(-3\cos^2 x \sin x\right)$ Using chain rule $z = \cos^3 x$

$= -9e^{\cos^3 x}\cos^2 x \sin x$

1.31 Differentiate $y = ae^{\sin bx}$ with respect to x

$$\frac{dy}{dx} = \left(abe^{\sin bx}\right)\cos bx = abe^{\sin bx}\cos bx \qquad \text{Using chain rule } z = \sin bx$$

1.32 Differentiate $y = \ln\left(x^2\right)$ with respect to x

$$\frac{dy}{dx} = \frac{1}{x^2}\left(2x\right) = \frac{2}{x} \qquad \text{Using chain rule } z = x^2$$

1.33 Find $\dfrac{dy}{dx}$ if $y = \tan^3 x + \ln(1 + x^2)$

$$\frac{dy}{dx} = 3\tan^2 x\sec^2 + \frac{1}{1+x^2}\times 2x = 3\tan^2 x\sec^2 + \frac{2x}{1+x^2}$$

Using the chain rule for chain rule $z = 1 + x^2$ and $s = \tan(x)$

1.34 Differentiate $y = \ln\left(x + \sqrt{1 + x^2}\right)$ with respect to x

Let $z = x + \sqrt{1 + x^2}$

Then $y = \ln z$

And $\dfrac{dy}{dz} = \dfrac{1}{z} = \dfrac{1}{x + \sqrt{1+x^2}}$

And $\dfrac{dz}{dx} = 1 + \dfrac{1}{2}\left(1 + x^2\right)^{-\frac{1}{2}}\times 2x = 1 + \dfrac{x}{\sqrt{1+x^2}}$

$$= \frac{\sqrt{1+x^2}+x}{\sqrt{1+x^2}}$$

$$\frac{dy}{dx} = \frac{dy}{dz}\cdot\frac{dz}{dx} = \left(\frac{1}{x + \sqrt{1+x^2}}\right)\times\frac{\sqrt{1+x^2}+x}{\sqrt{1+x^2}}$$

$$= \frac{\sqrt{1+x^2}+x}{\sqrt{1+x^2}\left(x + \sqrt{1+x^2}\right)}$$

$$= \frac{1}{\sqrt{1+x^2}}$$

1.35 Differentiate $y = \sec^2 2x$ with respect to x

$$\frac{dy}{dx} = 2\sec 2x \times \frac{d}{dx}\left(\frac{1}{\cos 2x}\right) \qquad \text{Using chain rule } z = \sec 2x$$

$$= \frac{\left(2\sec 2x\right)\left(2\sin 2x\right)}{\cos^2 2x}$$

$$= 4\sec^2 2x\tan 2x$$

1.36 Differentiate $x^2 + y^2 = 2y$ with respect to x

Differentiating the complete function with respect to x we get:

$$2x + 2y \frac{dy}{dx} = 2 \frac{dy}{dx}$$

$$\Rightarrow 2 \frac{dy}{dx} - 2y \frac{dy}{dx} = 2x$$

$$\Rightarrow (1 - y) \frac{dy}{dx} = x \qquad \text{Rearranging}$$

$$\Rightarrow \frac{dy}{dx} = \frac{x}{1 - y}$$

1.37 Differentiate $10^{3x} = y$ with respect to x

Taking logs of both sides we get:

$3x \ln 10 = \ln y$

Differentiating both sides with respect to x we get:

$$3 \ln 10 = \frac{1}{y} \frac{dy}{dx}$$

$$\Rightarrow \frac{dy}{dx} = 3y \ln 10 \qquad \text{Rearranging}$$

Hence $\frac{dy}{dx} = (3 \ln 10) 10^{3x}$ \qquad Substituting for $y = 10^{3x}$

MODULE D2.

CALCULUS D2 DIFFERENTIATION OF A PRODUCT

INTRODUCTION

Let $y = Cu$ where u is a function of x and C is a constant. Then:

$$y + \delta y = C(u + \delta u)$$

Where δy and δu are the increments in x and u respectively that correspond to an increment of δx in x

Then $Cu + \delta y = C(u + \delta u) = Cu + C\delta u$ Substituting for $y = Cu$ and expanding bracket

$$\Rightarrow \delta y = C\delta u$$

$$\Rightarrow \frac{\delta y}{\delta x} = C\frac{\delta u}{\delta x} \qquad \text{Dividing both sides by } \delta x$$

In the limit as $\delta x \to 0$ we then have:

$$\frac{dy}{dx} = C\frac{du}{dx}$$

Thus the differential coefficient of a constant multiplied by a function of x is equal to the constant multiplied by the differential coefficient of the function.

Now let $y = uv$ where u and v are given functions of x. If $\delta y, \delta u$ and δv are the increments in y, u and v respectively that correspond to an increment δx in x then:

$$y + \delta y = (u + \delta u)(v + \delta v)$$

$$= uv + u\delta v + v\delta u + \delta u\delta v \qquad \text{Expanding brackets}$$

$$\Rightarrow \delta y = uv + u\delta v + v\delta u + \delta u\delta v - uv \qquad \text{Substituting for } y = uv$$

$$= u\delta v + v\delta u + \delta u\delta v \qquad \text{Simplifying}$$

$$\Rightarrow \frac{\delta y}{\delta x} = u\frac{\delta v}{\delta x} + v\frac{\delta u}{\delta x} + \frac{\delta u\delta v}{\delta x} \qquad \text{Dividing both sides by } \delta x$$

$$= v\frac{\delta u}{\delta x} + (u + \delta u)\frac{\delta v}{\delta x} \qquad \text{Rearranging}$$

As $\delta x \to 0$, $\frac{\delta u}{\delta x}, \frac{\delta v}{\delta x}$ and $\frac{\delta y}{\delta x}$ tend to $\frac{du}{dx}, \frac{dv}{dx}$ and $\frac{dy}{dx}$ respectively hence:

$$\frac{dy}{dx} = u\frac{dv}{dx} + v\frac{du}{dx}$$

Thus the differential coefficient of the product of two functions of x is equal to the second function multiplied by the differential coefficient of the first plus the first function multiplied by differential coefficient of the second.

Similarly it can be shown that if $y = uvw$ then

$$\frac{dy}{dx} = vw\frac{du}{dx} + uw\frac{dv}{dx} + uv\frac{dw}{dx}$$

This result may be extended to the products of any finite number of functions by differentiating each function separately and multiplying by the remaining functions and then adding the results together.

Example (i)

Find $\dfrac{dy}{dx}$ if $y = x^3 \sin x$

Let $u = x^3$

And $v = \sin x$

Then $\dfrac{dy}{dx} = u\dfrac{dv}{dx} + v\dfrac{du}{dx}$ Standard formula for differentiating products

$\qquad = x^3 \times \cos x + \sin x \times 3x^2$

$\qquad = x^3 \cos x + 3x^2 \sin x$

Example (ii)

Find $\dfrac{d}{dx}\left(\sin x \cos x \right)$

Let $u = \sin x$

And $v = \cos x$

Then $\dfrac{dy}{dx} = u\dfrac{dv}{dx} + v\dfrac{du}{dx}$ Standard formula for differentiating products

$\qquad = \sin x \times \left(-\sin x \right) + \cos x \times \cos x$

$\qquad = -\sin x^2 + \cos^2 x$

$\qquad = \cos^2 x - \sin x^2$

$\qquad = \cos 2x$ Since $\cos^2 x - \sin x^2 = \cos 2x$

Example (iii)

Find $\dfrac{dy}{dx}$ when $y = 2x^2 e^x \cos x$

Let $u = 2e^x$; $v = x^2$; $w = \cos x$

Then $\dfrac{dy}{dx} = vw\dfrac{du}{dx} + uw\dfrac{dv}{dx} + uv\dfrac{dw}{dx}$

$vw\dfrac{du}{dx} = x^2 \cos x \times 2e^x = 2x^2 e^x \cos x$

$uw\dfrac{dv}{dx} = 2e^x \cos x \times 2x = 4xe^x \cos x$

$uv\dfrac{dw}{dx} = 2e^x x^2 \times \left(-\sin x \right) = -2x^2 e^x \sin x$

Hence $\dfrac{dy}{dx} = 2x^2 e^x \cos x + 4xe^x \cos x - 2x^2 e^x \sin x$ Adding the previous results together

$\qquad = 2xe^x \left(x \cos x + 2\cos x - x \sin x \right)$ Rearranging & simplifying

WORKED SOLUTIONS

1.1 Differentiate $y = (2x+3)(3x+1)$ with respect to x

Let $u = (2x+3) \Rightarrow \dfrac{du}{dx} = 2$

Let $v = (3x+1) \Rightarrow \dfrac{dv}{dx} = 3$

$\dfrac{d}{dx}(uv) = u\dfrac{dv}{dx} + v\dfrac{du}{dx}$ Standard formula for differentiating products

$= 3(2x+3) + 2(3x+1)$

Note that we could have simply multiplied the two bracketed term together I.e.

$y = (2x+3)(3x+1) = 6x^2 + 11x + 3$ and differentiated this to achieve the same result.

1.2 Find $\dfrac{dy}{dx}$ when $y = (x+2)^2 (2x-4)^3$

Let $u = (x+2)^2 \Rightarrow \dfrac{du}{dx} = 2(x+2) \times 1 = 2(x+2)$

Let $v = (2x-4)^3 \Rightarrow \dfrac{dv}{dx} = 3(2x-4)^2 \times 2 = 6(2x-4)^2$

$\dfrac{d}{dx}(uv) = u\dfrac{dv}{dx} + v\dfrac{du}{dx}$ Using chain rule

$= 6(x+2)^2 (2x-4)^2 + 2(x+2)(2x-4)^3$

1.3 Find $\dfrac{dy}{dx}$ when $y = \dfrac{(x+1)^3}{(x-2)}$

We can re-write $\dfrac{(x+1)^3}{(x-2)}$ as $(x+1)^3 (x-2)^{-1}$

Let $u = (x+1)^3 \Rightarrow \dfrac{du}{dx} = 3(x+1)^2 \times 1 = 3(x+1)^2$

Let $v = (x-2)^{-1} \Rightarrow \dfrac{dv}{dx} = -(x-2)^{-2} \times 1 = -(x-2)^{-2}$

$\dfrac{d}{dx}(uv) = u\dfrac{dv}{dx} + v\dfrac{du}{dx}$

$= (x+1)^3 \left(-(x-2)^{-2} \right) + (x-2)^{-1} 3(x+1)^2$

$= 3\dfrac{(x+1)^2}{(x-2)} - \dfrac{(x-1)^3}{(x-2)^2}$ Rearranging & Simplifying

1.4 Differentiate $y = \dfrac{(2x-4)^2}{(3x+4)^3}$ with respect to x

Just as before we can re-write $\dfrac{(2x-4)^2}{(3x+4)^3}$ as $(2x-4)^2 (3x+4)^{-3}$

Let $u = (2x-4)^2 \Rightarrow \dfrac{du}{dx} = 2(2x-4) \times 2 = 4(2x-4)$

Let $v = (3x+4)^{-3} \Rightarrow \dfrac{dv}{dx} = -3(3x+4)^{-4} \times 3 = -9(3x+4)^{-4}$

$\dfrac{d}{dx}(uv) = u\dfrac{dv}{dx} + v\dfrac{du}{dx}$

$\qquad = (2x-4)^2\left(-9(3x+4)^{-4}\right) + (3x+4)^{-3}\,4(2x-4)$

$\qquad = -9(2x-4)(3x+4)^{-4} + 4(3x+4)^{-3}(2x-4)$

$\qquad = \dfrac{4(2x-4)}{(3x+4)^3} - \dfrac{9(2x-4)^2}{(3x+4)^4}$ Rearranging & simplifying

1.4 Differentiate $y = \sin x \tan x$ with respect to x

We can re-write $y = \sin x \tan x$ as $y = \sin x \dfrac{\sin x}{\cos x} = \dfrac{\sin^2 x}{\cos x}$

$\qquad\qquad = \dfrac{1 - \cos^2 x}{\cos x} = \dfrac{1}{\cos x} - \cos x = \sec x - \cos x$

Hence $\dfrac{dy}{dx} = \sec x \tan x - (-\sin x)$ $\dfrac{d}{dx}(\sec x) = \sec x \tan x$ (standard form)

$\qquad\quad = \sec x \tan x + \sin x$ Simplifying

1.5 Differentiate $\left(4 - 3x^5\right)^6$ with respect to x

Let $y = \left(4 - 3x^5\right)^6$

Then $\dfrac{dy}{dx} = 6\left(4 - 3x^5\right)^5 \times \left(-15x^4\right)$

$\qquad\quad = -90x^4\left(4 - 3x^5\right)^5$ Simplifying

1.6 Differentiate $y = \left(x^3 - 5x + 7\right)^4 (2x+3)^3$ with respect to x

Let $u = \left(x^3 - 5x + 7\right)^4 \Rightarrow \dfrac{du}{dx} = 4\left(x^3 - 5x + 7\right)^3 \times \left(3x^2 - 5\right) = 4\left(x^3 - 5x + 7\right)^3\left(3x^2 - 5\right)$

Let $v = (2x+3)^3 \Rightarrow \dfrac{dv}{dx} = 3(2x+3)^2 \times 2 = 6(2x+3)^2$

$\dfrac{d}{dx}(uv) = u\dfrac{dv}{dx} + v\dfrac{du}{dx}$

$\dfrac{dy}{dx} = \left(x^3 - 5x + 7\right)^4 \times 6(2x+3)^2 + (2x+3)^3 \times 4\left(x^3 - 5x + 7\right)^3\left(3x^2 - 5\right)$

$\qquad = 6\left(x^3 - 5x + 7\right)^4 (2x+3)^2 + 4(2x+3)^3\left(x^3 - 5x + 7\right)^3\left(3x^2 - 5\right)$ Simplifying

1.7 Differentiate $x^2 \sin x$ with respect to x

Let $y = x^2 \sin x$

Let $u = x^2 \Rightarrow \dfrac{dy}{du} = 2x$

Let $v = \sin x \Rightarrow \dfrac{dv}{dx} = \cos x$

$\dfrac{d}{dx}(uv) = u\dfrac{dv}{dx} + v\dfrac{du}{dx}$

$\qquad\qquad = x^2 \cos x + 2x \sin x$

1.8 Differentiate $x^4 \sin 3x$ with respect to x

Let $y = x^4 \sin 3x$

Let $u = x^4 \Rightarrow \dfrac{du}{dx} = 4x^3$

Let $v = \sin 3x \Rightarrow dv = 3\cos 3x$

$\dfrac{d}{dx}(uv) = u\dfrac{dv}{dx} + v\dfrac{du}{dx}$

$\qquad\qquad = 3x^4 \cos 3x + \sin 3x \times 4x^3$

$\qquad\qquad = 3x^4 \cos 3x + 4x^3 \sin 3x \qquad\qquad\qquad\qquad$ Simplifying

1.9 Differentiate $e^{2x}\cos 3x$ with respect to x

Let $y = e^{2x} \cos 3x$

Let $u = e^{2x} \Rightarrow \dfrac{du}{dx} = 2e^{2x}$

Let $v = \cos 3x \Rightarrow \dfrac{dv}{dx} = -3\sin 3x$

$\dfrac{d}{dx}(uv) = u\dfrac{dv}{dx} + v\dfrac{du}{dx}$

$\qquad\qquad = e^{2x}(-3\sin 3x) + (\cos 3x)(2e^{2x})$

$\qquad\qquad = e^{2x}(2\cos 3x - 3\sin 3x) \qquad\qquad\qquad\qquad$ Simplifying

1.10 Differentiate $ae^{bx}\sin cx$ with respect to x where a and b are constants.

Let $y = ae^{bx}\sin 3x$

Let $u = ae^{bx} \Rightarrow \dfrac{du}{dx} = abe^{bx}$

Let $v = \sin 3x \Rightarrow \dfrac{dv}{dx} = 3\cos 3x$

$\dfrac{d}{dx}(uv) = u\dfrac{dv}{dx} + v\dfrac{du}{dx}$

$= (ae^{bx})(3\cos 3x) + (\sin 3x)(abe^{bx})$

$ae^{bx}(3\cos 3x + b\sin 3x) \qquad\qquad\qquad\qquad$ Simplifying

1.11 Differentiate $3ae^{bx}\sin^2 cx$ with respect to x where a, b and c are constants

Let $y = 3ae^{bx}\sin^2 cx$

Let $u = 3ae^{bx} \Rightarrow \dfrac{du}{dx} = 3abe^{bx}$

Let $v = \sin^2 cx \Rightarrow \dfrac{dv}{dx} = 2c \sin cx \cos cx$

$\dfrac{d}{dx}(uv) = u\dfrac{dv}{dx} + v\dfrac{du}{dx}$

$\qquad = \left(3ae^{bx}\right)\left(2c\sin cx \cos cx\right) + \left(\sin^2 cx\right)\left(3abe^{bx}\right)$

$\qquad = 6ace^{bx}c\sin cx \cos cx + 3abe^{bx}\sin^2 cx$

1.12 Differentiate $y = x^2 \ln(x^2)\sin x$ with respect to x

Note that this function is a product of three terms so the strategy in this case is to differentiate the product $x^2 \ln\left(x^2\right)$ first using the chain rule. Then we differentiate $x^2 \ln\left(x^2\right)$ as one product and $\sin x$ as the other using the chain rule a second time as shown.

Let $u = x^2 \Rightarrow \dfrac{du}{dx} = 2x$

Let $v = \ln\left(x^2\right) \Rightarrow \dfrac{dv}{dx} = \dfrac{1}{x^2} \times 2x = \dfrac{2}{x}$

$\dfrac{d}{dx}(uv) = u\dfrac{dv}{dx} + v\dfrac{du}{dx}$

$\qquad = x^2 \times \dfrac{2}{x} + \ln\left(x^2\right) \times 2x$

$\qquad = 2x + 2x\ln\left(x^2\right)$

Now let $u = x^2 \ln\left(x^2\right) \Rightarrow \dfrac{du}{dx} = 2x + 2x\ln\left(x^2\right)$ From above result

And let $v = \sin x$

Now using the chain rule again:

$\dfrac{d}{dx}(uv) = u\dfrac{dv}{dx} + v\dfrac{du}{dx}$

$\qquad = x^2 \ln\left(x^2\right)\cos x + \sin x\left(2x + 2x\ln\left(x^2\right)\right)$

$\qquad = x^2 \ln\left(x^2\right)\cos x + 2x\sin x\left(1 + \ln\left(x^2\right)\right)$ Simplifying

Note: This technique is often useful when differentiating three or more products. However you can still use the standard method we obtained previously to differentiate this function if you prefer as follows.

Let $u = x^2 \Rightarrow \dfrac{du}{dx} = 2x$

$v = \ln\left(x^2\right) \Rightarrow \dfrac{dv}{dx} = \dfrac{1}{x^2} \times 2x = \dfrac{2}{x}$

$w = \sin x \Rightarrow \dfrac{dw}{dx} = \cos x$

Now using $\dfrac{dy}{dx} = vw\dfrac{du}{dx} + uw\dfrac{dv}{dx} + uv\dfrac{dw}{dx}$ we obtain

$$\frac{dy}{dx} = 2x\ln(x^2)\sin x + x^2\sin x \times \frac{2}{x} + x^2\ln(x^2)\cos x$$

$$= 2x\ln(x^2)\sin x + 2x\sin x + x^2\ln(x^2)\cos x$$

$$x^2\ln(x^2)\cos x + 2x\sin x(1 + \ln(x^2)) \qquad \text{Yielding the same result as before}$$

1.13 Differentiate $y = x^2\ln(x^2)\sin^2 x$ with respect to x

This function also has three products which we will tackle in the same way as the previous question.

Let $u = x^2\ln(x^2) \Rightarrow \dfrac{du}{dx} = 2x + 2x\ln(x^2)$ $\qquad$ Using the result from the previous question

Let $v = \sin^2 x \Rightarrow \dfrac{dv}{dx} = 2\sin x\cos = \sin 2x$

$$\frac{d}{dx}(uv) = u\frac{dv}{dx} + v\frac{du}{dx}$$

$$= x^2\ln(x^2)\sin 2x + \sin^2 x(2x + 2x\ln(x^2))$$

$$= x^2\ln(x^2)\sin 2x + 2x\sin^2 x(1 + \ln(x^2))$$

1.14 Find $\dfrac{d}{dx}(x^2\ln(\sin(x)))$

Let $y = x^2\ln(\sin(x))$

Let $u = x^2 \Rightarrow \dfrac{du}{dx} = 2x$

Let $v = \ln(\sin x) \Rightarrow \dfrac{dv}{dx} = \dfrac{1}{\sin x} \times \cos x = \dfrac{\cos x}{\sin x}$

$$\frac{d}{dx}(uv) = u\frac{dv}{dx} + v\frac{du}{dx}$$

$$= x^2 \times \frac{\cos x}{\sin x} + \ln(\sin x) \times 2x$$

$$= x^2\cot x + 2x\ln(\sin x) \qquad \text{Simplifying}$$

1.15 Differentiate $y = \sin^2 x\cos^2 x$ with respect to x

Let $u = \sin^2 x \Rightarrow \dfrac{du}{dx} = 2\sin x\cos x = \sin 2x$

Let $v = \cos^2 x \Rightarrow \dfrac{dv}{dx} = \cos x \times -2\sin x = -2\sin x\cos x = -\sin 2x$

$$\frac{d}{dx}(uv) = u\frac{dv}{dx} + v\frac{du}{dx}$$

$$= \sin^2 x \times -\sin 2x + \cos^2 x \times \sin 2x$$

$$= -\sin 2x\sin^2 x + \sin 2x\cos^2 x$$

$$= \sin 2x\cos^2 x - \sin 2x\sin^2 x \qquad \text{Rearranging}$$

1.16 Differentiate $y = \sin^3 x\ln(\cos^2 x)$ with respect to x

Let $u = \sin^3 x \Rightarrow \dfrac{du}{dx} = 3\sin^2 x \times \cos x = 3\sin^2 x\cos x$

Let $v = \ln\left(\cos^2 x\right) \Rightarrow \dfrac{dv}{dx} = \dfrac{1}{\cos^2 x} \times \left(-2\cos x \sin x\right) = -2\tan x$

$\dfrac{d}{dx}(uv) = u\dfrac{dv}{dx} + v\dfrac{du}{dx}$

$= \sin^3 x \times \left(-2\tan x\right) + \ln\left(\cos^2 x\right) \times 3\sin^2 x \cos x$

$= -2\tan x \sin^3 x + 3\sin^2 x \cos x \ln\left(\cos^2 x\right)$

$3\sin^2 x \cos x \ln\left(\cos^2 x\right) - 2\tan x \sin^3 x$ Rearranging

1.17 Differentiate $y = e^{3x}\ln\left(\sqrt{x}\right)$ with respect to x

Let $u = e^{3x} \Rightarrow \dfrac{du}{dx} = 3e^{3x}$

Let $v = \ln\left(x^{\frac{1}{2}}\right) \Rightarrow \dfrac{dv}{dx} = \dfrac{1}{x^{\frac{1}{2}}} \times \left(\dfrac{1}{2}x^{-\frac{1}{2}}\right) = \dfrac{1}{2x}$

$\dfrac{d}{dx}(uv) = u\dfrac{dv}{dx} + v\dfrac{du}{dx}$

$= e^{3x} \times \dfrac{1}{2x} + \ln\left(\sqrt{x}\right) \times 3e^{3x}$

$= e^{3x}\left(\dfrac{1}{2x} + 3\ln\left(\sqrt{x}\right)\right)$ Rearranging

1.18 Differentiate $y = e^{3x}\ln\left(\sqrt{x}\right)\cos^4 x$ with respect to x

This is another example of differentiating three products

Let $u = e^{3x}\ln\left(\sqrt{x}\right) \Rightarrow \dfrac{du}{dx} == e^{3x}\left(\dfrac{1}{2x} + 3\ln\left(\sqrt{x}\right)\right)$ From previous question

Let $v = \cos^4 x \Rightarrow \dfrac{dv}{dx} = 4\cos^3 x \times \left(-\sin x\right) = -4\cos^3 x \sin x$

$\dfrac{d}{dx}(uv) = u\dfrac{dv}{dx} + v\dfrac{du}{dx}$

$= e^{3x}\ln\left(\sqrt{x}\right) \times \left(-4\cos^3 x \sin x\right) + \cos^4 x \times e^{3x}\left(\dfrac{1}{2x} + 3\ln\left(\sqrt{x}\right)\right)$

$= e^{3x}\cos^4 x\left(\dfrac{1}{2x} + 3\ln\left(\sqrt{x}\right)\right) - 4e^{3x}\cos^3 x \sin x \ln\left(\sqrt{x}\right)$ Rearranging

1.19 Differentiate $y = x^2 \sin x \cos x$ with respect to x

Just as before we will attack the $x^2 \sin x$ product first.

Let $u = x^2 \Rightarrow \dfrac{du}{dx} = 2x$

Let $v = \sin x \Rightarrow \dfrac{du}{dx} = \cos x$

$\dfrac{d}{dx}(uv) = u\dfrac{dv}{dx} + v\dfrac{du}{dx}$

$$= x^2 \times \cos x + \sin x \times 2x = x^2 \cos x + 2x \sin x$$

Now let $u = x^2 \sin x \Rightarrow \dfrac{du}{dx} = x^2 \cos x + 2x \sin x$ From above result

And Let $v = \cos x \Rightarrow \dfrac{dv}{dx} = -\sin x$

$$\frac{d}{dx}(uv) = u\frac{dv}{dx} + v\frac{du}{dx}$$

$$= x^2 \sin x \times (-\sin x) + \cos x \times (x^2 \cos x + 2x \sin x)$$

$$= -x^2 \sin^2 x + \cos x (x^2 \cos x + 2x \sin x)$$

$$= \cos x (x^2 \cos x + 2x \sin x) - x^2 \sin^2 x$$ Rearranging

In many cases there is an alternative approach to solving a problem. If we look again at this question we can tackle it as follows:

We can re-write $y = x^2 \sin x \cos x$ as

$$y = x^2 \times \frac{1}{2}\sin 2x = \frac{1}{2}x^2 \sin 2x$$ Since $2\sin x \cos x = \sin 2x \Rightarrow \sin x \cos x = \frac{1}{2}\sin 2x$

Now we can proceed as before using the chain rule:

Let $u = \dfrac{1}{2}x^2 \Rightarrow \dfrac{du}{dx} = x$

Let $v = \sin 2x \Rightarrow \dfrac{dv}{dx} = 2\cos 2x$

$$\frac{d}{dx}(uv) = u\frac{dv}{dx} + v\frac{du}{dx}$$

$$= \frac{1}{2}x^2 \times 2\cos 2x + \sin 2x \times x$$

$$= x^2 \cos 2x + x \sin 2x$$ Simplifying

By writing $\cos 2x = \cos^2 x - \sin^2 x$ and $\sin 2x$ as $2\sin x \cos x$, we obtain the same result as before.

1.20 Differentiate $y = x^2 \sin^2 x \cos^2 x$ with respect to x

Again we will differentiate the product $x^2 \sin^2 x$ first as follows:

Let $u = x^2 \Rightarrow \dfrac{du}{dx} = 2x$

Let $v = \sin^2 x \Rightarrow \dfrac{dv}{dx} = 2\sin x \cos x$

$$\frac{d}{dx}(uv) = u\frac{dv}{dx} + v\frac{du}{dx}$$

$$= x^2 \times 2\sin x \cos x + \sin^2 x \times 2x$$

$$= 2x^2 \sin x \cos x + 2x \sin^2 x$$ Rearranging

Now let $u = x^2 \sin^2 x \Rightarrow \dfrac{du}{dx} = 2x^2 \sin x \cos x + 2x \sin^2 x$ From above result

And let $v = \cos^2 x \Rightarrow \dfrac{dv}{dx} = 2\cos x \times (-\sin x) = -2\sin x \cos x$

$$\frac{d}{dx}(uv) = u\frac{dv}{dx} + v\frac{du}{dx}$$

$$= x^2\sin^2 x \times (-2\sin x\cos x) + \cos^2 x\left(2x^2\sin x\cos x + 2x\sin^2 x\right)$$

$$= \cos^2 x\left(2x^2\sin x\cos x + 2x\sin^2 x\right) - 2x^2\sin^3 x\cos x \qquad \text{Simplifying}$$

MODULE D3.

CALCULUS D3 DIFFERENTIATION OF QUOTIENTS

INTRODUCTION

Let $y = \dfrac{u}{v}$ where u and v are given functions of x

Now $y = \dfrac{u}{v} \Rightarrow u = yv$. Hence we can use the rule for differentiating the product yv

$$\frac{du}{dx} = v\frac{dy}{dx} + y\frac{dv}{dx}$$

$$\Rightarrow \frac{dy}{dx} = \frac{1}{v}\frac{du}{dx} - \frac{y}{v}\frac{dv}{dx}$$

$$\Rightarrow \frac{dy}{dx} = \frac{1}{v}\frac{du}{dx} - \frac{u}{v^2}\frac{dv}{dx} \qquad \text{Substituting for } y = \frac{u}{v}$$

$$= \frac{v\dfrac{du}{dx} - u\dfrac{dv}{dx}}{v^2} \qquad \text{Rearranging}$$

Hence the differential coefficient of the quotient of two functions of x is equal to the denominator multiplied by the differential coefficient of the numerator minus the numerator multiplied by the differential coefficient of the denominator, all divided by the square of the denominator.

Example (i)

Find $\dfrac{dy}{dx}$ when $y = \dfrac{\sin x}{x^2}$

Let $u = \sin x \Rightarrow \dfrac{du}{dx} = \cos x$

Let $v = x^2 \Rightarrow \dfrac{dv}{dx} = 2x$

$$\frac{dy}{dx} = \frac{v\dfrac{du}{dx} - u\dfrac{dv}{dx}}{v^2} \qquad \text{Using rule for differentiating quotients}$$

$$= \frac{x^2 \times \cos x - \sin x \times 2x}{x^4}$$

$$= \frac{\cos x}{x^2} - \frac{2\sin x}{x^3} \qquad \text{Rearranging}$$

Example (ii)

Find $\dfrac{dy}{dx}$ when $y = \dfrac{1 - x^2}{1 + x}$

Let $u = 1 - x^2 \Rightarrow \dfrac{du}{dx} = -2x$

Let $v = 1 + x \Rightarrow \dfrac{dv}{dx} = 1$

$$\frac{dy}{dx} = \frac{(1+x) \times (-2x) - (1-x^2) \times 1}{(1+x^2)}$$ Using rule for differentiating quotients

$$= -\frac{2x}{1+x} - \frac{(1-x^2)}{1+x^2}$$ Rearranging

WORKED SOLUTIONS

1.1 Find $\dfrac{dy}{dx}$ when $y = \dfrac{1-x}{1+x}$

Let $u = 1-x \Rightarrow \dfrac{du}{dx} = -1$

Let $v = 1+x \Rightarrow \dfrac{dv}{dx} = 1$

$$\frac{dy}{dx} = \frac{v\dfrac{du}{dx} = u\dfrac{dv}{dx}}{v^2}$$ Using rule for differentiating quotients

$$= \frac{(1+x) \times (-1) - (1-x) \times 1}{(1+x)^2} = \frac{-2}{(1+x)^2}$$

1.2 Differentiate $\dfrac{2x^2 - 3}{(1-x^2)}$ with respect to x

Let $y = \dfrac{2x^2 - 3}{1-x^2}$

Let $u = 2x^2 - 3 \Rightarrow \dfrac{du}{dx} = 4x$

Let $v = 1-x^2 \Rightarrow \dfrac{dv}{dx} - 2x$

$$\frac{dy}{dx} = \frac{(1-x^2) \times 4x - (2x^2 - 3) \times (-2x)}{(1-x^2)^2}$$ Using Rule for differentiating quotients

$$= \frac{4x - 4x^3 + 4x^3 - 6x}{(1-x^2)^2}$$

$$= \frac{-2x}{(1-x^2)^2}$$ Simplifying

1.3 Differentiate $\dfrac{x^2 + 5}{(4x^2 - 3)}$ with respect to x

Let $y = \dfrac{x^2 + 5}{(4x^2 - 3)}$

Let $u = x^2 + 5 \Rightarrow \dfrac{du}{dx} = 2x$

Let $v = 4x^2 - 3 \Rightarrow \dfrac{dv}{dx} = 8x$

$$\frac{dy}{dx} = \frac{\left(4x^2 - 3\right) \times 2x - \left(x^2 + 5\right) \times 8x}{\left(4x^2 - 3\right)^2}$$

Using Rule for differentiating quotients

$$= \frac{8x^3 - 6x - 8x^3 - 40x}{\left(4x^2 - 3\right)^2}$$

$$= \frac{-46x}{\left(4x^2 - 3\right)^2}$$

Simplifying

1.4 Differentiate $\left(\dfrac{2x}{1-x}\right)^2$ with respect to x

Let $y = \left(\dfrac{2x}{(1-x)}\right)^2 = \dfrac{4x^2}{(1-x)^2}$

Expanding

Let $u = 4x^2 \Rightarrow \dfrac{du}{dx} = 8x$

Let $v = (1-x)^2 \Rightarrow \dfrac{dv}{dx} = 2x - 2$

$$\frac{dy}{dx} = \frac{(1-x)^2 \times 8x - 4x^2 \times (2x - 2)}{(1-x)^4}$$

Using Rule for differentiating quotients

$$= \frac{8x\left(1 - 2x + x^2\right) - 8x^3 + 8x^2}{(1-x)^4}$$

$$= \frac{8x - 16x^2 + 8x^3 - 8x^3 + 8x^2}{(1-x)^4}$$

$$= \frac{8x - 8x^2}{(1-x)^4} = \frac{8x(1-x)}{(1-x)^4} = \frac{8x}{(1-x)^3}$$

Simplifying

1.5 Differentiate $\dfrac{\sqrt{x}}{1 + 3\sqrt{x}}$ with respect to x

Let $y = \dfrac{\sqrt{x}}{1 + 3\sqrt{x}}$

Let $u = \sqrt{x} = x^{\frac{1}{2}} \Rightarrow \dfrac{du}{dx} = \dfrac{1}{2}x^{-\frac{1}{2}}$

Let $v = 1 + 3\sqrt{x} = 1 + 3x^{\frac{1}{2}} \Rightarrow \dfrac{dv}{dx} = \dfrac{3}{2}x^{-\frac{1}{2}}$

$$\frac{dy}{dx} = \frac{\left(1 + 3x^{\frac{1}{2}}\right) \times \left(\dfrac{1}{2}x^{-\frac{1}{2}}\right) - x^{\frac{1}{2}}\left(\times \dfrac{3}{2}x^{-\frac{1}{2}}\right)}{\left(1 + 3\sqrt{x}\right)^2}$$

Using Rule for differentiating quotients

$$= \frac{\frac{1}{2}x^{-\frac{1}{2}} + \frac{3}{2} - \frac{3}{2}}{\left(1+3\sqrt{x}\right)^2}$$

$$= \frac{\frac{1}{2}x^{-\frac{1}{2}}}{\left(1+3\sqrt{x}\right)^2}$$

$$= \frac{1}{2\sqrt{x}\left(1+3\sqrt{x}\right)^2} \qquad \text{Simplifying}$$

1.6 Find $\dfrac{d}{dx}\left(\dfrac{1-x}{1+x}\right)^2$

Let $y = \left(\dfrac{1-x}{1+x}\right)^2 = \dfrac{(1-x)^2}{(1+x)^2}$

Let $u = (1-x)^2 \Rightarrow \dfrac{du}{dx} = 2(1-x)\times(-1) = 2x-2$

Let $v = (1+x)^2 \Rightarrow \dfrac{dv}{dx} = 2(1+x)\times1 = 2x+2$

$$\frac{dy}{dx} = \frac{(1+x)^2\times(2x-2)-(1-x)^2\times(2x+2)}{(1+x)^4} \qquad \text{Using Rule for differentiating quotients}$$

$$= \frac{\left(1+2x+x^2\right)(2x-2)-\left(1-2x+x^2\right)(2x+2)}{(1+x)^4}$$

$$= \frac{2x+4x^2+2x^3-2-4x-2x^2-\left(2x-4x^2+2x^3+2-4x+2x^2\right)}{(1+x)^4}$$

$$= \frac{2x+4x^2+2x^3-2-4x-2x^2-2x+4x^2-2x^3-2+4x-2x^2}{(1+x)^4}$$

$$= \frac{4x^2-4}{(1+x)^4} = \frac{4\left(x^2-1\right)}{(1+x)^4} = \frac{4(x+1)(x-1)}{(1+x)^4} \qquad \text{Rearranging}$$

$$= \frac{4(x-1)}{(1+x)^3}$$

1.7 Find $\dfrac{dy}{dx}$ when $y = \dfrac{\cos x}{1+x}$

Let $u = \cos x \Rightarrow \dfrac{du}{dx} = -\sin x$

Let $v = 1+x \Rightarrow \dfrac{dv}{dx} = 1$

$$\frac{dy}{dx} = \frac{(1+x)\times(-\sin x)-\cos x\times1}{(1+x)^2} \qquad \text{Using Rule for differentiating quotients}$$

$$= \frac{-\sin x(1+x) - \cos x}{(1+x)^2}$$

Rearranging

1.8 Differentiate $\dfrac{\sin^2 x}{(3x+2)^3}$ with respect to x

Let $y = \dfrac{\sin^2 x}{(3x+2)^3}$

Let $u = \sin^2 x \Rightarrow \dfrac{du}{dx} = 2\sin x \cos x$

Let $v = (3x+2)^3 \Rightarrow \dfrac{dv}{dx} = 3(3x+2)^2 \times 3 = 9(3x+2)^2$

$$\frac{dy}{dx} = \frac{(3x+2)^3 \times 2\sin x \cos - \sin^2 x \times 9(3x+2)^2}{(3x+2)^6}$$

Using Rule for differentiating quotients

$$= \frac{2(3x+2)^3 \sin x \cos x - 9(3x+2)^2 \sin^2 x}{(3x+2)^6}$$

$$= \frac{2\sin x \cos x}{(3x+2)^3} - \frac{9\sin^2 x}{(3x+2)^4}$$

Simplifying

1.9 Differentiate $\dfrac{\sin x \cos x}{x^2 + 2x + 2}$ with respect to x

Let $y = \dfrac{\sin x \cos x}{x^2 + 2x + 2}$

Let $u = \sin x \cos x \Rightarrow \dfrac{du}{dx} = -\sin^2 x + \cos^2 x$

Using differentiation of a product rule

Let $v = x^2 + 2x + 2 \Rightarrow \dfrac{dv}{dx} = 2x + 2$

$$\frac{dy}{dx} = \frac{(x^2 + 2x + 2) \times (-\sin^2 x + \cos^2 x) - \sin x \cos x \times (2x+2)}{(x^2 + 2x + 2)^2}$$

Using Rule for differentiating quotients

$$= \frac{(x^2 + 2x + 2)(\cos^2 x - \sin^2 x) - 2(x+2)\sin x \cos x}{(x^2 + 2x + 2)^2}$$

$$= \frac{(x^2 + 2x + 2)\cos 2x - (x+2)\sin 2x}{(x^2 + 2x + 2)^2}$$

Since $(\cos^2 x - \sin^2 x) = \cos 2x$ and

$2\sin x \cos x = \sin 2x$

1.10 Find $\dfrac{d}{dx}\left(\dfrac{\cos^2 x}{3\sin^2 x} \right)$

Let $y = \dfrac{\cos^2 x}{3\sin^3 x}$

Let $u = \cos^2 x \Rightarrow \dfrac{du}{dx} = -2\cos x \sin x$

Let $v = 3\sin^3 x \Rightarrow \dfrac{dv}{dx} = 9\sin^2 x \cos x$

$\dfrac{dy}{dx} = \dfrac{3\sin^3 x \times (-2\cos x \sin x) - \cos^2 x \times 9\sin^2 x \cos x}{9\sin^6 x}$ ⸺ Using Rule for differentiating quotients

$= \dfrac{-6\sin^4 x \cos x - 9\sin^2 x \cos^3 x}{9\sin^6 x}$ ⸺ Simplifying

$= \dfrac{-2\sin^2 x \cos x - 3\cos^3 x}{3\sin^4 x}$ ⸺ Dividing top & bottom by $\sin^2 x$

$= \dfrac{-\cos x(2\sin^2 x + 3\cos^2 x)}{3\sin^4 x}$ ⸺ Simplifying

$= \dfrac{-\cos x(2 - 2\cos^2 x + 3\cos^2 x)}{3\sin^4 x}$ ⸺ Substituting for $\sin^2 x = 1 - \cos^2 x$

$= \dfrac{-\cos x(\cos^2 x + 2)}{3\sin^4 x}$

1.11 Prove that $\dfrac{dy}{dx} = 4\tan x \sec^2 x$ when $y = \dfrac{1 + \sin^2 x}{1 - \sin^2 x}$

Let $u = 1 + \sin^2 x \Rightarrow \dfrac{du}{dx} = 2\sin x \cos x$

Let $v = 1 - \sin^2 x \Rightarrow \dfrac{dv}{dx} = -2\sin x \cos x$

$\dfrac{dy}{dx} = \dfrac{(1 - \sin^2 x) \times 2\sin x \cos x - (1 + \sin^2 x) \times (-2\sin x \cos x)}{(1 - \sin^2 x)^2}$ ⸺ Using Rule for differentiating quotients

$= \dfrac{2\sin x \cos x - 2\sin^3 x \cos x + 2\sin x \cos x + 2\sin^3 x \cos x}{(1 - \sin^2 x)^2}$ ⸺ Expanding brackets

$= \dfrac{4\sin x \cos x}{(1 - \sin^2 x)^2}$ ⸺ Simplifying

$= \dfrac{4\sin x \cos x}{(\cos^2 x)^2} = \dfrac{4\sin x \cos x}{\cos^4 x} = \dfrac{4\sin x}{\cos^3 x}$ ⸺ Substituting for $1 - \sin^2 x = \cos^2 x$

$= \dfrac{4\sin x}{\cos} \times \dfrac{1}{\cos^2 x} = 4\tan x \sec^2 x$ ⸺ Rearranging

1.12 Find $\dfrac{d}{dx}\left(\dfrac{2x}{\sqrt{x^2 - 3}}\right)$

Let $u = 2x \Rightarrow \dfrac{du}{dx} = 2$

Let $v = \sqrt{x^2 - 3} = \left(x^2 - 3\right)^{\frac{1}{2}} \Rightarrow \dfrac{dv}{dx} = \dfrac{1}{2}\left(x^2 - 3\right)^{-\frac{1}{2}} \times 2x = x\left(x^2 - 3\right)^{-\frac{1}{2}}$

$\dfrac{dy}{dx} = \dfrac{\left(x^2 - 3\right)^{\frac{1}{2}} \times 2 - 2x \times x\left(x^2 - 3\right)^{-\frac{1}{2}}}{x^2 - 3}$ Using Rule for differentiating quotients

$= \dfrac{2\left(x^2 - 3\right)^{\frac{1}{2}} - 2x^2\left(x^2 - 3\right)^{-\frac{1}{2}}}{x^2 - 3}$

$= \dfrac{2\left(x^2 - 3\right) - 2x^2}{\left(x^2 - 3\right)^{\frac{3}{2}}}$ Multiplying top and bottom by $\left(x^2 - 3\right)^{\frac{1}{2}}$

$= \dfrac{2x^2 - 6 - 2x^2}{\left(x^2 - 3\right)^{\frac{3}{2}}}$

$= \dfrac{-6}{\left(x^2 - 3\right)^{\frac{3}{2}}}$ Simplifying

1.13 Differentiate $\dfrac{x^2}{\left(x^3 - 3x + 2\right)^{\frac{3}{2}}}$ with respect to x

Let $u = x^2 \Rightarrow \dfrac{du}{dx} = 2x$

Let $v = \left(x^3 - 3x + 2\right)^{\frac{3}{2}} \Rightarrow \dfrac{dv}{dx} = \dfrac{3}{2}\left(x^3 - 3x + 2\right)^{\frac{1}{2}} \times \left(3x^2 - 3\right)$

$\dfrac{dy}{dx} = \dfrac{\left(x^3 - 3x + 2\right)^{\frac{3}{2}} \times 2x - x^2 \times \dfrac{3}{2}\left(3x^2 - 3\right)\left(x^3 - 3x + 2\right)^{\frac{1}{2}}}{\left(x^3 - 3x + 2\right)^3}$ Using Rule for differentiating quotients

$= \dfrac{2x\left(x^3 - 3x + 2\right)^{\frac{3}{2}} - \dfrac{3x^2}{2}\left(3x^2 - 3\right)\left(x^3 - 3x + 2\right)^{\frac{1}{2}}}{\left(x^3 - 3x + 2\right)^3}$ Simplifying

$= \dfrac{4x\left(x^3 - 3x + 2\right) - 3x^2\left(3x^2 - 3\right)}{2\left(x^3 - 3x + 2\right)^{\frac{5}{2}}}$ Multiplying top and bottom by $2\left(x^3 - 3x + 2\right)^{-\frac{1}{2}}$

$= \dfrac{2x}{\left(x^3 - 3x + 2\right)^{\frac{3}{2}}} - \dfrac{3x^2\left(3x^2 - 3\right)}{2\left(x^3 - 3x + 2\right)^{\frac{5}{2}}}$ Rearranging & simplifying

1.14　Differentiate $\sqrt{\dfrac{3x+2}{2x-3}}$ with respect to x

Let $y = \sqrt{\dfrac{3x+2}{2x-3}}$

Let $u = (3x+2)^{\frac{1}{2}} \Rightarrow \dfrac{du}{dx} = \dfrac{1}{2}(3x+2)^{-\frac{1}{2}} \times 3 = \dfrac{3}{2}(3x+2)^{-\frac{1}{2}}$

Let $v = (2x-3)^{\frac{1}{2}} \Rightarrow \dfrac{dv}{dx} = \dfrac{1}{2}(2x-3)^{-\frac{1}{2}} \times 2 = (2x-3)^{-\frac{1}{2}}$

$$\dfrac{dy}{dx} = \dfrac{(2x-3)^{\frac{1}{2}} \times \dfrac{3}{2}(3x+2)^{-\frac{1}{2}} - \dfrac{1}{2}(3x+2)^{\frac{1}{2}} \times (2x-3)^{-\frac{1}{2}}}{2x-3}$$

$$= \dfrac{3(2x-3)^{\frac{1}{2}}(3x+2)^{-\frac{1}{2}} - (3x+2)^{\frac{1}{2}}(2x-3)^{-\frac{1}{2}}}{2x-3} \qquad \text{Rearranging}$$

$$= \dfrac{3(2x-3)(3x+2)^{-\frac{1}{2}} - (3x+2)^{\frac{1}{2}}}{(2x-3)^{\frac{3}{2}}} \qquad \text{Multiplying top and bottom by } (2x-3)^{\frac{1}{2}}$$

$$= \dfrac{3(2x-3)(3x+2)^{-\frac{1}{2}}}{(2x-3)^{\frac{3}{2}}} - \dfrac{(3x+2)^{\frac{1}{2}}}{(2x-3)^{\frac{3}{2}}} \qquad \text{Splitting into two fractions}$$

$$= \dfrac{3(2x-3)}{(2x-3)^{\frac{3}{2}}(3x+2)^{\frac{1}{2}}} - \dfrac{(3x+2)^{\frac{1}{2}}}{(2x-3)^{\frac{3}{2}}} \qquad \text{Rearranging}$$

$$= \dfrac{3}{(2x-3)^{\frac{1}{2}}(3x+2)^{\frac{1}{2}}} - \dfrac{(3x+2)^{\frac{1}{2}}}{(2x-3)^{\frac{3}{2}}} \qquad \text{Dividing top and bottom of fraction by } 2x-3$$

$$= \dfrac{3}{\sqrt{2x-3}\sqrt{3x+2}} - \dfrac{\sqrt{3x+2}}{(2x-3)^{\frac{3}{2}}} \qquad \text{Simplifying}$$

1.15　Differentiate $\dfrac{\sqrt{(x+3)^5}}{3x^2}$ with respect to x

Let $y = \dfrac{\sqrt{x+3}}{3x^2}$

Let $u = \sqrt{(x+3)^5} = (x+3)^{\frac{5}{2}} \Rightarrow \dfrac{du}{dx} = \dfrac{5}{2}(x+3)^{\frac{3}{2}} \times 1 = \dfrac{5}{2}(x+3)^{\frac{3}{2}}$

Let $v = 3x^2 \Rightarrow \dfrac{du}{dx} = 3x$

$$\dfrac{dy}{dx} = \dfrac{3x^2 \times \frac{15}{2}(x+3)^{\frac{3}{2}} - (x+3)^{\frac{5}{2}} \times 6x}{9x^4} \qquad \text{By rule for differentiating quotients}$$

$$= \frac{\frac{15}{2}x^2(x+3)^{\frac{3}{2}} - 6x(x+3)^{\frac{5}{2}}}{9x^4} \qquad \text{Rearranging}$$

$$= \frac{15x^2(x+3)^{\frac{3}{2}} - 12x(x+3)^{\frac{5}{2}}}{18x^4} \qquad \text{Multiplying top and bottom by 2}$$

$$= \frac{15x^2(x+3)^{\frac{3}{2}}}{18x^4} - \frac{12x(x+3)^{\frac{5}{2}}}{18x^4} \qquad \text{Splitting into two fractions}$$

$$= \frac{5(x+3)^{\frac{3}{2}}}{6x^2} - \frac{2(x+3)^{\frac{5}{2}}}{3x^3} \qquad \text{Simplifying}$$

1.16 Differentiate $\dfrac{x^3 e^x}{(1+x^2)}$ with respect to x

Let $y = \dfrac{x^3 e^x}{(1+x^2)}$

Let $u = x^3 e^x \Rightarrow \dfrac{du}{dx} = x^3 \times e^x + e^x \times 3x^2 = x^3 e^x + 3x^2 e^x$ $\qquad$ By rule for differentiating products

Let $v = 1 + x^2 \Rightarrow \dfrac{dv}{dx} = 2x$

$$\frac{dy}{dx} = \frac{(1+x) \times (x^3 e^x + 3x^2 e^x) - x^3 e^x \times 2x}{(1+x)^2} \qquad \text{By rule for differentiating quotients}$$

$$= \frac{(1+x)(x^3 e^x + 3x^2 e^x) - 2x^4 e^x}{(1+x)^2} \qquad \text{Simplifying}$$

$$= \frac{(1+x)(x^3 e^x + 3x^2 e^x)}{(1+x)^2} - \frac{2x^4 e^x}{(1+x)^2} \qquad \text{Splitting into two fractions}$$

$$= \frac{x^2 e^x(x+3)}{1+x^2} - \frac{2x^4 e^x}{(1+x^2)^2} \qquad \text{Simplifying}$$

1.17 Differentiate $\dfrac{e^{x^2}}{\sin x}$ with respect to x

Let $u = e^{x^2} \Rightarrow \dfrac{du}{dx} = 2x \times e^{x^2} = 2x e^{x^2}$ $\qquad$ By rule for differentiating products

Let $v = \sin x \Rightarrow \dfrac{du}{dx} = \cos x$

$$\frac{dy}{dx} = \frac{\sin x \times 2x e^{x^2} - e^{x^2} \times \cos x}{\sin^2 x} \qquad \text{By Rule for differentiating quotients}$$

$$= \frac{2xe^{x^2}\sin x - e^{x^2}\cos x}{\sin^2 x} \qquad \text{Rearranging}$$

$$= \frac{2xe^{x^2} - e^{x^2}\cot x}{\sin x} \qquad \text{Dividing top \& bottom by } \sin x$$

$$= 2xe^{x^2}\csc x - e^{x^2}\cot x \csc x \qquad \text{Simplifying}$$

$$= e^{x^2}\left(2x\csc x - \cot x \csc x\right) \qquad \text{Simplifying}$$

1.18 Find $\dfrac{d}{dx}\left(\dfrac{3x^2 e^x}{\cos^2 x}\right)$

Let $y = \dfrac{3x^2 e^x}{\cos^2 x}$

Let $u = 3x^2 e^x \Rightarrow \dfrac{du}{dx} = e^x \times 6x + 3x^2 e^x = 6xe^x + 3x^2 e^x$ By rule for differentiating products

Let $v = \cos^2 x \Rightarrow \dfrac{dv}{dx} = 2\cos x \times \sin x = 2\sin x \cos x$

$$\frac{dy}{dx} = \frac{\cos^2 x \times \left(6xe^x + 3x^2 e^x\right) - 3x^2 e^x \times 2\sin x \cos x}{\cos^4 x} \qquad \text{By Rule for differentiating quotients}$$

$$= \frac{\left(6xe^x + 3x^2 e^x\right)\cos^2 x - 6x^2 e^x \sin x \cos x}{\cos^4 x} \qquad \text{Rearranging}$$

$$= \frac{6xe^x + 3x^2 e^x}{\cos^2 x} + \frac{6x^2 e^x \sin x}{\cos^3 x} \qquad \text{Splitting into separate fractions and simplifying}$$

$$= \frac{6xe^x + 3x^2 e^x}{\cos^2 x} + \frac{6x^2 e^x \sin x}{\cos x \cos^2 x} \qquad \text{Rearranging}$$

$$= 3xe^x\left(2 + x\right)\sec^2 x + 6x^2 e^x \tan x \sec^2 x \qquad \text{Simplifying}$$

1.19 Find $\dfrac{d}{dx}\left(\dfrac{2x^3 e^x}{\sin x \cos^2 x}\right)$

Let $y = \dfrac{2x^3 e^x}{\sin x \cos^2 x}$

Let $u = 2x^3 e^x \Rightarrow \dfrac{du}{dx} = e^x \times 6x^2 + 2x^3 \times e^x$ By Rule for differentiating products

$$= 6x^2 e^x + 2x^3 e^x \qquad \text{Rearranging}$$

Let $v = \sin x \cos^2 x$

$$\Rightarrow \frac{dv}{dx} = \cos^2 x \times \cos x + \sin x \times 2\cos x \times (-\sin x) \qquad \text{By Rule for differentiating products}$$

$$= \cos^3 x - 2\sin^2 x \cos x \qquad \text{Rearranging}$$

$$\frac{dy}{dx} = \frac{\sin x \cos^2 x \times \left(6x^2 e^x + 2x^3 e^x\right) - 2x^3 e^x \times \left(\cos^3 x - 2\sin^2 x \cos x\right)}{\sin^2 x \cos^4 x} \qquad \text{By Rule for differentiating quotients}$$

$$= \frac{\left(6x^2 e^x + 2x^3 e^x\right)\sin x \cos^2 x - 2x^3 e^x\left(\cos^3 x - 2\sin^2 x \cos x\right)}{\sin^2 x \cos^4 x} \qquad \text{Rearranging}$$

$$= \frac{\left(6x^2e^x + 2x^3e^x\right)\sin x \cos^2 x}{\sin^2 x \cos^2 x} - \frac{2x^3e^x \cos^3 x}{\sin^2 x \cos^2 x} + \frac{4x^3e^x \sin^2 x \cos x}{\sin^2 x \cos^4 x} \qquad \text{Splitting into fractions}$$

$$= \left(6x^2e^x + 2x^3e^x\right)\csc x \sec^2 x - 2x^3e^x \sec x \csc^2 x + 4x^3e^x \sec^3 x \qquad \text{Simplifying}$$

1.20   Find $\dfrac{d}{dx}\left(\dfrac{x^2 \sin x \ln\left(x^3\right)}{\cos^2 x}\right)$

Let $y = \dfrac{x^2 \sin x \ln\left(x^3\right)}{\cos^2 x}$

The numerator contains a triple product so we need to split it into two parts first as demonstrated in previous questions as follows:

Let $u = x^2 \sin x \Rightarrow \dfrac{du}{dx} = \sin x \times 2x + x^2 \times \cos x = 2x\sin x + x^2 \cos x$

Let $v = \ln\left(x^3\right) \Rightarrow \dfrac{dv}{dx} = 3x^2 \times \dfrac{1}{x^3} = \dfrac{3}{x}$

Then $\dfrac{d}{dx}\left(x^2 \sin x \ln\left(x^3\right)\right) = \ln\left(x^3\right) \times \left(2x\sin x + x^2 \cos x\right) + x^2 \sin x \times \dfrac{3}{x}$

$$= \left(2x\sin x + x^2 \cos x\right)\ln\left(x^3\right) + 3x\sin x \qquad \text{Simplifying}$$

Now let $u = x^2 \sin x \ln\left(x^3\right)$

$$\Rightarrow \frac{du}{dx} = \left(2x\sin x + x^2 \cos x\right)\ln\left(x^3\right) + 3x\sin x \qquad \text{From above result}$$

And let $v = \cos^2 x \Rightarrow \dfrac{dv}{dx} = 2\cos x \times \left(-\sin x\right) = -2\sin x \cos x$

Hence

$$\frac{dy}{dx} = \frac{\cos^2 x \times \left[\left(2x\sin x + x^2 \cos x\right)\ln\left(x^3\right) + 3x\sin x\right] - x^2 \sin x \ln\left(x^3\right) \times \left(-2\sin x \cos x\right)}{\cos^4 x} \qquad \begin{array}{l}\text{By rule for}\\\text{differentiating}\\\text{quotients}\end{array}$$

$$= \frac{\cos^2 x \left[\left(2x\sin x + x^2 \cos x\right)\ln\left(x^3\right) + 3x\sin x\right] + 2x^2 \ln\left(x^3\right)\sin^2 x \cos x}{\cos^4 x}$$

$$= \frac{\left[\left(2x\sin x + x^2 \cos x\right)\ln\left(x^3\right) + 3x\sin x\right]}{\cos^2 x} + \frac{2x^2 \ln\left(x^3\right)\sin^2 x}{\cos^3 x} \qquad \text{Splitting into 2 fractions}$$

$$= \sec^2 x \left[\left(2x\sin x + x^2 \cos x\right)\ln\left(x^3\right) + 3x\sin x\right] + 2x^2 \ln\left(x^3\right)\tan^2 x \sec x \qquad \text{Simplifying}$$

1.21   Find $\dfrac{d}{dx}\left(\dfrac{x^2 \ln\left(\cos x\right)}{\sin^3 3x}\right)$

Let $y = \dfrac{x^2 \ln\left(\cos x\right)}{\sin^3 3x}$

Let $u = x^2 \ln\left(\cos x\right)$

$$\Rightarrow \frac{du}{dx} = \ln\left(\cos x\right) \times 2x + x^2 \times \frac{1}{\cos x} \times \left(-\sin x\right) \qquad \text{By rule for differentiating products}$$

$$= 2x\ln\left(\cos x\right) - x^2 \tan x \qquad \text{Simplifying}$$

Let $v = \sin^3 3x \Rightarrow \dfrac{dv}{dx} = 3\sin^2 3x \times 3\cos 3x$ By Rule for differentiating products

$$= 9\sin^2 3x \cos 3x$$ Simplifying

$$\frac{dy}{dx} = \frac{\sin^3 3x \times \left(2x\ln(\cos x) - x^2 \tan x\right) - x^2 \ln(\cos x) \times 9\sin^2 3x \cos 3x}{\sin^6 3x}$$ By rule for differentiating quotients

$$= \frac{\left(2x\ln(\cos x) - x^2 \tan x\right)}{\sin^3 3x} - \frac{9x^2 \ln(\cos x)\cos 3x}{\sin^4 3x}$$ Splitting into separate fractions

$$= \frac{2x\ln(\cos x)}{\sin^3 3x} - \frac{x^2 \tan x}{\sin^3 3x} - \frac{9x^2 \ln(\cos x)\cos 3x}{\sin^4 3x}$$ Rearranging

$$= 2x\ln(\cos x)\csc^3 3x - x^2 \tan 3x \csc^3 3x - 9x^2 \ln(\cos x)\cot 3x \csc^3 3x$$ Simplifying

1.22   Find $\dfrac{\ln(\cos x)\tan^2 x}{\sin^3 3x \cos 4x}$

Let $y = \dfrac{\ln(\cos x)\tan^2 x}{\sin^3 3x \cos 4x}$

Here we have a product of two functions in the numerator and a product of two functions in the denominator. In order to proceed we have to differentiate each of these first.

Dealing with the numerator first:

Let $z = \ln(\cos x)\tan^2 x$

$$\Rightarrow \frac{dz}{dx} = \tan^2 x \times \left(\frac{1}{\cos x} \times -\sin x\right) + \ln(\cos x) + \ln(\cos x) \times 2\tan x \times \frac{1}{\cos^2 x}$$

By Rule for differentiating products

$$= -\tan^2 x \tan x + 2\ln(\cos(x))\tan x \sec^2 x$$

$$= -\tan^3 x + 2\ln(\cos(x))\tan x \sec^2 x$$

$$= 2\ln(\cos(x))\sec^2 x \tan x - \tan^3 x$$ Simplifying

Now we will deal with the denominator

Now let $z = \sin^3 3x \cos 4x$

$$\Rightarrow \frac{dz}{dx} = \cos 4x \times \left(3\sin^2 3x \times 3\cos 3x\right) + \sin^3 3x \times (-4\sin 4x)$$

By Rule for differentiating products

$$= 9\sin^2 3x \cos 4\cos 3x - 4\sin^3 3x \sin 4x$$ Simplifying

Now let $u = \ln(\cos x)\tan^2 x$

$$\Rightarrow \frac{du}{dx} = 2\ln(\cos(x))\sec^2 x \tan x - \tan^3 x$$ From previous result

And let $v = \sin^3 3x \cos 4x$

$$\Rightarrow \frac{dv}{dx} = 9\sin^2 3x \cos 4\cos 3x - 4\sin^3 3x \sin 4x$$ From previous result

Hence $\dfrac{dy}{dx} = \dfrac{\sin^3 3x \cos 4x \times \left(2\ln(\cos(x))\sec^2 x \tan x - \tan^3 x\right)}{\sin^6 3x \cos^2 4x}$

Using rule for differentiating quotients

132

$$-\frac{\ln\left(\cos x\right)\tan^2 x \times \left(9\sin^2 3x\cos 4\cos 3x - 4\sin^3 3x\sin 4x\right)}{\sin^6 3x\cos^2 4x}$$

$$= 2\ln(\cos x)\sec^2 x\sec 4x\tan x\csc^3 3x - \sec 4x\tan^3 x\csc^3 3x$$

$$-9\cot 3x\ln\left(\cos x\right)\sec 4x\tan^2 x\csc^3 3x + 4\ln\left(\cos x\right)\sec 4x\tan^2 x\tan 4x\csc^3 3x$$

Admittedly this question is quite involved but it does give good practice in differentiating products and quotients and in manipulating and rationalising of trigonometric identities.

MODULE D4.

CALCULUS D4 DIFFERENTIATION OF IMPLICIT FUNCTIONS

INTRODUCTION

So far we have been examining explicit functions where the dependent variable is expressed explicitly in terms of the independent variable - for example $y = 6x^2 + 3x + 3$ is an explicit function where the dependent variable y is expressed explicitly in terms of the independent variable x. However, the equation $x + y + \cos y = 6$ cannot be expressed in this way as y is expressed implicitly in terms of x. In other words, we cannot rearrange this equation so that we arrive at an equation y = terms containing x or a constant.

To solve an implicit equation, we differentiate it term by term as shown in the following examples:

1.1 Find $\dfrac{dy}{dx}$ when $x^2 + y^2 = 2x$

Differentiating with respect to x we get:

$$\frac{d}{dx}\left(2x^2\right) + \frac{d}{dy}\left(y^2\right) \times \frac{dy}{dx} = \frac{d}{dx}\left(2x\right)$$ Using rule for differentiating a function of a function

$$\Rightarrow 2x + 2y\frac{dy}{dx} = 2$$

$$2y\frac{dy}{dx} = 2 - 2x$$ Rearranging

$$\text{Hence } \frac{dy}{dx} = \frac{1-x}{y}$$ Rearranging

Note that in this case it is possible to express this as an explicit function since we can say that

$$y^2 = 2x - x^2$$ y^2 is expressed explicitly in terms of x

$$\Rightarrow y = \sqrt{2x - x^2} = \left(2x - x^2\right)^{\frac{1}{2}}$$

$$\Rightarrow \frac{dy}{dx} = \tfrac{1}{2}\left(2x - x^2\right)^{-\frac{1}{2}} \times \left(2 - 2x\right)$$

$$= \frac{1-x}{\sqrt{2x - x^2}} = \frac{1-x}{y}$$ Since $y = \sqrt{2x - x^2}$

1.2 Find $\dfrac{dy}{dx}$ when $2x^2 - 3y^3 + 4x - 5y = 0$

Differentiating we get:

$$4x - 9y^2\frac{dy}{dx} + 4 - 5\frac{dy}{dx} = 0$$

$$\Rightarrow 9y^2\frac{dy}{dx} + 5\frac{dy}{dx} = 4x + 4$$ Rearranging

$$\Rightarrow \frac{dy}{dx}\left(9y^2 + 5\right) = 4x + 4$$ Rearranging

$$\Rightarrow \frac{dy}{dx} = \frac{4x+4}{(9y^2+5)} = \frac{4(x+1)}{(9y^2+5)} \qquad \text{Rearranging and simplifying}$$

1.3 Find $\frac{dy}{dx}$ when $3x^2 + 7xy + 9y^2 = 6$

Differentiating we get:

$$6x + (y \times 7) + \left(7x\frac{dy}{dx}\right) \Rightarrow +18y\frac{dy}{dx} = 0 \qquad \text{Using rule for differentiating products for } 7xy$$

$$\Rightarrow 7x\frac{dy}{dx} + 18y\frac{dy}{dx} = -6x - 7y \qquad \text{Rearranging}$$

$$\Rightarrow \frac{dy}{dx}(7x + 18y) = -(6x + 7y) \qquad \text{Rearranging}$$

$$\Rightarrow \frac{dy}{dx} = \frac{-(6x+7y)}{(7x+18y)} \qquad \text{Rearranging and simplifying}$$

1.4 Find $\frac{dy}{dx}$ when $x^3 + y^2 = 6xy + 4$

Differentiating we get:

$$3x^2 + 2y\frac{dy}{dx} = (y \times 6) + \left(6x \times \frac{dy}{dx}\right) \qquad \text{Using rule for differentiating products for } 6xy$$

$$\Rightarrow 3x^2 + 2y\frac{dy}{dx} = 6y + 6x\frac{dy}{dx} \qquad \text{Simplifying}$$

$$\Rightarrow 2y\frac{dy}{dx} - 6x\frac{dy}{dx} = 6y - 3x^2 \qquad \text{Rearranging}$$

$$\Rightarrow \frac{dy}{dx}(2y - 6x) = 6y - 3x^2 \qquad \text{Rearranging}$$

$$\Rightarrow \frac{dy}{dx} = \frac{6y - 3x^2}{(2y - 6x)} \qquad \text{Rearranging}$$

1.5 Find $\frac{dy}{dx}$ when $x + y = \cos x = 3$

Differentiating we get:

$$1 + \frac{dy}{dx} - \sin y\frac{dy}{dx} = 0$$

$$\Rightarrow \frac{dy}{dx}(1 - \sin y) = -1 \qquad \text{Rearranging}$$

$$\Rightarrow \frac{dy}{dx} = \frac{-1}{1 - \sin y} = \frac{1}{\sin y - 1} \qquad \text{Rearranging}$$

1.6 Find $\frac{dy}{dx}$ when $\sqrt{x} + \sqrt{y} = 4$

$$\sqrt{x} + \sqrt{y} = x^{\frac{1}{2}} + y^{\frac{1}{2}}$$

Hence we can re-write $\sqrt{x} + \sqrt{y} = 4$ as $x^{\frac{1}{2}} + y^{\frac{1}{2}} = 4$ $\qquad$ Rearranging

Differentiating we get:

$$\tfrac{1}{2}x^{-\frac{1}{2}} + \tfrac{1}{2}y^{-\frac{1}{2}}\frac{dy}{dx} = 0$$

$$\Rightarrow \frac{dy}{dx} = \frac{-\tfrac{1}{2}x^{-\frac{1}{2}}}{\tfrac{1}{2}y^{-\frac{1}{2}}} \qquad\qquad\qquad\text{Rearranging}$$

$$= -\frac{\sqrt{y}}{\sqrt{x}} = -\sqrt{\frac{y}{x}} \qquad\qquad\qquad\text{Rearranging}$$

1.7 Find $\dfrac{dr}{d\theta}$ when $r^2\cos\theta = a$ where a is a constant

Differentiating we get:

$$\cos\theta \times 2r\frac{dr}{d\theta} + r^2 \times (-\sin\theta) = 0$$

$$\Rightarrow 2r\cos\theta\frac{dr}{d\theta} = r^2\sin\theta \qquad\qquad\text{Rearranging}$$

$$\Rightarrow \frac{dr}{d\theta} = \frac{r^2\sin\theta}{2r\cos} \qquad\qquad\qquad\text{Rearranging}$$

$$= \frac{r}{2}\tan\theta \qquad\qquad\qquad\qquad\text{Simplifying}$$

1.8 Find $\dfrac{dy}{dx}$ when $y^2 - \sin 3x = 5$

Differentiating we get:

$$2y\frac{dy}{dx} - 3\cos 3x = 0$$

$$\Rightarrow 2y\frac{dy}{dx} = 3\cos 3x \qquad\qquad\qquad\text{Rearranging}$$

$$\Rightarrow \frac{dy}{dx} = \frac{3\cos 3x}{2y} \qquad\qquad\qquad\text{Simplifying}$$

Note: We could also say in this case that $y^2 = \sin 3x + 5 \Rightarrow y = \sqrt{\sin 3x + 5}$ and differentiate this to yield the same result.

1.9 Find $\dfrac{dy}{dx}$ when $\dfrac{x^2}{a^2} + \dfrac{y^2}{b^2} = 1$ where a and b are constants

Differentiating we get:

$$\frac{2x}{a^2} + \frac{2y}{b^2}\frac{dy}{dx} = 0$$

$$\Rightarrow \frac{2y}{b^2}\frac{dy}{dx} = -\frac{2x}{2a^2}$$

$$\Rightarrow \frac{dy}{dx} = -\frac{2b^2 x}{2a^2 y} = \frac{b^2 x}{a^2 y} \qquad\qquad\text{Rearranging and simplifying}$$

1.9 Find $\dfrac{dy}{dx}$ when $ax^2 + 2gxy + by^2 = 1$ where a, b and g are constants

Differentiating we get:

$$2ax + (y \times 2g) + \left(2gx \times \frac{dy}{dx} \right) + \left(2by \times \frac{dy}{dx} \right) = 0 \qquad$$ Using rule for differentiating products for $2gxy$

$$\Rightarrow 2gx\frac{dy}{dx} + 2by\frac{dy}{dx} = -2ax - 2gy \qquad$$ Rearranging

$$\Rightarrow \frac{dy}{dx}(2by + 2gx) = -(2ax + 2gy) \qquad$$ Rearranging

$$\Rightarrow \frac{dy}{dx} = \frac{-(2ax + 2gy)}{(2by + 2gx)}$$

$$= -\frac{(ax + gy)}{(by + gx)} \qquad$$ Simplifying

1.10 Find $\dfrac{dy}{dx}$ when $x^3 + y^3 = 3axy$ where a is a constant

Differentiating we get:

$$3x^2 + 3y^2\frac{dy}{dx} = 3ay + 3ax\frac{dy}{dx} \qquad$$ Using rule for differentiating products for $3axy$

$$\Rightarrow 3y^2\frac{dy}{dx} - 3ax\frac{dy}{dx} = 3ay - 3x^2 \qquad$$ Rearranging

$$\Rightarrow \frac{dy}{dx}(y^2 - ax) = (ay - x^2) \qquad$$ Rearranging

$$\Rightarrow \frac{dy}{dx} = \frac{ay - x^2}{y^2 - ax} \qquad$$ Rearranging and simplifying

1.11 If $xy = ax^2 + b$ prove that $x^2\dfrac{d^2y}{dx^2} = 2y$ where a and b are constants

Differentiating we get:

$$(y \times 1) + \left(x \times \frac{dy}{dx} \right) = 2ax \qquad$$ Using rule for differentiating products for xy

$$\Rightarrow x\frac{dy}{dx} = 2ax - y$$

Differentiating again we get:

$$\left(\frac{dy}{dx} \times 1 \right) + \left(x \times \frac{d^2y}{dx^2} \right) = 2a - \frac{dy}{dx}$$

$$\frac{dy}{dx} + x\frac{d^2y}{dx^2} = 2a - \frac{dy}{dx} \qquad$$ Simplifying

$$\Rightarrow x\frac{d^2y}{dx^2} = 2a - 2\frac{dy}{dx} \qquad$$ Rearranging

$$= 2a - 2\left(\frac{2ax - y}{x} \right) \qquad$$ Since $\dfrac{dy}{dx} = 2ax - y$ from above result

$$\therefore x^2 \frac{d^2y}{dx^2} = x\left[2a - 2\left(\frac{2ax-y}{x}\right)\right]$$
$$= 2ax - 2ax - 2y = 2y$$

1.12 Find $\frac{dy}{dx}$ when $x^2 + xy^2 + y^3 = 3$

Differentiating we get:

$$2x + \left(x \times 2y\frac{dy}{dx}\right) + \left(y^2 \times 1\right) + 3y^2\frac{dy}{dx} = 0 \qquad \text{Using rule for differentiating products for } xy^2$$

$$\Rightarrow 2x + 2xy\frac{dy}{dx} + y^2 + 3y^2\frac{dy}{dx} = 0 \qquad \text{Simplifying}$$

$$\Rightarrow 2xy\frac{dy}{dx} 3y^2\frac{dy}{dx} = -2x - y^2 \qquad \text{Rearranging}$$

$$\Rightarrow \frac{dy}{dx}\left(2xy + 3y^2\right) = -\left(2x + y^2\right) \qquad \text{Rearranging}$$

$$\Rightarrow \frac{dy}{dx} = \frac{-\left(2x + y^2\right)}{2xy + 3y^2} \qquad \text{Rearranging}$$

1.13 Find $\frac{dy}{dx}$ when $x = ye^x$

Differentiating we get:

$$1 = ye^x + e^x\frac{dy}{dx} \qquad \text{Using rule for differentiating products for } ye^x$$

$$\Rightarrow e^x\frac{dy}{dx} = 1 - ye^x \qquad \text{Rearranging}$$

$$\Rightarrow \frac{dy}{dx} = \frac{1 - ye^x}{e^x} \qquad \text{Rearranging}$$

Note that in this case it is possible to express $x = ye^x$ as an explicit function since we can say that:

$$y = \frac{x}{e^x} = xe^{-x} \qquad \text{Rearranging to change RHS to a product instead of a quotient}$$

Differentiating we get:

$$\frac{dy}{dx} = \left(x \times \left(-e^{-x}\right)\right) + \left(e^{-x} \times 1\right) \qquad \text{Using rule for differentiating products}$$

$$= -xe^{-x} + e^{-x} \qquad \text{Simplifying}$$

We can rearrange this result to yield the same result as before as follows:

$$\frac{dy}{dx} = -\frac{x}{e^x} + \frac{1}{e^x} = -y + \frac{1}{e^x} \qquad \text{Substituting for } y = \frac{x}{e^x}$$

$$= \frac{1 - ye^x}{e^x} \qquad \text{Rearranging}$$

1.14 Find $\frac{dy}{dx}$ when $y^2 - 2y\sqrt{1 + x^3} + x^3 - 2 = 0$

Differentiating term by term we get:

$$2y\frac{dy}{dx} - \left(2y \times \left[\frac{1}{2}(1+x^3)^{-\frac{1}{2}} \times 3x^2\right] + \left[(1+x^3)^{\frac{1}{2}} \times 2\frac{dy}{dx}\right]\right) + 2x = 0$$

$$2y\frac{dy}{dx} + 3x^2y(1+x^3)^{-\frac{1}{2}} + 2\frac{dy}{dx}(1+x^3)^{\frac{1}{2}} + 2x = 0$$

$$2y\frac{dy}{dx} + 2\frac{dy}{dx}(1+x^3)^{\frac{1}{2}} + 2x + 3x^2y(1+x^3)^{-\frac{1}{2}} = 0$$

$$2y\frac{dy}{dx}(1+(1+x^3)^{\frac{1}{2}}) = -2x - 3x^2y(1+x^3)^{-\frac{1}{2}}$$

$$\frac{dy}{dx} = \frac{-2x - 3x^2y(1+x^3)^{-\frac{1}{2}}}{2y}$$

$$= \frac{-2x - 3x^2y}{2y\left(\sqrt{1+x^3}\right)} = \frac{-x(2+3xy)}{2y\left(\sqrt{1+x^3}\right)}$$

1.15 If $y^2 - 2y\sqrt{1+x^2} + x^2 = 0$ show that $\frac{dy}{dx} = \frac{x}{\sqrt{1+x^2}}$

Taking $u = 2y$ and $v = \sqrt{1+x^2}$ and differentiating term by term we get:

$$2y \times \frac{dy}{dx} - \left(2y \times \frac{1}{2} \times (1+x^2)^{-\frac{1}{2}} \times 2x + \sqrt{1+x^2} \times 2\frac{dy}{dx}\right) + 2x = 0 \qquad \text{Using product rule}$$

$$\Rightarrow 2y\frac{dy}{dx} - \frac{2xy}{\sqrt{1+x^2}} - 2\sqrt{1+x^2}\frac{dy}{dx} + 2x = 0 \qquad \text{Simplifying}$$

$$y\frac{dy}{dx} - \frac{xy}{\sqrt{1+x^2}} - \sqrt{1+x^2}\frac{dy}{dx} + x = 0 \qquad \text{Dividing throughout by 2}$$

$$\Rightarrow y\frac{dy}{dx} + \sqrt{1+x^2}\frac{dy}{dx} = \frac{2xy}{\sqrt{1+x^2}} + 2x \qquad \text{Rearranging}$$

$$\Rightarrow \frac{dy}{dx}\left(y + \sqrt{1+x^2}\right) = \frac{xy}{\sqrt{1+x^2}} + 1 \qquad \text{Factorising LHS}$$

$$\Rightarrow \frac{dy}{dx}\left(y + \sqrt{1+x^2}\right) = \frac{xy + x\sqrt{1+x^2}}{\sqrt{1+x^2}} \qquad \text{Rearranging RHS}$$

$$= \frac{x\left(y + \sqrt{1+x^2}\right)}{\left(\sqrt{1+x^2}\right)\left(y + \sqrt{1+x^2}\right)} \qquad \text{Dividing both sides by } \left(y + \sqrt{1+x^2}\right)$$

$$= \frac{x}{\sqrt{1+x^2}} \qquad \text{Dividing top \& bottom by } \left(y + \sqrt{1+x^2}\right)$$

1.16 Find $\frac{dy}{dx}$ when y is given by $x^2\sin y - y\cos x = 0$

In this case we have two products so we will have to use the product rule for each of them

Firstly we differentiate the product $x^2 \sin y$ using the product rule

Let $u = x^2$ and $v = \sin y$

Then $\dfrac{d}{dx}\left(x^2 \sin y\right) = x^2 \times \cos y \dfrac{dy}{dx} + \sin y \times 2x = 2x\sin y + x^2 \cos y \dfrac{dy}{dx}$

Now we do the same with the second product $y \cos x$

Let $u = y$ and $v = \cos x$

Giving $\dfrac{d}{dx}\left(y\cos x\right) = \left(y \times -\sin x + \cos x \times \dfrac{dy}{dx}\right) = -y\sin x + \cos x \dfrac{dy}{dx}$

Hence $2x\sin y + x^2 \cos y \dfrac{dy}{dx} - \left(-y\sin x + \cos x \dfrac{dy}{dx}\right) = 0$

$\Rightarrow 2x\sin y + x^2 \cos y \dfrac{dy}{dx} + y\sin x - \cos x \dfrac{dy}{dx} = 0$ Simplifying & Rearranging

$\Rightarrow x^2 \cos y \dfrac{dy}{dx} - \cos x \dfrac{dy}{dx} = -y\sin x - 2x\sin y$ Rearranging

$\Rightarrow \dfrac{dy}{dx}\left(x^2 \cos y - \cos x\right) = -y\sin x - 2x\sin y$ Factorising LHS

$\therefore \dfrac{dy}{dx} = \dfrac{-y\sin x - 2x\sin y}{x^2 \cos y - \cos x} = \dfrac{2x\sin y + y\sin x}{\cos x - x^2 \cos y}$ Dividing both sides by

$x^2 \cos y - \cos x$ and multiplying top & bottom by -1

1.18 Find $\dfrac{dy}{dx}$ when y is given by $x\cos y - y^2 \sin x = 0$

As before we have two products so we will have to use the product rule for each of them

Let $u = x$ and $v = \cos y$

Giving $\dfrac{d}{dx}\left(x\cos y\right) = -x\sin \times y \dfrac{dy}{dx} + \cos y \times 1 = -x\sin y \dfrac{dy}{dx} + \cos y$

Now let $u = y^2$ and $v = \sin x$

Giving $\dfrac{d}{dx}\left(y^2 \sin x\right) = y^2 \times \cos x + \sin x \times 2y\dfrac{dy}{dx} = y^2 \cos x + 2y\sin x \dfrac{dy}{dx}$

$\therefore -x\sin y \dfrac{dy}{dx} + \cos y - \left(y^2 \cos x + 2y\sin x \dfrac{dy}{dx}\right) = 0$

$\Rightarrow -2y\sin x \dfrac{dy}{dx} - x\sin y \dfrac{dy}{dx} = y^2 \cos x - \cos y$ Rearranging

$\Rightarrow -\left(2y\sin x \dfrac{dy}{dx} - x\sin y \dfrac{dy}{dx}\right) = y^2 \cos x - \cos y$

Hence $\dfrac{dy}{dx}\left(2y\sin x + x\sin y\right) = \cos y - y^2 \cos x$ Factorising LHS

And $\dfrac{dy}{dx} = \dfrac{\cos y - y^2 \cos x}{2y\sin x + x\sin y}$ Dividing both sides by $2y\sin x + x\sin y$

MODULE D5.

CALCULUS D5 MAXIMA AND MINIMA

INTRODUCTION

Let us assume that the graph of $y = f(x)$ is as shown in Fig 5.1. The points **A, B** and **C** are called **turning points or stationary points** As the value of x increases, the value of y will increase until the point **A** is reached then decrease from point **A** to point **B** and then increase again from point **B** to point **C** after which the value of y decreases again.

However, at point points **A, B** and **C**, the value of y is neither increasing nor decreasing. Frequently, these points are called **maximum** or **minimum** values, **maxima** at A and C and **minimum** at point **B**.

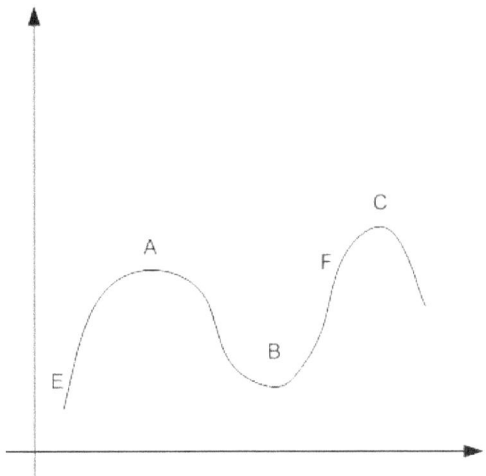

Fig 5.1 Maximum and minimum points

It should be noted, however, that a maximum or minimum value is the greatest or the least value in the **neighbourhood** we are examining. For example there are points at **E** where the values of y are less than the minimum value at **B**. Similarly, there are points at **F** that have greater values of y than the maximum value at **A.**

This is also illustrated in Fig 5.2 below.

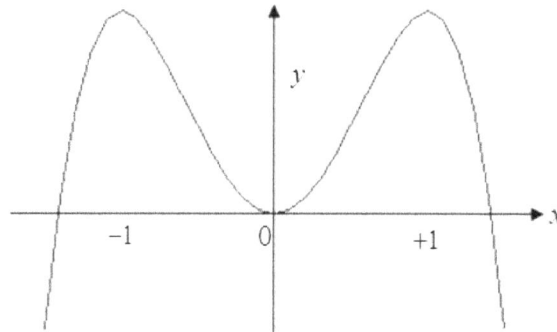

Fig 5.2 Graph of $2x^2 - x^4$

This graph shows that there are maximum values where $x = \pm 1$ and these do represent the greatest values of the function but the value of the function where $x = 0$ is not a minimum value since there are points where the value of y is are less than that at $x = 0$

We can see from the graph that where the derivative (slope or gradient) of the graph is positive, the value of the function (y) is increasing and where it is negative it is decreasing.

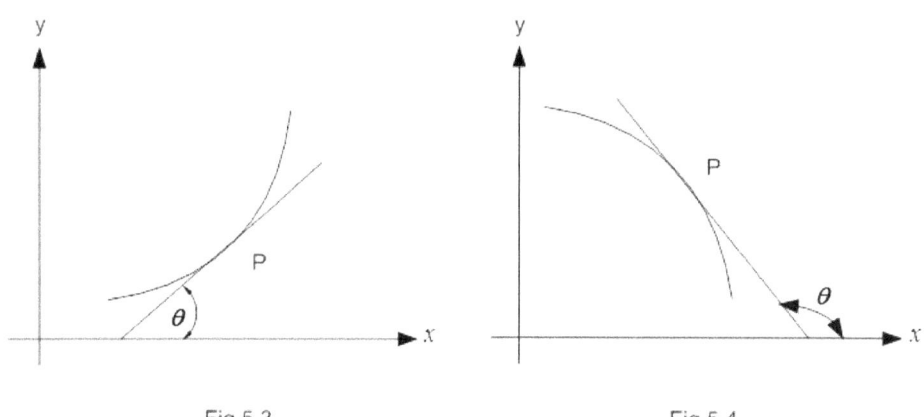

Fig 5.3   Fig 5.4

This is illustrated in Figs 5.3 and 5.4 which show the graphs of functions that are increasing and decreasing respectively as the independent variable x increases. In Fig 5.3, the tangent at the point **P** makes and acute angle θ with the x axis. Since the tangent of an acute angle is positive and since $\dfrac{dy}{dx} = \tan\theta$ then the derivative $\left(\dfrac{dy}{dx}\right)$ will also be positive.

In Fig 5.4 the angle θ is obtuse; the tangent of θ at the point **P** will be negative which means that the derivative $\left(\dfrac{dy}{dx}\right)$ will also be negative.

At points at **A** and **C** in Fig 5.1, the tangent to the curves at these points will be parallel to the x axis, hence the tangents will make a zero angle with the x axis. Therefore at such points the value of the function is neither increasing nor decreasing

$$\text{Therefore } \frac{dy}{dx} = 0$$

We can now introduce some rules for determining the positions of turning points and distinguish between maximum and minimum values as follows:

(i) At a **turning point** $\frac{dy}{dx} = 0$

(ii) At a point where a **maximum** value occurs, $\frac{dy}{dx}$ changes from positive to negative as the value of x is just less and just greater respectively than the value of x at the turning point

(iii) At a point where a **minimum** value occurs, $\frac{dy}{dx}$ changes from negative to positive as the value of x is just less and just greater respectively than the value of x at the turning point.

Example (i) Find the turning points for the function $y = 2x^2 - 4x$ and determine whether it is a minimum or a maximum value of $y = 2x^2 - 4x$

$$\frac{dy}{dx} = 4x - 4$$

Since $\frac{dy}{dx} = 0$ at a turning point we have:

$$4x - 4 = 0$$

Hence $x = 1$ at this point

To determine whether this point is a minimum or a maximum, we have to examine the value of the function just before and just after this point as follows:

When $x = 0.9, \frac{dy}{dx} = 4 \times 0.9 - 4 = 3.6 - 4 = -0.4$

When $x = 1.1, \frac{dy}{dx} = 4 \times 1.1 - 4 = +0.4$

The derivative $\left(\frac{dy}{dx}\right)$ changes from a negative value to a positive value so the point

$x = 1, y = -2$ is a minimum value of the function. This result can be seen graphically in Fig 5.5

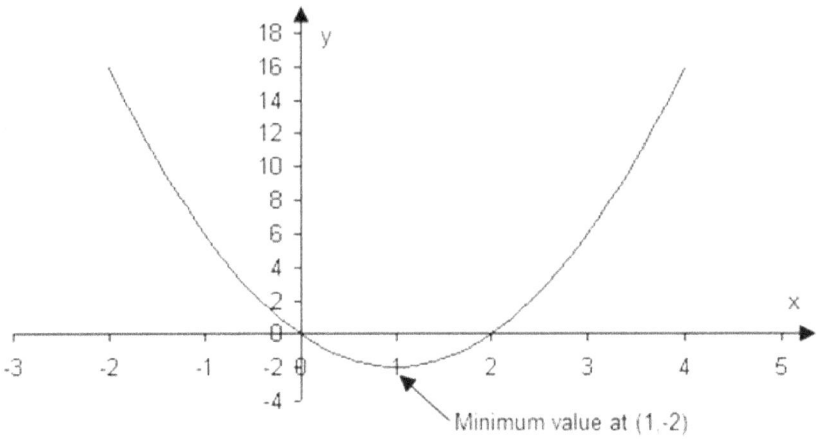

Fig 5 5 Graph of $y = 2x^2 - 4x$

So far we have been using the first derivative $\left(\dfrac{dy}{dx}\right)$ to determine the behaviour of functions.

However, we can also use the second derivative $\left(\dfrac{d^2y}{dx^2}\right)$ of a function to gain further

information. The second derivative $\left(\dfrac{d^2y}{dx^2}\right)$ means $\dfrac{d}{dx}\left(\dfrac{dy}{dx}\right)$. In other words, we differentiate

a function first to obtain $\dfrac{dy}{dx}$ (first derivative) and then differentiate this result again to obtain

$\dfrac{d^2y}{dx^2}$ (second derivative).

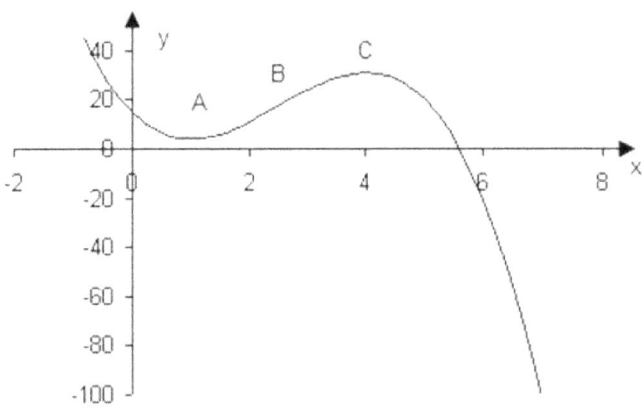

Fig 5.6 Graph of $y = -2x^3 + 15x^2 - 24x + 15$

In Fig 5.6 we can see that when a function is passing through a minimum value as at the point **A** in the graph, the sign of $\frac{dy}{dx}$ changes from negative to positive. In other words $\frac{dy}{dx}$ is increasing and consequently $\frac{d^2y}{dx^2}$ is **positive**.

When the function passes through a maximum as at the point **C** in the graph, the sign of $\frac{dy}{dx}$ changes from positive to negative and consequently $\frac{d^2y}{dx^2}$ is **negative.**

This allows us to determine whether turning points found by finding $\frac{dy}{dx}$, occurs at a maximum or minimum value.

$$\frac{d^2y}{dx^2} \textit{ positive indicates a minimum value}$$

$$\frac{d^2y}{dx^2} \textit{ negative indicates a maximum value}$$

Between point **A** and **C** on the graph there must be point **B** where the curve begins to change direction. On the left of point **B**, the curve is concave. After the point **B**, the curve changes to convex. A point like this is called a ***point of inflection***.

At this point, the gradient of the function stops increasing and starts to decrease. In Fig 5.6 the minimum value precedes the maximum value. However, if a maximum values preceded a minimum value, the reverse is true.

Hence, at a point of inflection we have:

$$\frac{d^2y}{dx^2} = 0$$

These result make it very much easier to sketch the graph of cubic functions or of any other integral function if $\frac{dy}{dx} = 0$ and $\frac{d^2y}{dx^2} = 0$ are solvable.

SUMMARY OF PROCEDURE
Turning Points, Maxima and Minima and Points of Inflection

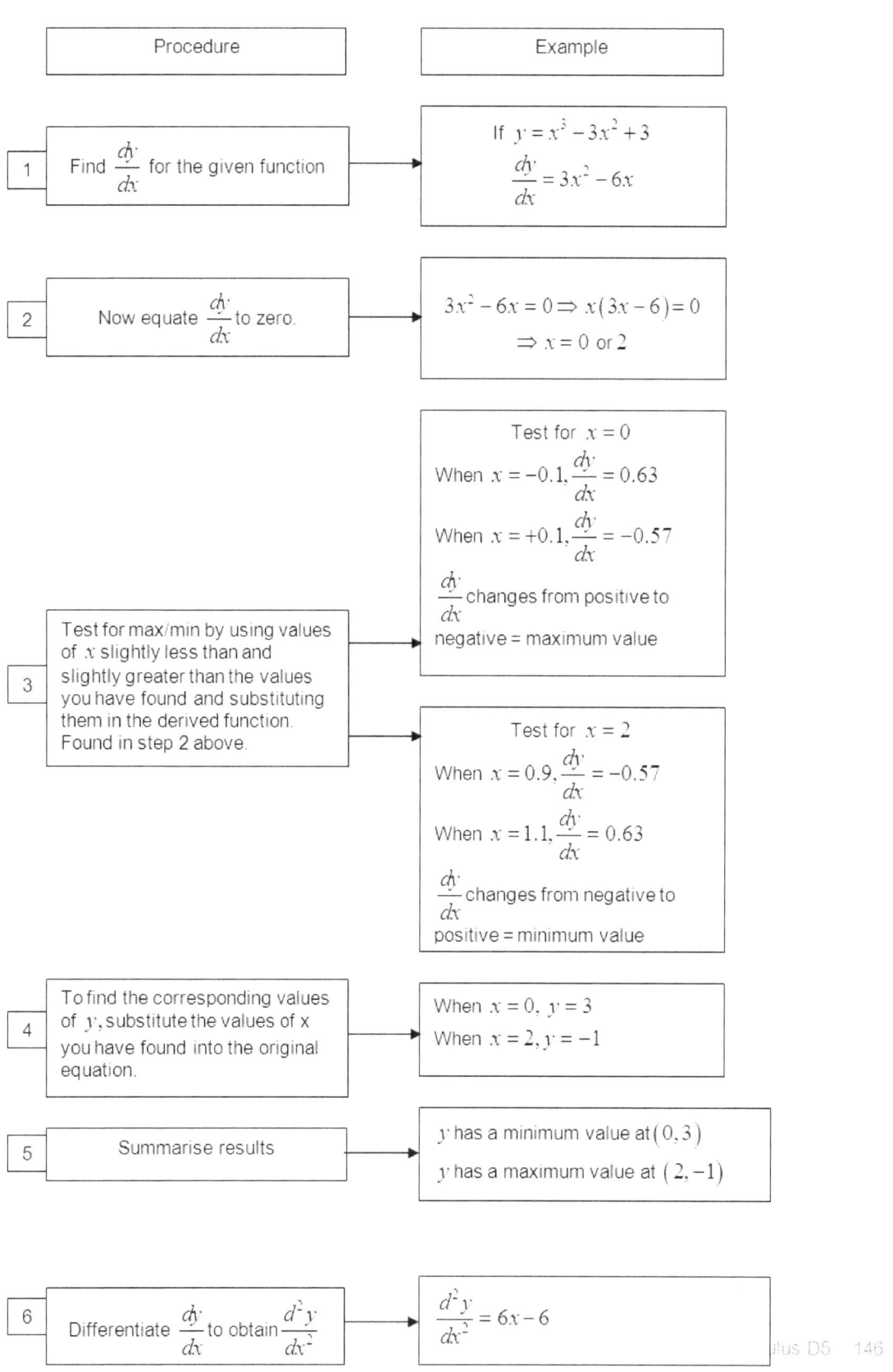

Procedure	Example

1 Find $\dfrac{dy}{dx}$ for the given function

If $y = x^3 - 3x^2 + 3$

$$\dfrac{dy}{dx} = 3x^2 - 6x$$

2 Now equate $\dfrac{dy}{dx}$ to zero.

$$3x^2 - 6x = 0 \Rightarrow x(3x - 6) = 0$$
$$\Rightarrow x = 0 \text{ or } 2$$

3 Test for max/min by using values of x slightly less than and slightly greater than the values you have found and substituting them in the derived function. Found in step 2 above.

Test for $x = 0$

When $x = -0.1, \dfrac{dy}{dx} = 0.63$

When $x = +0.1, \dfrac{dy}{dx} = -0.57$

$\dfrac{dy}{dx}$ changes from positive to negative = maximum value

Test for $x = 2$

When $x = 0.9, \dfrac{dy}{dx} = -0.57$

When $x = 1.1, \dfrac{dy}{dx} = 0.63$

$\dfrac{dy}{dx}$ changes from negative to positive = minimum value

4 To find the corresponding values of y, substitute the values of x you have found into the original equation.

When $x = 0$, $y = 3$
When $x = 2$, $y = -1$

5 Summarise results

y has a minimum value at $(0, 3)$
y has a maximum value at $(2, -1)$

6 Differentiate $\dfrac{dy}{dx}$ to obtain $\dfrac{d^2y}{dx^2}$

$$\dfrac{d^2y}{dx^2} = 6x - 6$$

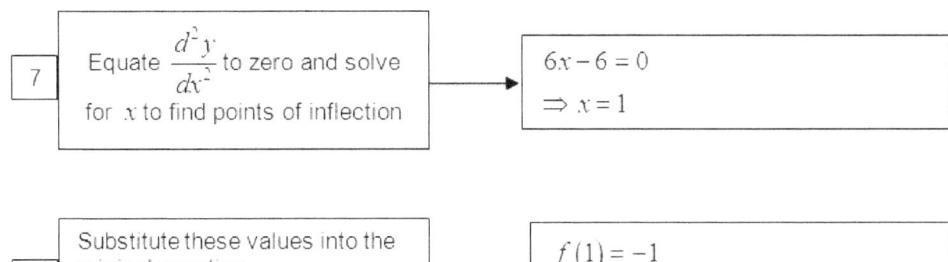

| 7 | Equate $\dfrac{d^2y}{dx^2}$ to zero and solve for x to find points of inflection | → | $6x - 6 = 0$
$\Rightarrow x = 1$ |

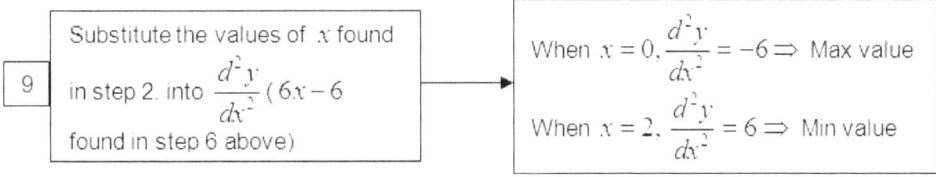

| 8 | Substitute these values into the original equation $y = x^3 - 3x^2 + 3$ to obtain coordinates of POI | → | $f(1) = -1$
$\Rightarrow$ POI at $(1,1)$ |

Alternative method to find maximum and minimum values

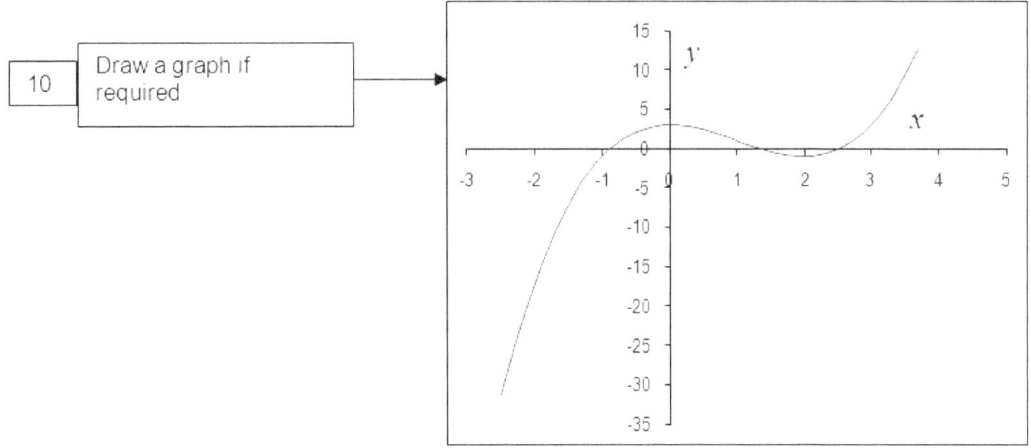

| 9 | Substitute the values of x found in step 2. into $\dfrac{d^2y}{dx^2}$ ($6x - 6$ found in step 6 above) | → | When $x = 0$, $\dfrac{d^2y}{dx^2} = -6 \Rightarrow$ Max value
When $x = 2$, $\dfrac{d^2y}{dx^2} = 6 \Rightarrow$ Min value |

| 10 | Draw a graph if required | → |

To illustrate these techniques we will now use them to show how the graph in Fig 5.6 may be sketched:

When $x = 0$, $y = 15$

$$\frac{dy}{dx} = -6x^2 + 30x - 24$$

And $\dfrac{d^2y}{dx^2} = -12x^2 + 30$

Therefore turning points will occur where:

$-6x^2 + 30x - 24 = 0$

$\Rightarrow 6x^2 - 30x + 24 = 0$

$\Rightarrow 3x^2 - 15x + 24 = 0$

$\Rightarrow (3x - 3)(x - 4) = 0$

Hence $x = 1$ or 4

Also, when $x = 1$, $\dfrac{d^2 y}{dx^2} = 18$

And when $x = 4$, $\dfrac{d^2 y}{dx^2} = -18$

This means that since the sign of $\dfrac{d^2 y}{dx^2}$ is positive when $x = 1$ and negative when $x = 4$, the first value gives a minimum value and the second a maximum value

Substituting these values in $y = -2x^3 + 15x^2 - 24x + 15$ gives $y = 4$ and 31 respectively.

Hence the turning point are at $(1, 4)$ and $(4, 36)$

And points of inflection will occur when:

$-12x^2 + 30 = 0$

$\Rightarrow 12x - 30 = 0$

$\Rightarrow x = 2\frac{1}{2} \Rightarrow y = 17.5$

We are now in a position to draw up a table based on these results as follows:

	Intersection with axes		Min	Max	Point of inflection	Other points		
x	0	15	1	4	2.5	-1	5	6
y	15	0	4	31	17.5	56	20	-21

WORKED SOLUTIONS

1.1 Find the maximum or minimum values of the function $2x^2 + 4x + 5$

Let $y = 2x^2 + 4x + 5$

Then $\dfrac{dy}{dx} = 4x + 4$

The turnings points occur when $\dfrac{dy}{dx} = 0$

I.e. when $4x + 4 = 0$

$\Rightarrow x = -1$

When $x = -1$, $y = 3$

Hence the turning point is $(-1, 3)$

And $\dfrac{d^2 y}{dx^2} = 4$

Since $\dfrac{d^2y}{dx^2}$ is positive, this indicates a minimum value

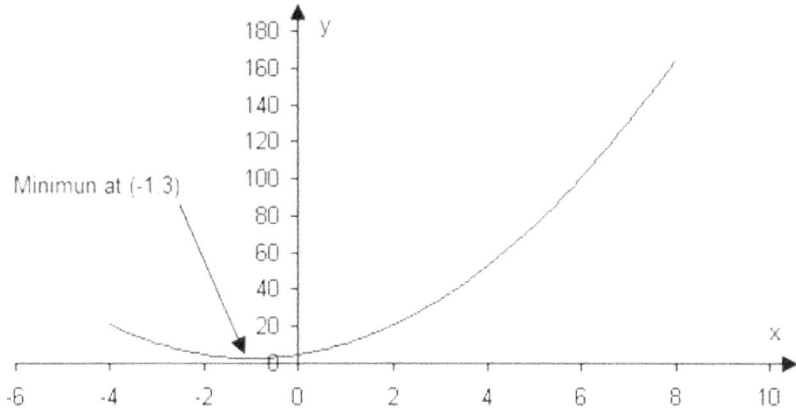

Fig 5 7 Graph of $y = 2x^2 + 4x + 5$

1 2 Find the maximum or minimum values of the function $2x^3 - 15x^2 + 5$

Let $y = 2x^3 - 15x^2 + 5$

Then $\dfrac{dy}{dx} = 6x^2 - 30x$

The turnings points occur when $\dfrac{dy}{dx} = 0$

I.e. when $6x^2 - 30x = 0$

$\Rightarrow x(2x - 10) = 0$

Note. Do not make the mistake of saying $6x^2 - 30x = 0$ is the same as $6x - 30 = 0 \Rightarrow x = 5$ since this reasoning neglects the possibility that $x = 0$ (which in this case, it does)

$\Rightarrow x = 0$ Or $x = 5$
When $x = 0$, $y = 5$
When $x = 5$, $y = -120$

$\dfrac{d^2y}{dx^2} = 12x - 30$

When $x = 0$, $\dfrac{d^2y}{dx^2} = -30$ which indicates a maximum value

When $x = 5$, $\dfrac{d^2y}{dx^2} = 30$ which indicates a maximum value

Also $\dfrac{d^2y}{dx^2} = 0$ for a point of inflection

Hence a point of inflection occurs when $12x - 30 = 0 \Rightarrow x = 2\frac{1}{2}$

When $x = 2\frac{1}{2}$, $y = -57\frac{1}{2}$

Hence the turning points are $(0, 5)$ maximum and $(5, -120)$ minimum with a point of inflection at $\left(2\frac{1}{2}, -57\frac{1}{2} \right)$

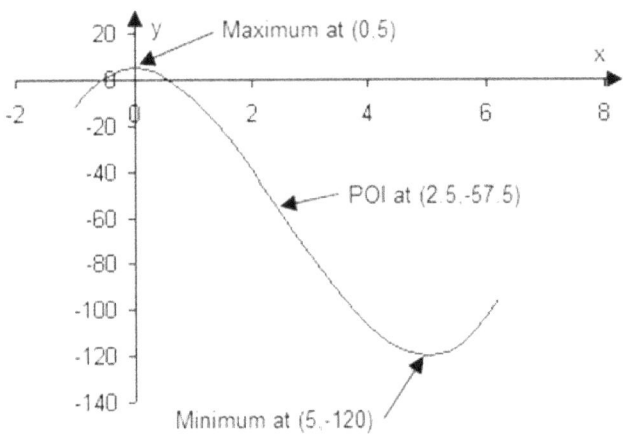

Maximum at (0,5)

POI at (2.5,-57.5)

Minimum at (5,-120)

Fig 5 8 Graph of $y = 2x^3 - 15x^2 + 5$

1.3 Find the maximum or minimum values of the function $x^3 - 3x^2 - 9x + 10$

Let $y = x^3 - 3x^2 - 9x + 10$

Then $\dfrac{dy}{dx} = 3x^2 - 6x - 9$

The turnings points occur when $\dfrac{dy}{dx} = 0$

I.e. when $3x^2 - 6x - 9 = 0$

$\Rightarrow x^2 - 2x - 3 = 0$

$\Rightarrow (x - 3)(x + 1) = 0$

$\Rightarrow x = 3$ Or $x = -1$

When $x = 3, y = -17$

When $x = -1, y = 15$

$\dfrac{d^2 y}{dx^2} = 6x - 6$

$\dfrac{d^2 y}{dx^2} = 0$ for a point of inflection

Hence a point of inflection occurs when $6x - 6 = 0 \Rightarrow x = 1$

When $x = 1, y = -1$

Hence the turning points are $(3, -17)$ and $(-1, 15)$ and a point of inflection occurs at $(1, -1)$

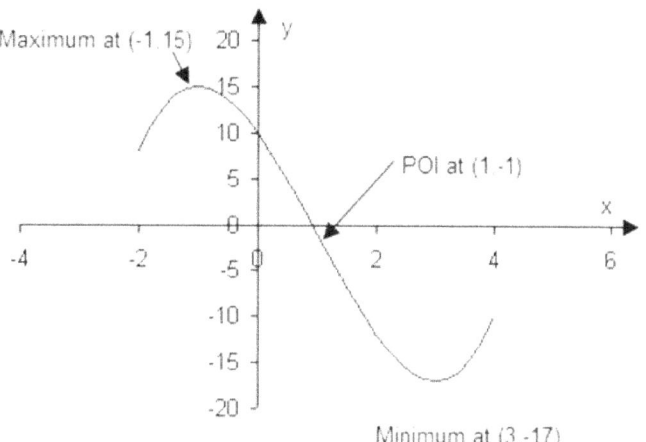

Maximum at (-1,15)

POI at (1,-1)

Minimum at (3,-17)

Fig 5.9 Graph of $y = x^3 - 3x^2 - 9x + 10$

1.4 Find the maximum or minimum values of the function $\dfrac{(x-1)(x-2)}{x}$

Let $y = \dfrac{(x-1)(x-2)}{x}$

$= \dfrac{x^2 - 3x + 2}{x}$

Then $\dfrac{dy}{dx} = \dfrac{x \times (2x - 3) - (x^2 - 3x + 2) \times 1}{x^2}$

$= \dfrac{2x^2 - 3x - x^2 + 3x - 2}{x^2}$ Expanding brackets

$= \dfrac{x^2 - 2}{x^2}$ Simplifying

$= 1 - \dfrac{2}{x^2}$ Rearranging

The turnings points occur when $\dfrac{dy}{dx} = 0$

I.e. when $1 - \dfrac{2}{x^2} = 0$

$\Rightarrow x^2 = 2 \Rightarrow x = \pm\sqrt{2}$

When $x = \sqrt{2}$, $y = 2\sqrt{2} - 3$

When $x = -\sqrt{2}$, $y = -2\sqrt{2} - 3$

Hence the turning points are $\left(\sqrt{2}, 2\sqrt{2} - 3\right)$ and $\left(-\sqrt{2}, -2\sqrt{2} - 3\right)$

$\dfrac{d^2y}{dx^2} = \dfrac{d}{dx}\left(1 - 2x^{-2}\right) = 4x^{-3}$

$\dfrac{d^2y}{dx^2} = 0$ for a point of inflection

I.e. when $4x^{-3} = 0 \Rightarrow x = 0$

Hence $\dfrac{d^2y}{dx^2}$ is neither positive or negative

However, we can clearly see from the graph of the function that a minimum value occurs at $\left(-\sqrt{2}, -2\sqrt{2} - 3\right)$ and a maximum value at $\left(\sqrt{2}, 2\sqrt{2} - 3\right)$ as shown.

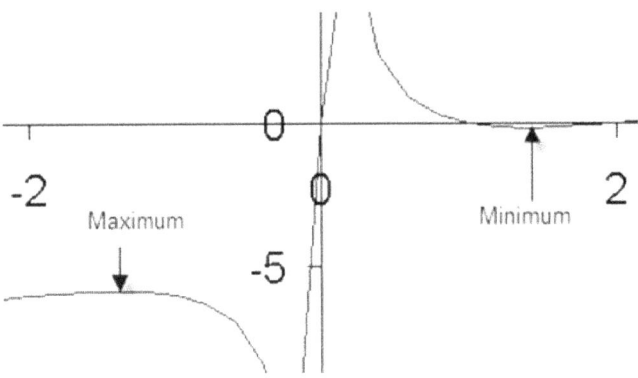

Fig 5.10a Enlarged view of turning points for $y = \dfrac{(x-1)(x-2)}{x}$

1.5 Find the maximum or minimum values of the function $y = (x - 2)(x - 3)^2$

$$y = (x-2)(x-3)^2$$
$$= (x-2)(x^2 - 6x + 9)$$
$$= x^3 - 6x^2 + 9x - 2x^2 + 12x - 18 \qquad \text{Expanding brackets}$$
$$= x^3 - 8x^2 + 21x - 18 \qquad \text{Simplifying}$$
$$\frac{dy}{dx} = 3x^2 - 16x + 21$$

The turnings points occur when $\dfrac{dy}{dx} = 0$

I.e. when $3x^2 - 16x + 21 = 0$
$$\Rightarrow (x-3)(3x-7) = 0$$
$$\Rightarrow x = 3 \text{ Or } x = \frac{7}{3}$$

When $x = 3, y = 0$

When $x = \dfrac{7}{3}, y = \dfrac{4}{27}$

$$\frac{d^2y}{dx^2} = 6x^2 - 16$$

$\dfrac{d^2y}{dx^2} = 0$ at a point of inflection

Hence points of inflection occurs when $6x^2 - 16 = 0$
$$\Rightarrow x = \pm\sqrt{\frac{8}{3}}$$

Therefore $\dfrac{d^2y}{dx^2}$ is negative when $x < \sqrt{\dfrac{8}{3}}$

And positive when $-\sqrt{\dfrac{8}{3}} < x < \sqrt{\dfrac{8}{3}}$

And negative when $x > \sqrt{\dfrac{8}{3}}$

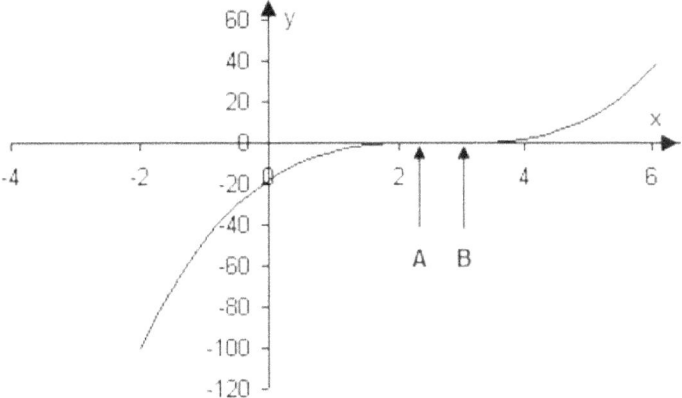

Fig 5.11 Graph of $y = (x-2)(x-3)^2$ where $\mathbf{A} = \left(\dfrac{7}{3}, \dfrac{4}{27}\right)$ and $\mathbf{B} = (3,0)$

1.6 Sketch the graph of function $y = 4x^3 - 9x^2 + 12x + 6$ and show that it has a point of inflection but does not have any turning points.

$\dfrac{dy}{dx} = 12x^2 - 18x + 12$

The turnings points occur when $\dfrac{dy}{dx} = 0$

I.e. when $12x^2 - 18x + 12 = 0$

This equation has no real roots and hence no solution as can be verified by completing the square or using the formula for the solution of quadratic equations. This is illustrated in Fig 5.12. which shows the curve does not intersect the x axis. (Note that this is the graph of the derivative)

Hence the function $y = 4x^3 - 9x^2 + 12x + 6$ has no turning points. This is further demonstrated in Fig 5.13 below.

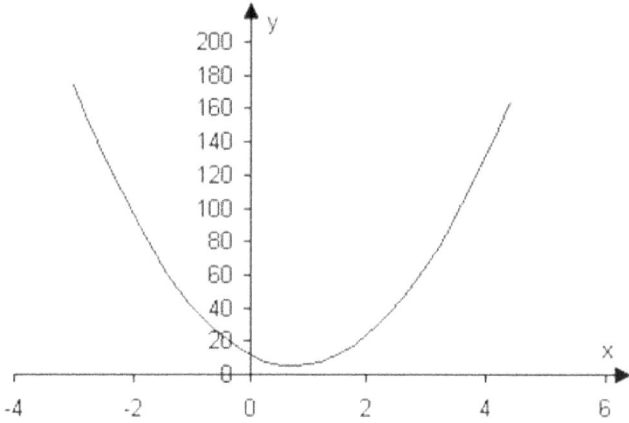

Fig 5.12 $y = 12x^2 - 18x + 12$ (Derivative)

Points of inflection occur where $\dfrac{d^2y}{dx^2} = 0$

$\dfrac{d^2y}{dx^2} = 24x - 18$

Hence there is point of inflection when $24x - 18 = 0$

I.e. when $x = \dfrac{4}{3}$

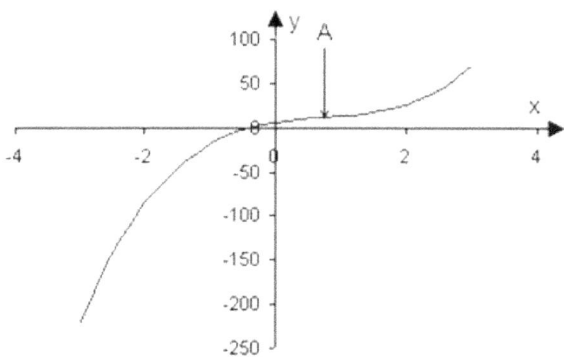

Fig 5.13 Graph of $y = 4x^3 - 9x^2 + 12x + 6$ showing point of inflection at **A**

1.7 Prove that if the sum of two numbers is given, the product of these two numbers will be greatest when the two numbers are equal

Let a be the sum of the two numbers
Let x be one of the numbers
Then $a - x$ must be the other number
Let y be equal to the product of the two numbers

Then we require $y = x(a - x)$ to be a maximum for this proposition to be true

This function will have a maximum/minimum value when $\dfrac{dy}{dx} = 0$

I.e. when $\frac{d}{dx}\left(ax - x^2\right) = 0$

Now $\frac{dy}{dx} = a - 2x$

So a maximum/minimum value will occur when $a - 2x = 0$

$\Rightarrow x = \frac{a}{2}$

Now when $x < \frac{a}{2}$, $(a - 2x)$ is positive

And when $x > \frac{a}{2}$, $(a - 2x)$ is negative

Hence the sign of $\frac{dy}{dx}$ changes from positive to negative as x passes through the value $\frac{a}{2}$

Therefore, the value of the function $ax - x^2$ is a maximum when $x = \frac{a}{2}$

Now if $x = \frac{a}{2}$, the other number must also be $\frac{a}{2}$

Therefore the product of the two numbers is a maximum when the two numbers are equal.

Just to show there is more than one way of cracking a nut, this problem may also be solved algebraically as follows:

The product $ax - x^2$ may be expressed as $-\left(x^2 - ax\right)$ Rearranging

$= -\left(x^2 - ax + \left(\frac{ax}{2}\right)^2 - \left(\frac{ax}{2}\right)^2\right)$ Note the trick of adding and subtracting $\left(\frac{ax}{2}\right)^2$

$= -\left(\left(x - \frac{ax}{2}\right)^2 - \left(\frac{ax}{2}\right)^2\right)$ Rearranging

$= \left(\frac{ax}{2}\right)^2 - \left(x - \frac{ax}{2}\right)^2$ Rearranging

Since a complete square cannot be negative, the minimum value of $\left(x - \frac{ax}{2}\right)^2$ is zero.

$\Rightarrow$ The maximum value of $\left(\frac{ax}{2}\right)^2 - \left(x - \frac{ax}{2}\right)^2$ is $\left(\frac{ax}{2}\right)^2$

Therefore the product must be a maximum value when $x = \pm \frac{ax}{2}$

1.8 A sphere is to be machined down into a cylinder involving the least possible wastage of material. Determine the dimensions of the cylinder and find the ratio of its volume with that of the sphere.

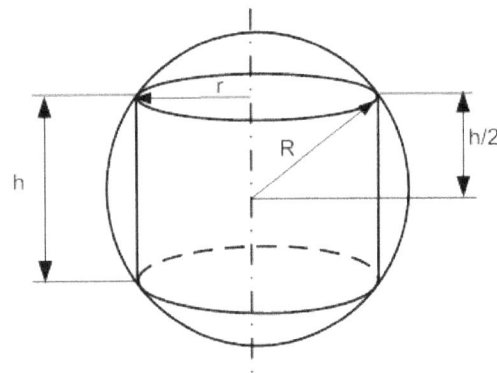

Fig 5.14

Let R = the radius of the sphere
Let V = the volume of the sphere
Let r = the radius of the cylinder
Let h = the height of the cylinder
Let v = the volume of the cylinder

Observations

1. Since the volume of the sphere is fixed, its radius R must also be a fixed constant value.

2. In order for the loss of material to be a minimum, the volume of the cylinder must be a maximum.

3. The volume of the cylinder is dependent on two variables; the radius r and the height h. We can only use on variable when differentiating so we need to eliminate one of them before we can proceed which we can achieve by establishing a relationship between them as follows:

Solution

From the right angled triangle in Fig 5.14

$$r^2 = R^2 - \left(\frac{h}{2}\right)^2 = R^2 - \frac{h^2}{4} \qquad (5.1)$$

The volume of the cylinder is $\pi r^2 h$, hence the volume of the cylinder becomes:

$$v = \pi h \left(R^2 - \frac{h^2}{4} \right)$$

$$= \pi R^2 h - \frac{\pi h^3}{4}$$

Now the condition for a maximum or minimum value for v occurs when $\dfrac{dv}{dx} = 0$

$$\frac{dv}{dh} = \frac{d}{dh}\left(\pi R^2 h - \frac{\pi h^3}{4} \right)$$

$$= \pi R^2 - 3\pi \frac{h^2}{4} \qquad\qquad \text{Remember that } R \text{ is a constant value}$$

$$= R^2 - \tfrac{3}{4} h^2 \qquad\qquad (5.2)\qquad \text{Dividing throughout by } \pi$$

Hence a maximum/minimum value occurs when:

$$R^2 - \tfrac{3}{4} h^2 = 0$$

$$\Rightarrow \frac{3}{4} h^2 = R^2$$

$$\Rightarrow h = \sqrt{\frac{4}{3} R^2} = \frac{2R}{\sqrt{3}}$$

Substituting for h in 5.1 we obtain:

$$r^2 = R^2 - \frac{1}{4}\left(\frac{2R}{\sqrt{3}}\right)^2$$

$$= R^2 - \frac{R^2}{3}$$

$$= \frac{2}{3} R^2$$

$$\therefore r = R\sqrt{\frac{2}{3}}$$

Using 5.2 $\dfrac{d^2v}{dh^2} = -\dfrac{3}{2} h$ indicating a maximum value

Hence the volume of the cylinder is a maximum value when:

$$h = \frac{2R}{\sqrt{3}} \text{ and } r = R\sqrt{\frac{2}{3}}$$

Note, in this case, the test for a maximum/minimum value is not really necessary since the two extreme cases would be a long thin rod or a thin flat disc which would represent a very small volume. Hence the shape of the cylinder of maximum value would lie between them.

However, the proof has been established here for the sake of completeness and to illustrate its use in other problems where this intuitive deduction may not be as obvious.

The volume of the cylinder is $\pi r^2 h = \pi \left(R\sqrt{\dfrac{2}{3}} \right)^2 \dfrac{2R}{\sqrt{3}} = \dfrac{4R^3}{3\sqrt{3}}$

And the volume of the sphere is $\dfrac{4}{3} \pi R^3$

Hence the ratio of the volume of the cylinder to that of the sphere is:

$$\frac{4\pi R^3}{3\sqrt{3}} : \frac{4}{3} \pi R^3 = \frac{1}{\sqrt{3}} : 1 \text{ Or } 1 : \sqrt{3} \qquad \text{Simplifying and multiplying both sides by } \sqrt{3}$$

1.9 A cylindrical tank, open at one end is to be made from thin sheet metal to holds a given volume V. Prove that the area of the material used will be a minimum when the height of the tank is equal to its radius.

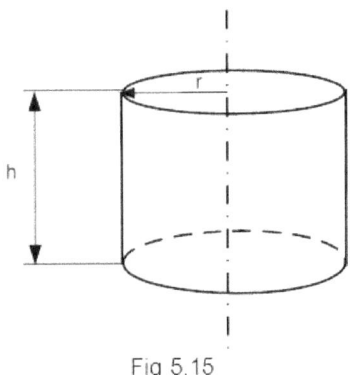

Fig 5.15

Let r = radius of cylinder
Let h = the height of cylinder
Let V = the volume of the cylinder
Let S = the surface area of the cylinder

Observations

1. The size of the piece of sheet material is irrelevant since we are looking for the required relationship $(r = h)$ in any size of sheet material.

2. Note that one end of the cylinder is open – only one end is closed.

3. It is the surface area of the sheet material that we need to be a minimum. The volume V is fixed.

4. There are two variables r and h. We can only use on variable when differentiating so we need to eliminate one of them before we can proceed which we can achieve by establishing a relationship between them as follows:

Solution
The volume of the cylinder $V = \pi r^2 h$
The surface area of the cylinder $S = \pi r^2 + 2\pi rh$

$$= \pi r^2 + \frac{2V}{r}$$

Now the surface area will be a maximum/minimum when $\dfrac{dS}{dr} = 0$

$$\frac{dS}{dr} = 0$$

$$\frac{dS}{dr} = 2\pi r - 2Vr^{-2} = 2\pi r - \frac{2V}{r^2}$$

$$= \frac{2\pi}{r^2}\left(r^3 - \frac{V}{\pi}\right)$$

Rearranging

Now this is zero when $r^3 = \dfrac{V}{\pi}$

Negative if $r^3 < \dfrac{V}{\pi}$

And positive if $r^3 > \dfrac{V}{\pi}$ indicating this value of r makes S a minimum

Hence when $r^3 = \dfrac{V}{\pi}$, $V = \pi r^3 = \pi r^2 h$

Hence $r = h$ Dividing both sides by πr^2

1.10 Find the maximum volume of a cone that is inscribed in a sphere of radius R and determine its proportion of the sphere.

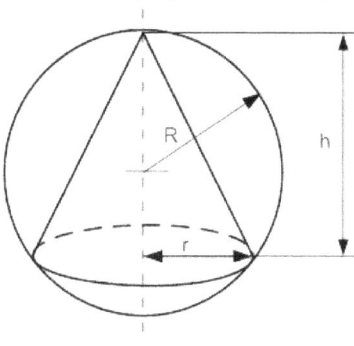

Fig 5.16   Fig 5.17

Fig 5.17 shows a pictorial sketch of this arrangement and Fig 5.17 shows a slice through the centre of both the sphere and cone.

Let r = base radius of cone
Let h = height of cone
Let v = volume of cone
Let R = radius of sphere
Let V = volume of sphere

Observations

1.   the radius of the sphere (R) is fixed

2.   We want the volume of the cone to be a maximum.

3.   There are two variables r and h. We can only use one variable when differentiating so we need to eliminate one of them before we can proceed which we can achieve by establishing a relationship between them as follows:

The volume of the cone is $v = \dfrac{1}{3}\pi r^2 h$ (5.3)

With reference to Fig 5.17:

$h \times (2R - h) = r \times r$ From the property of intersecting chords of a circle

Or $h(2R - h) = r^2$ (5.4)

Hence $v = \dfrac{1}{3}\pi h^2 (2R - h)$ Substituting for r^2 from (5.4) in (5.3)   (5.5)

Then $\dfrac{dv}{dh} = \dfrac{1}{3}\pi (4Rh^2 - 3h^3)$ Remember R is fixed

A maximum/minimum value occurs when $\dfrac{dv}{dx} = 0$

I.e. when $\dfrac{1}{3}\pi\left(4Rh-3h^2\right)=0$

$\Rightarrow h=0 \text{ Or } \dfrac{4R}{3}$

And h is positive for all values in between these.

Since there is a turning point at $\dfrac{4R}{3}$ it must be a maximum value

Substituting this result into (5.5) we get:

$$v=\frac{1}{3}\pi\left(2R\left(\frac{4R}{3}\right)^2-\left(\frac{4R}{3}\right)^3\right)$$

$$=\frac{1}{3}\pi\left(\frac{32R^3}{9}-\frac{64R^3}{27}\right)$$

$$=\frac{32\pi R^3}{81}$$

The volume of a sphere is $V=\dfrac{4}{3}\pi R^3$

Therefore the maximum volume of the cone as a proportion of the sphere is $\dfrac{v}{V}$

$$=\frac{\dfrac{32\pi R^3}{81}}{\dfrac{4}{3}\pi R^3}=\frac{32}{108}=\frac{8}{27}$$

1.11 An open rectangular box is to be made from 12 cm x 8cm by cutting square areas out of the corners and then folding up the edges. Determine the dimensions of the box so that it has the maximum possible volume.

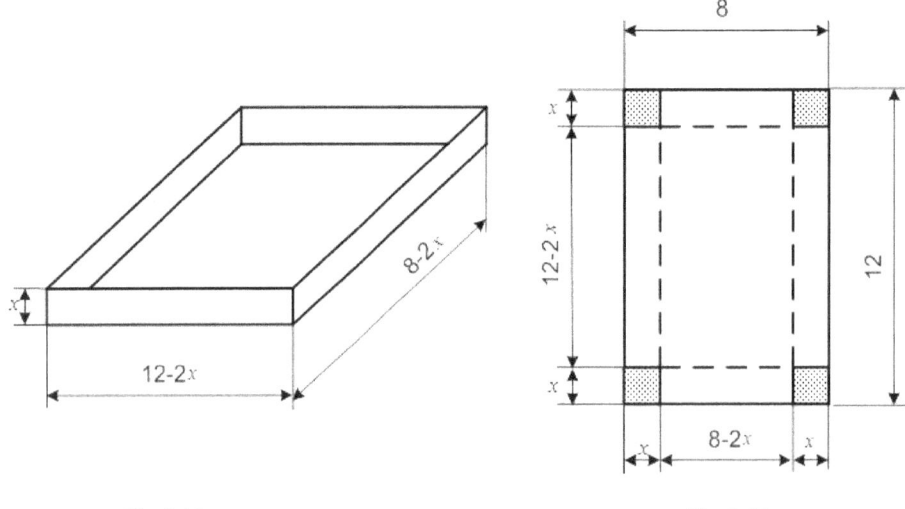

Fig 5.18   Fig 5.19

Fig 5.18 shows a pictorial view of the box and Fig 5.19 shows the sheet metal plate with the corners that have to cut out (shaded).

<u>Observations</u>

1. Once the corners have been removed, the four sides are folded so that they are at right angles to the rest of the plate.

2. The volume of the resulting box has to be a maximum.

3. In this case, we only have one variable - x, hence we only need to express the volume of the box in terms of the dimensions shown in Fig 5.18

Let V = volume of box
Let x = height of box

Then
$$V = x \times (8 - 2x) \times (12 - 2x)$$
$$= x(96 - 40x + 4x^2)$$
$$= 4x^3 - 40x^2 + 96x$$
$$= x^3 - 10x^2 - 24x$$
$$\frac{dV}{dx} = 3x^2 - 20x + 24$$

Now the volume of the box will have maximum /minimum value when $\frac{dV}{dx} = 0$

I.e. when $3x^2 - 20x + 24 = 0$

Unfortunately, this will not factorise so we will use the formula for solving a quadratic equation as follows:

$$x = \frac{-b \pm \sqrt{b^2 - 4ac}}{2a}$$
Where $a = 3$
$b = -20$
$c = 24$
Hence $x = \frac{20 \pm \sqrt{400 - 4.3.24}}{6}$

$$= \frac{20 \pm \sqrt{112}}{6} = 1.57 \text{ Or } 5.1$$

Now the value of 5.1 gives a minimum value of the function $V = x^3 - 10x^2 - 24x$ but we can immediately see that it is clearly not a solution to the problem. This is because, the sheet metal plate is only 8 cm wide and we cannot fold up the sheet 5.1 cm all the way around!

We can also see that when $x = 0, V = 0$ and when $x = 4$, V is also zero. Clearly neither of these values is acceptable as the box would have no depth. Therefore the value of $x = 1.57$ cm gives the maximum value we require.

We could also apply the standard test with $\frac{d^2V}{dx^2} = 0$ as used in previous questions.

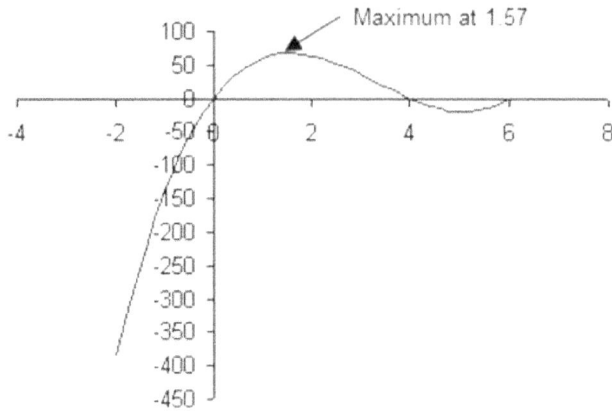

Fig 5.20 Graph of $3x^2 - 20x + 24$ showing a maximum value at 1.57

1.12 A parcel delivery service stipulates that the sum of the length and girth of a parcel must not exceed 80 cm. If a cylindrical parcel is to be sent, find its dimensions for maximum volume.

Observations

1. The girth of a cylinder is its circumference

2. The volume of the cylinder has to be a maximum

3. We have two variables in this problem – the radius and height of the cylinder. We have to reduce this to one variable before we can proceed differentiate.

Let the radius of the cylinder $= r$
Let the height of the cylinder $= h$
Let the Volume of the cylinder $= v$
Let the maximum length + girth of the parcel $= l = 80$

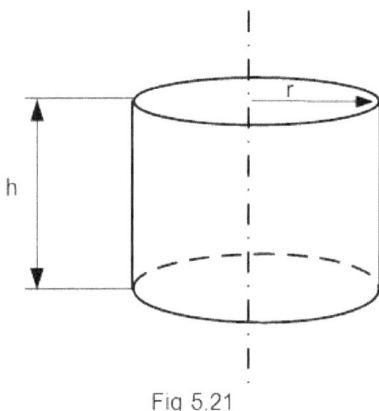

Fig 5.21

Then
$$h + 2\pi r = 80$$
$$\Rightarrow h = 80 - 2\pi r$$
$$v = \pi r^2 h \qquad\qquad \text{Volume of a cylinder}$$

$$v = \pi r^2 (80 - 2\pi r) \qquad\qquad \text{Substituting for } h = 80 - 2\pi r$$
$$= 80\pi r^2 - 2\pi^2 r^3$$

Then $\dfrac{dv}{dr} = 160\pi r - 6\pi^2 r^2$

The volume of the cylinder will have maximum /minimum value when $\dfrac{dv}{dr} = 0$

I.e. when $160\pi r - 6\pi^2 r^2 = 0$

Or when $r(80 - 3\pi r) = 0$

$\Rightarrow r = 0$ Or $r = \dfrac{80}{3\pi}$ cm

When $r = 0, v = 0$ and $v = 0$ when $r = \dfrac{40}{\pi}$ hence r must have a maximum value when

$r = \dfrac{80}{3\pi} \Rightarrow h = 80 - 2\pi \times \dfrac{80}{3\pi} = 80 - \dfrac{160}{3} = \dfrac{80}{3}$

Hence the dimensions for the maximum volume of the cylinder are:

$r = \dfrac{80}{3\pi}$ cm and $h = \dfrac{80}{3}$ cm

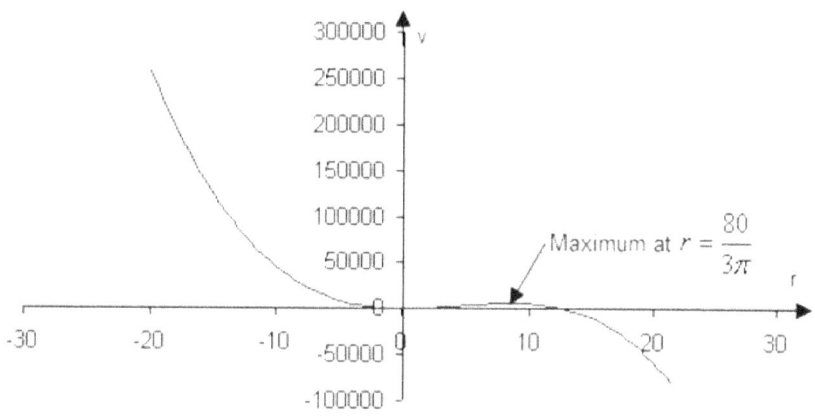

Fig 5.22 Graph of $v = \pi r^2 (80 - 2\pi r)$

1.13 Find the stationary values and the points of inflection of the function $y = x + \dfrac{1}{x}$

$\dfrac{dy}{dx} = 1 - x^{-2}$

Turning points or stationary values occur where $\dfrac{dy}{dx} = 0$

I.e. when $1 - x^{-2} = 0$

$\Rightarrow x = \pm 1$

When $x = 1, y = 2$

When $x = -1, y = -2$

Therefore stationary values occur at $(1,2)$ and $(-1,-2)$

Points of inflection occur where $\dfrac{d^2y}{dx^2} = 0$

I.e. where $2x^{-3} = 0$

$\qquad \Rightarrow x = 0$

When $x < 0, \dfrac{d^2y}{dx^2}$ is negative indicating a maximum value.

When $x > 0, \dfrac{d^2y}{dx^2}$ is positive indicating a minimum value.

Hence the point $(-1, -2)$ is a maximum value

And the point $(1, 2)$ is a minimum value.

This is illustrated in fig 5.23 below

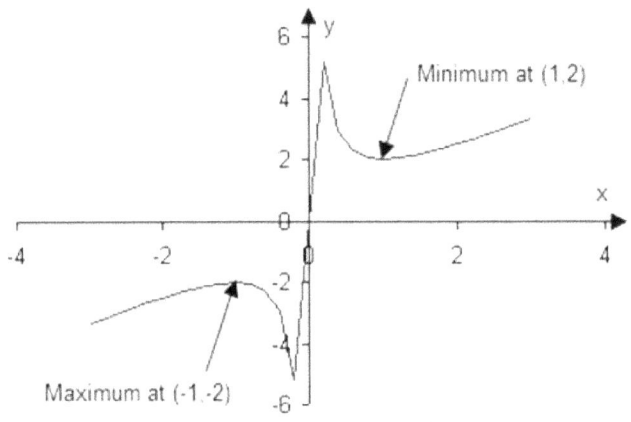

Fig 5.23 Graph of $y = x + \dfrac{1}{x}$

1.14 Find the stationary values of the function $y = xe^x$ and the points of inflection.

$\dfrac{dy}{dx} = \dfrac{d}{dx}\left(xe^x\right)$

$\qquad = e^x + xe^x = e^x\left(1 + x\right)$ Using rule for differentiating a product

The stationary point occur where $\dfrac{dy}{dx} = 0$

I.e. where $e^x\left(1 + x\right) = 0$

This has only one real solution where $x = -1$

When $x = -1, y = 0.368$

$\dfrac{d^2y}{dx^2} = \dfrac{d}{dx}\left(e^x + xe^x\right)$

$\qquad = e^x + e^x + xe^x$

$\qquad = e^x\left(x + 2\right)$

Points of inflection occur where $\dfrac{d^2y}{dx^2} = 0$

I.e. where $e^x\left(x + 2\right) = 0$

This has only one real solution where $x = -2$

When $x = -2$, $y = 0.0338$

Hence there is a maximum value at $x = -2$, $y = -0.368$ and point of inflection occurs at $x = -2$, $y = -0.271$

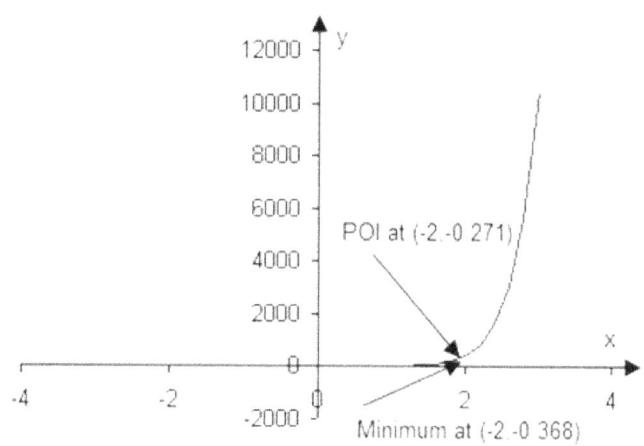

Fig 5.24 Graph of $y = xe^x$

1.15 Find the stationary values of the function $y = x^2 - 4x - 2$

$$\frac{dy}{dx} = 2x - 4$$

Turning points or stationary values occur where $\frac{dy}{dx} = 0$

I.e. where $2x - 4 = 0$

$$\Rightarrow x = 2$$

When $x = 2$, $y = 1$

Hence a stationary point occurs at $(2, -6)$

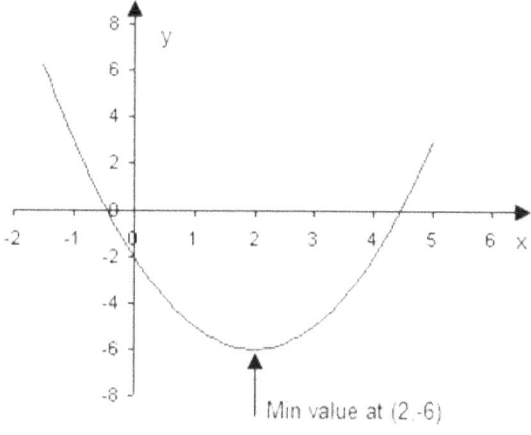

Fig 5 25 Graph of $y = x^2 - 4x - 2$

By substituting values of x in $\frac{dy}{dx}$ $(= 2x - 4)$ that are slightly less and slightly greater than 2

(say 1.9 and 2.1), the gradient changes from negative (-0.2) to positive $(+0.41)$ respectively indicating a stationary minimum value at this point.

We can also clearly see from the graph that $(2, -6)$ is a minimum value of y.

1.16 Find the turning point of the function $y = (x-1)(x-2)(x-4)$ and sketch the graph of this curve.

There are two ways we can solve this problem as follows:

This function will be zero when $x - 1 = 0, x - 2 = 0$ and $x - 4 = 0$

I.e. when $x = 1, x = 2$ and $x = 4$

The curve will therefore cut the x axis at these values. Assuming that the curve is continuous, there must be turning points between:

$x = 1$ and $x = 2$ and

$x = 2$ and $x = 4$

Now $(x-1)(x-2)(x-4) = x^3 - 7x^2 - 14x - 8$ Expanding brackets

$\frac{dy}{dx} = 3x^2 - 14x - 14$

To find the actual values of these turning points we require $\frac{dy}{dx} = 0$

I.e. when $3x^2 - 14x - 14 = 0$

Which leads to the solution $x = 3.215$ and $x = 1.451$ Using formula for quadratic equations

Once again we can confirm algebraically that $(1.451, 0.631)$ represents a maximum value by examining values either side of $x = 1.451$, (say 1.4 and 1.6). By substituting these values in

$\frac{dy}{dx} = 3x^2 - 14x - 14$ we obtain

$f(1.4) = +0.28$

$f(1.6) = -0.72$

As the gradient changes from positive to negative, this shows that $(1.451, 0.631)$ is a maximum value.

Similarly, we can examine the point $(3.215, -2.113)$ by examining values either side of 3.215, say (3.1 and 3.3)

$f(3.1) = -0.57$ Remember, $f(x)$ is the same as $\frac{dy}{dx}$

$f(3.3) = +0.47$

As the gradient changes from negative to positive, this shows that $(3.215, -2.113)$ is a minimum value.

Fig 5.26 shows the graph $y = (x-1)(x-2)(x-4)$ (or $y = x^3 - 7x^2 - 14x - 8$) confirming these results.

Points of inflection occur where $\frac{d^2y}{dx^2} = 0$

$$\frac{d^2y}{dx^2} = 6x - 14$$

Hence a point of inflection occur where $6x - 14 = 0 \Rightarrow x = \frac{7}{3}$

If we substitute $x = 3.215$ in $\frac{d^2y}{dx^2} = 6x - 14$

We obtain $\frac{d^2y}{dx^2} = +5.29$ indicating a minimum value at $(3.215, -2.113)$

If we substitute $x = 1.451$ in $\frac{d^2y}{dx^2} = 6x - 14$

We obtain $\frac{d^2y}{dx^2} = -5.29$ Indicating a maximum value at $(1.451, 0.631)$

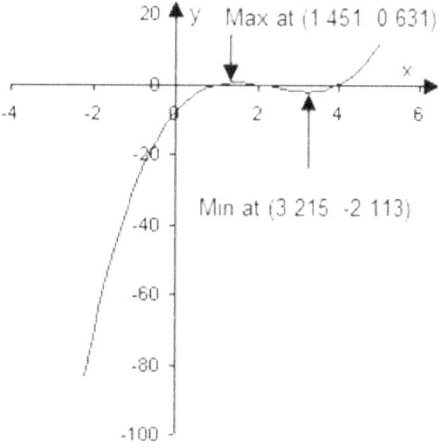

Fig 5 26 Graph of $y = (x-1)(x-2)(x-4)$

1.17 Find the turning points of the function $y = \sin x + \cos x$ ($\cos x \neq 0$)

$$\frac{dy}{dx} = \cos x - \sin x$$

Turning points occur where $\frac{dy}{dx} = 0$

I.e. when $\cos x - \sin x = 0$

$\Rightarrow \tan x = 1$

$\Rightarrow x = \frac{\pi}{4} + k\pi$ (where k is an integer) since both $\sin x$ and $\cos x$ are

continuous functions and have a period of 2π as can be seen in Fig 5.28.

Since the function is continuously differentiable and bounded (because both $\sin x$ and $\cos x$ are bounded) then $\left|\sqrt{2}\right|$ represents the upper and lower bound of the function.

In this case will only consider $k = 0$ and $k = 1$

$\Rightarrow x = \dfrac{\pi}{4}$ and $x = \dfrac{5\pi}{4}$

When $x = \dfrac{\pi}{4}$, $y = \sqrt{2}$

When $x = \dfrac{5\pi}{4}$, $y = -\sqrt{2}$

Hence the turning points are $\left(\dfrac{\pi}{4}, \sqrt{2}\right)$ and $\left(\dfrac{5\pi}{4}, -\sqrt{2}\right)$

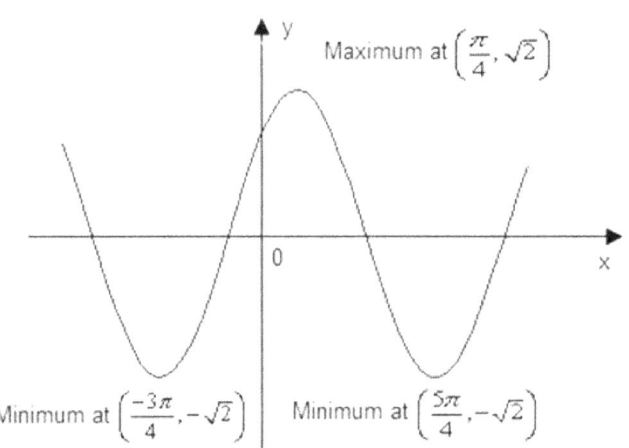

Fig 5 28 Graph of $y = \sin x + \cos x$

Since the function is bounded by $\left|\sqrt{2}\right|$, we can see that $\left(\dfrac{\pi}{4}, \sqrt{2}\right)$ is a maximum value and $\left(\dfrac{5\pi}{4}, -\sqrt{2}\right)$ is a minimum value.

1.18 A measure of the stiffness of a rectangular steel beam is proportional to $\dfrac{bd^3}{12}$ where b is the breadth of the beam and d is the depth of the beam. The perimeter of the cross section of the beam is 50 cm. Determine the bread and depth of the beam having the greatest stiffness

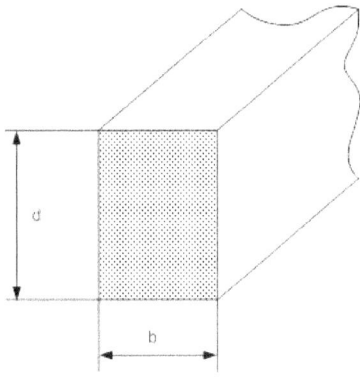

Fig 5.29

Observations

1. The perimeter of the beam is the total distance around the outside of the cross section.

2. The stiffness of the beam has to be a maximum

3. We have two variables in this problem – the breadth and depth of the beam. Again, we have to reduce this to one variable before we can proceed to differentiate.

Let S = stiffness of beam
Let P = perimeter of the cross section
Then $P = 2b + 2d$ and

$$S = \frac{bd^3}{12}$$

We are given that the perimeter is 50 cm
Hence $P = 50$
$\Rightarrow 2b + 2d = 50$
$\Rightarrow b = 25 - d$

Substituting for $b = 25 - d$ in $S = \dfrac{bd^3}{12}$ we obtain

$$S = \frac{(25 - d)d^3}{12} = \frac{25d^3 - d^4}{12}$$

Now we can differentiate S with respect to d to find the maximum value of S

$$\frac{dS}{dd} = \frac{75d^2 - 4d^3}{12}$$

Turning points occur where $\dfrac{dS}{dd} = 0$

I.e. where $\dfrac{75d^2 - 4d^3}{12} = 0$

$$\Rightarrow d = \frac{75}{4} = 18.75 \, \text{cm}$$

Substituting for $d = 18.75$ in $b = 25 - d$ we obtain
$25 - 18.75 = 6.25 \, \text{cm}$

Hence, the turning point is at $(18.75, 6.25)$

We can confirm this is a maximum value by examining values of d that are slightly less than and slightly greater than 18.75 which shows the corresponding values of S changing from positive to negative or by testing with $\dfrac{d^2 S}{dd^2}$ both of which indicate a maximum value of S at this point.

This is confirmed in the graph of the function shown in Fig 5.30

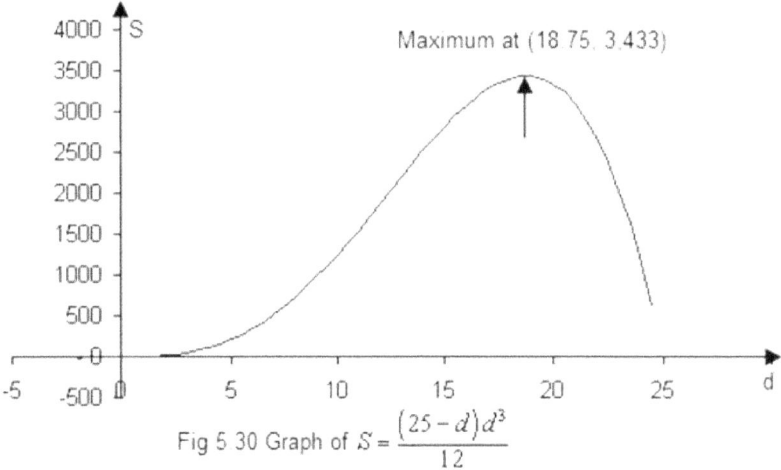

Fig 5 30 Graph of $S = \dfrac{(25-d)d^3}{12}$

1.19 A body is projected vertically upward with an initial velocity of $100\ m\ \text{sec}^{-1}$. The height reached after time t seconds is given by the formula $S = 100t - 5t^2$. Find the maximum height reached by the body and the time taken to reach this height.

Observations

1. We have to find the turning points of the function $S = 100t - 5t^2$ to establish the time taken to reach the maximum height

2. We then need to substitute the time we have found in this equation to establish the maximum height reached.

Turning point occur when $\dfrac{dS}{dt} = 0$

Differentiating the function given we obtain

$$\frac{dS}{dt} = 100 - 10t$$

Hence turning points occur when $100 - 10t = 0$

$$\Rightarrow t = 10\,\text{secs}$$

Substituting this in $S = 100t - 5t^2$, we obtain:

$S = 1,000 - 500 = -500\ \text{m}$ Height reached after 10 seconds

The negative sign indicates the body has travelled upwards.

Fig 5.1 shows the graph of $S = 100t - 5t^2$

This clearly shows that both the distance and time are the same when the body falls back to the starting position again.

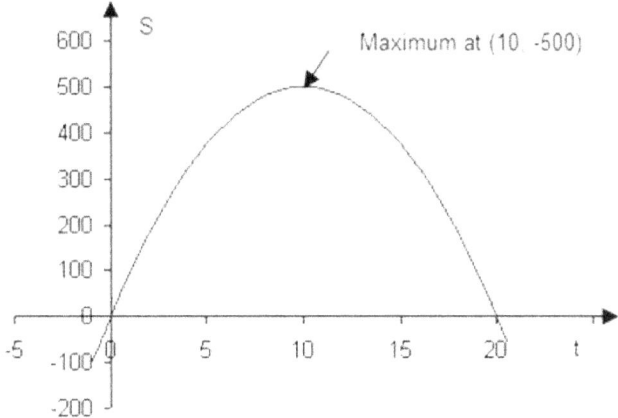

Fig 5.31 Graph of $S = 100t - 5t^2$

1.20 Find the maximum and minimum values of the function $y = (x+1)(x-3)^2$ and the points of inflexion.

$$\frac{dy}{dx} = (x-3)^2 \times 1 + (x+1) \times 2(x-3) \times 1 \qquad \text{Rule for differentiating a product}$$

$$= (x-3)^2 + 2(x+1)(x-3)$$

$$= (x-3)((x-3) + 2(x+1))$$

$$= (x-3)(3x-1)$$

Turning points occur when $\frac{dy}{dx} = 0$

I.e. when $(x-3)(3x-1) = 0$

$$\Rightarrow x = 3 \text{ and } x = \frac{1}{3}$$

When $x = 3, y = 0$ Substituting for $x = 3$ in $y = (x+1)(x-3)^2$

When $x = \frac{1}{3}, y = \frac{256}{27}$ Substituting for $x = \frac{1}{3}$ in $y = (x+1)(x-3)^2$

$$\frac{d^2y}{dx^2} = \frac{d}{dx}((x-3)(3x-1))$$

$$= \frac{d}{dx}(3x^2 - 10x + 3) \qquad \text{Rearranging}$$

$$= 6x - 10$$

Points of inflection occur where $\frac{d^2y}{dx^2} = 0$

I.e. When $6x - 10 = 0$

$$\Rightarrow x = \frac{5}{3}$$

When $x = \dfrac{5}{3}$, $y = \dfrac{128}{27}$ Substituting for $x = \dfrac{5}{3}$ in $y = (x+1)(x-3)^2$

Substituting for $x = 3$ in $\dfrac{d^2 y}{dx^2} = 6x - 10$ gives $+8$ indicating a minimum value at this point

Substituting for $x = \dfrac{1}{3}$ in $\dfrac{d^2 y}{dx^2} = 6x - 10$ gives -8 indicating a maximum value at this point

Hence there are turning points at $(3, 0)$ minimum and at $\left(\dfrac{1}{3}, \dfrac{256}{27} \right)$ maximum

And a point of inflection occurs at $\left(\dfrac{5}{3}, \dfrac{128}{27} \right)$

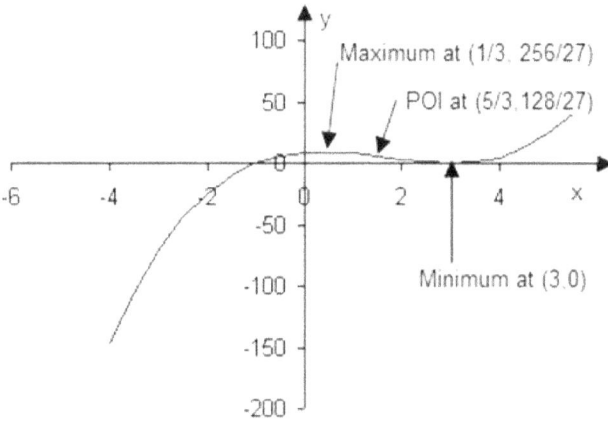

Fig 5.32 Graph of $y = (x+1)(x-3)^2$

MODULE D6.

CALCULUS D6 DIFFERENTIATION OF INVERSE FUNCTONS

INTRODUCTION

Firstly a point of clarification. You will often see the notation $\sin^{-1}\theta$ which is used on scientific calculators to save space. However this notation is also used for the reciprocal of a quantity. E.g. x^{-1} means $\dfrac{1}{x}$ whereas $\sin^{-1}\theta$ means the sine whose angle is θ. So that if $y = \sin^{-1}\theta$ then $\theta = \sin y$. $\sin^{-1}\theta$ **does not mean** $\dfrac{1}{\sin\theta}$. To avoid this ambiguity it is preferable to use $\arcsin\theta$ instead of $\sin^{-1}\theta$.

1. Differentiate $\sin^{-1}\left(\dfrac{1-x^2}{1+x^2}\right)$ with respect to x

Let $y = \sin^{-1}\left(\dfrac{1-x^2}{1+x^2}\right)$

Let $u = \dfrac{1-x^2}{1+x^2}$ so that $u = \sin y$

Then $\dfrac{du}{dy} = \cos y \Rightarrow \dfrac{dy}{du} = \dfrac{1}{\cos y} = \dfrac{1}{\sqrt{1-\sin^2 y}}$

$= \dfrac{1}{\sqrt{1-u^2}}$ Since $\sin y = u$

$= \dfrac{1}{\sqrt{1-\left(\dfrac{1-x^2}{1+x^2}\right)^2}} = \dfrac{1}{\sqrt{1-\dfrac{1-2x^2-x^4}{\left(1+x^2\right)^2}}} = \dfrac{1}{\sqrt{\dfrac{4x^2}{\left(1+x^2\right)^2}}} = \dfrac{2x}{1+x^2}$ Simplifying

Hence $\dfrac{dy}{du} = \dfrac{1+x^2}{2x}$ Since $\dfrac{dy}{du} = \dfrac{1}{\dfrac{du}{dy}}$

And $\dfrac{du}{dx} = \dfrac{\left(1-x^2\right)\times(2x) - \left(1+x^2\right)\times-(2x)}{\left(1+x^2\right)^2}$ Using quotient rule

$= \dfrac{2x-2x^3+2x+2x^3}{\left(1+x^2\right)^2} = \dfrac{4x}{\left(1+x^2\right)^2}$ Simplifying

But $\dfrac{dy}{dx} = \dfrac{dy}{du}\times\dfrac{du}{dx} = \dfrac{1+x^2}{2x}\times\dfrac{4x}{\left(1+x^2\right)^2} = \dfrac{2}{1+x^2}$

2. If $y = \sin^{-1}\left(3x-4x^3\right)$, show that $\sqrt{1-x^2}\,\dfrac{dy}{dx} = \dfrac{3\sqrt{1-x^2}\left(1-4x^2\right)}{\sqrt{1-\left(3x-4x^3\right)^2}}$

Let $u = 3x - 4x^3$

Then $y = \sin^{-1}u \Rightarrow\Rightarrow u = \sin y$

So that $\dfrac{du}{dy} = \cos y \Rightarrow \dfrac{dy}{du} = \dfrac{1}{\cos y} = \dfrac{1}{\sqrt{1-\sin^2 y}}$

$= \dfrac{1}{\sqrt{1-u^2}} = \dfrac{1}{\sqrt{1-\left(3x-4x^3\right)^2}}$ Since $u = \sin y$

$\dfrac{du}{dx} = 3 - 12x^2$

$\dfrac{dy}{dx} = \dfrac{dy}{du} \times \dfrac{du}{dx} = \dfrac{1}{\sqrt{1-\left(3x-4x^3\right)^2}} \times \left(3-12x^2\right)$ Using chain rule

And $\sqrt{1-x^2}\,\dfrac{dy}{dx} = \dfrac{3\sqrt{1-x^2}\left(1-4x^2\right)}{\sqrt{1-\left(3x-4x^3\right)^2}}$

3. Find $\dfrac{dy}{dx}$ when $y = \sqrt{x}\,\sin^{-1}\left(\sqrt{x}\right)$

Let $u = \sqrt{x}$

Let $z = \sin^{-1}\left(\sqrt{x}\right) \Rightarrow \dfrac{dz}{dx} = \dfrac{1}{\sqrt{1-x}}$ From previous results

And let $u = \sqrt{x} = x^{\frac{1}{2}} \Rightarrow \dfrac{du}{dx} = \dfrac{1}{2}x^{-\frac{1}{2}} = \dfrac{1}{2\sqrt{x}}$

Then $\dfrac{dz}{dx} = \dfrac{dz}{du} \times \dfrac{du}{dx} = \dfrac{1}{\sqrt{1-x}} \times \dfrac{1}{2\sqrt{x}} = \dfrac{1}{2\sqrt{x}\sqrt{1-x}}$ Using chain rule

$\dfrac{dy}{dx} = \sqrt{x} \times \dfrac{1}{2\sqrt{x}\sqrt{1-x}} + \dfrac{1}{2\sqrt{x}} \times \sin^{-1}\sqrt{x}$ Using product rule

$= \dfrac{1}{2\sqrt{x}}\left(\sin^{-1}\sqrt{x} + \dfrac{\sqrt{x}}{\sqrt{1-x}}\right)$

4. Find $\dfrac{dy}{dx}$ if $y = \cos^{-1}\left(2x^2 - 1\right)$

Let $u = 2x^2 - 1$

Then $y = \cos^{-1} u \Rightarrow u = \cos y$

$\therefore \dfrac{du}{dy} = -\sin y \Rightarrow \dfrac{dy}{du} = \dfrac{1}{-\sin y}$

$= -\dfrac{1}{\sqrt{1-\cos^2 y}}$ Since $\sin^2 y = 1 - \cos^2 y$

$= -\dfrac{1}{\sqrt{1-u^2}}$ Since $y = \cos u$

$= -\dfrac{1}{\sqrt{1-\left(2x^2-1\right)^2}}$ Since $u = 2x^2 - 1$

$= -\dfrac{1}{\sqrt{1-\left(4x^4 - 4x^2 + 1\right)}}$ Expanding bracket

$$= -\frac{1}{\sqrt{4x^2 - 4x^4}}$$ Simplifying

$$= -\frac{1}{2x\sqrt{1 - x^2}}$$ Rearranging

And $\dfrac{du}{dx} = 4x$

$$\frac{dy}{dx} = \frac{dy}{du} \times \frac{du}{dx} = 4x \times \left(-\frac{1}{2x\sqrt{1 - x^2}}\right) = -\frac{2}{\sqrt{1 - x^2}}$$ Using chain rule

5. Differentiate $\sin^{-1}\left(\dfrac{x}{a}\right)$ with respect to x

(i) Let $y = \sin^{-1}\left(\dfrac{x}{a}\right)$

And let $u = \dfrac{x}{a}$

So that $y = \sin^{-1} u$

$\Rightarrow u = \sin y$

From previous results

$$\frac{du}{dy} = \cos y \Rightarrow \frac{dy}{du} = \frac{1}{\cos y} = \frac{1}{\sqrt{1 - \sin^2 y}}$$

$$= \frac{1}{\sqrt{1 - u^2}} = \frac{1}{\sqrt{1 - \left(\dfrac{x}{a}\right)^2}}$$ Substituting for $u = \dfrac{x}{a}$

$$= \frac{1}{\sqrt{1 - \dfrac{x^2}{a^2}}} = \frac{1}{\sqrt{\dfrac{a^2 - x^2}{a^2}}}$$

$$\frac{du}{dx} = \frac{1}{a}$$

$$\frac{dy}{dx} = \frac{dy}{du} \times \frac{du}{dx} = \frac{1}{a} \times \frac{1}{\sqrt{\dfrac{a^2 - x^2}{a^2}}}$$ Using chain rule

$$= \frac{1}{\sqrt{a^2 - x^2}}$$ Note this is a standard result

6. Differentiate $\cos^{-1}\left(\dfrac{x}{a}\right)$ with respect to x

This time we proceed as follows:

Let $y = \cos^{-1}\left(\dfrac{x}{a}\right)$

And let $u = \dfrac{x}{a}$

So that $y = \cos^{-1} u$

$\Rightarrow u = \cos y$

Exactly the same procedure as before gives $\dfrac{dy}{dx} = -\dfrac{1}{\sqrt{a^2 - x^2}}$ Note this is a standard result

7. Differentiate $\tan^{-1}\left(\dfrac{x}{a}\right)$ with respect to x

Let $y = \tan^{-1}\left(\dfrac{x}{a}\right)$

And let $u = \dfrac{x}{a}$

So that $y = \tan^{-1} u$

$\qquad \Rightarrow u = \tan y$

$\dfrac{du}{dy} = \sec^2 y \Rightarrow \dfrac{dy}{du} = \dfrac{1}{\sec^2 y}$

$\qquad = \dfrac{1}{1 + \tan^2 y} = \dfrac{1}{1 + u^2}$ Since $\tan y = u$

$\qquad = \dfrac{1}{1 + \left(\dfrac{x^2}{a^2}\right)} = \dfrac{1}{\dfrac{a^2 + x^2}{a^2}} = \dfrac{a^2}{a^2 + x^2}$ Substituting for $u = \dfrac{x}{a}$

As before $\dfrac{du}{dx} = \dfrac{1}{a}$

Hence $\dfrac{dy}{dx} = \dfrac{dy}{du} \times \dfrac{du}{dx} = \dfrac{a^2}{a^2 + x^2} \times \dfrac{1}{a} = \dfrac{a}{a^2 + x^2}$ Using chain rule

8. Differentiate $\cos^{-1}(1 - 3x)$ with respect to x

Let $y = \cos^{-1}(1 - 3x)$

Let $u = 1 - 3x$

Then $y = \cos^{-1} u \Rightarrow u = \cos y$

And $\dfrac{du}{dy} = -\sin y \Rightarrow \dfrac{dy}{du} = \dfrac{-1}{\sin y} = \dfrac{-1}{\sqrt{1 - \cos^2 y}} = \dfrac{-1}{\sqrt{1 - u^2}}$ From previous results

$\qquad = \dfrac{-1}{\sqrt{1 - (1 - 3x)^2}}$ Since $u = 1 - 3x$

$\qquad = \dfrac{-1}{\sqrt{1 - (1 - 6x + 9x^2)}}$ Expanding bracket

$\qquad = \dfrac{-1}{\sqrt{6x - 9x^2}}$

And $\dfrac{du}{dx} = -3$

$\dfrac{dy}{dx} = \dfrac{dy}{du} \times \dfrac{du}{dx} = \dfrac{-1}{\sqrt{6x - 9x^2}} \times -3$ Using chain rule

$\qquad = \dfrac{3}{\sqrt{6x - 9x^2}} = \dfrac{3}{\sqrt{3x(2 - 3x)}}$ Simplifying

9. Differentiate $\tan^{-1}\left(\dfrac{2x+1}{3}\right)$ with respect to x

Let $y = \tan^{-1}\left(\dfrac{2x+1}{3}\right)$

Let $u = \dfrac{2x+1}{3}$

Then $y = \tan^{-1} u$

Hence $u = \tan y$

$\dfrac{dy}{du} = \dfrac{1}{1+u^2}$ From previous results

$= \dfrac{1}{1+\left(\dfrac{2x+1}{3}\right)^2} = \dfrac{1}{\dfrac{9+4x^2+4x+1}{9}} = \dfrac{9}{2\left(2x^2+2x+5\right)}$ Substituting for $u = \dfrac{2x+1}{3}$

$\dfrac{du}{dx} = \dfrac{2}{3}$

$\dfrac{dy}{dx} = \dfrac{dy}{du} \times \dfrac{du}{dx} = \dfrac{9}{2\left(2x^2+2x+5\right)} \times \dfrac{2}{3}$ Using chain rule

$= \dfrac{3}{\left(2x^2+2x+5\right)}$

Note that we could have used the result of Q8 above for this questions where we established that

$\dfrac{d}{dx}\left(\tan^{-1}\left(\dfrac{x}{a}\right)\right) = \dfrac{a}{a^2+x^2}$ with $a = 3$ and $x = 2x+1$

10. Differentiate $\left(\tan^{-1} x\right)^2$ with respect to x

Let $y = \left(\tan^{-1} x\right)^2$

Let $u = \tan^{-1} x$
$\Rightarrow x = \tan y$

Then $y = \tan^{-1} u^2 \Rightarrow u^2 = \tan y$

And $2u \dfrac{du}{dy} = \sec^2 y$ See module 4. Differentiation of Implicit Functions

$\Rightarrow \dfrac{du}{dy} = \dfrac{\sec^2 y}{2u}$

$\Rightarrow \dfrac{dy}{dx} = \dfrac{2u}{\sec^2 y}$

$\dfrac{2u}{1+\tan^2 y}$

$= \dfrac{2\tan^{-1} x}{1+x^2}$ Since $u = \tan^{-1} x$ and $\tan y = x$

Alternatively we could tackle this problem as follows

$\dfrac{dy}{dx} = 2\tan^{-1} x \times \dfrac{d}{dx}\left(\tan^{-1} x\right)$

Using the result $\dfrac{d}{dx}\tan^{-1}\left(x/a\right)=\dfrac{a}{a^2+x^2}$ obtained in Q8 and putting $a=1$

$$\frac{dy}{dx}=2\tan^{-1}x\times\frac{1}{1+x^2}$$

$$=\frac{2\tan^{-1}x}{1+x^2}$$

This method shows the advantage of remembering and using standard results like this. It is clearly much simpler and quicker

11. Differentiate $\tan^{-1}\left(\dfrac{1+x}{1-x}\right)$ with respect to x

Let $y=\tan^{-1}\left(\dfrac{1+x}{1-x}\right)$

Let $u=\dfrac{1+x}{1-x}$

Then $y=\tan^{-1}u\Rightarrow u=\tan y$

And $\dfrac{du}{dy}=\sec^2 y\Rightarrow\dfrac{dy}{du}=\dfrac{1}{\sec^2 y}=\dfrac{1}{1+u^2}$ Standard result

$$=\frac{1}{1+\left(\dfrac{1+x}{1-x}\right)^2}=\frac{1}{1+\dfrac{1+2x+x^2}{(1-x)^2}}$$

$$=\frac{1}{\dfrac{1-2x+x^2+1+2x+x^2}{(1-x)^2}}=\frac{(1-x)^2}{2(1+x^2)}$$

$\dfrac{du}{dx}=\dfrac{(1-x)\times(1)-(1+x)\times(-1)}{(1-x)^2}=\dfrac{2}{(1-x)^2}$ Using quotient rule

$\dfrac{dy}{dx}=\dfrac{dy}{du}\times\dfrac{du}{dx}=\dfrac{(1-x)^2}{2(1+x^2)}\times\dfrac{2}{(1-x)^2}=\dfrac{1}{(1+x)^2}$ Using chain rule

Just as in the previous question, we can the result $\dfrac{d}{dx}\tan^{-1}\left(x/a\right)=\dfrac{a}{a^2+x^2}$ and put $a=1$ and

$x=\dfrac{1+x}{1-x}$ to give $\dfrac{dy}{du}=\dfrac{(1-x)^2}{2(1+x)^2}$ directly. Then $\dfrac{dy}{dx}=\dfrac{dy}{du}\times\dfrac{du}{dx}=\dfrac{(1-x)^2}{2(1+x^2)}\times\dfrac{2}{(1-x)^2}=\dfrac{1}{(1+x)^2}$ as before.

12. Differentiate $\tan^{-1}\left(\dfrac{2x}{1-x^2}\right)$ with respect to x

Just as in the previous question, we can use the result $\dfrac{d}{dx}\tan^{-1}\left(x/a\right)=\dfrac{a}{a^2+x^2}$ and put $a=1$ and

$x=\dfrac{2x}{1-x^2}$

Let $y = \tan^{-1}\left(\dfrac{2x}{1-x^2}\right)$

And let $u = \dfrac{2x}{1-x^2}$

Then $y = \tan^{-1} u$

$\Rightarrow u = \tan y$

$\Rightarrow \dfrac{du}{dy} = \sec^2 y \Rightarrow \dfrac{dy}{du} = \dfrac{1}{1 + \left(\dfrac{2x}{1-x^2}\right)^2}$ Using the above standard result

$= \dfrac{1}{\dfrac{1 - 2x^2 + x^4 + 4x^2}{\left(1-x^2\right)^2}} = \dfrac{1}{x^4 + 2x^2 + 1} = \dfrac{\left(1-x^2\right)^2}{\left(1+x^2\right)^2}$ Simplifying

$\dfrac{du}{dx} = \dfrac{\left(1-x^2\right) \times 2 - 2x \times (-2x)}{\left(1-x^2\right)^2}$ Using quotient rule

$= \dfrac{2 - 2x^2 + 4x^2}{\left(1-x^2\right)^2} = \dfrac{2x^2 + 2}{\left(1-x^2\right)^2}$ Simplifying

$= \dfrac{2\left(x^2 + 1\right)}{\left(1-x^2\right)^2}$ Simplifying

$\dfrac{dy}{dx} = \dfrac{dy}{du} \times \dfrac{du}{dx} = \dfrac{\left(1-x^2\right)^2}{\left(1+x^2\right)^2} \times \dfrac{2\left(x^2+1\right)}{\left(1-x^2\right)^2} = \dfrac{2}{1+x^2}$ Using chain rule and simplifying

13. Differentiate $\tan^{-1}\left(\sqrt{1-2x^2}\right)$ with respect to x

Let $y = \tan^{-1}\left(\sqrt{1-2x^2}\right)$

And $u = \sqrt{1-2x^2}$

Hence $y = \tan^{-1} u \Rightarrow u = \tan y$

$\therefore \dfrac{du}{dy} = \sec^2 y \Rightarrow \dfrac{dy}{du} = \dfrac{1}{1 + (1-2x^2)} = \dfrac{1}{2(1-x^2)}$ Using standard result

$\dfrac{du}{dx} = \dfrac{1}{2}\left(1-2x^2\right)^{-1/2} \times (-2x) = \dfrac{1}{2\sqrt{1-2x^2}} \times (-2x) = \dfrac{-x}{\sqrt{1-2x^2}}$

$\dfrac{dy}{dx} = \dfrac{dy}{du} \times \dfrac{du}{dx} = \dfrac{1}{2(1-x^2)} \times \dfrac{-x}{\sqrt{1-2x^2}}$ Using chain rule

$= \dfrac{-x}{(1-x^2)\sqrt{1-2x^2}}$ Simplifying

14. If $u = \theta^2 + \left(\sin^{-1}\theta\right)^2 - 2\theta\sqrt{1-\theta^2}\,\sin^{-1}\theta$ show that $\left(\sqrt{1-\theta^2}\right)\dfrac{du}{d\theta} = 4\theta^2\sin^{-1}\theta$

However, since this question uses the notation $\sin^{-1}\theta$, we will solve the problem using this.

To solve this problem, we have to deal with the two terms containing $\sin^{-1}\theta$ first.

(i). The term $\left(\sin^{-1}\theta\right)^2$

Let $z = \left(\sin^{-1}\theta\right)^2$

Then $\dfrac{dz}{d\theta} = 2\sin^{-1}\theta\,\dfrac{d}{d\theta}\left(\sin^{-1}\theta\right)$

Now let $t = \sin^{-1}\theta \Rightarrow \theta = \sin t$

Then $\dfrac{d\theta}{dt} = \cos t \Rightarrow \dfrac{dt}{d\theta} = \dfrac{1}{\cos t} = \dfrac{1}{\sqrt{1-\sin^2 t}}$ 　　Using $\cos^2 t = 1 - \sin^2 t$

$\qquad\qquad = \dfrac{1}{\sqrt{1-\theta^2}}$ 　　Reversing the substitution for $\theta = \sin t$. This is a standard result

Hence $\dfrac{dz}{d\theta} = \dfrac{2\sin^{-1}\theta}{\sqrt{1-\theta^2}}$

(ii) The term $2\theta\sqrt{1-\theta^2}\,\sin^{-1}\theta$

Here we have three products

Let $z = 2\theta\sqrt{1-\theta^2}\,\sin^{-1}\theta$

Let $w = \sqrt{1-\theta^2} \Rightarrow \dfrac{dw}{d\theta} = \dfrac{1}{2}\left(1-\theta^2\right)^{-\frac{1}{2}} \times \left(-2\theta\right) = -\theta\left(1-\theta^2\right)^{-\frac{1}{2}} = \dfrac{-\theta}{\sqrt{1-\theta^2}}$

Let $x = 2\theta \Rightarrow \dfrac{dx}{d\theta} = 2$

Let $y = \sin^{-1}\theta \Rightarrow \dfrac{dy}{d\theta} = \dfrac{1}{\sqrt{1-\theta^2}}$

$\dfrac{dz}{d\theta} = xy\dfrac{dw}{\theta} + wy\dfrac{dx}{d\theta} + wx\dfrac{dy}{d\theta}$ 　　Using product rule

$= 2\theta\sin^{-1}\theta\left(\dfrac{-\theta}{\sqrt{1-\theta^2}}\right) + \sqrt{1-\theta^2}\,\sin^{-1}\theta \times 2 + \sqrt{1-\theta^2} \times 2\theta \times \dfrac{1}{\sqrt{1-\theta^2}}$

$= 2\theta\sin^{-1}\theta\left(\dfrac{-\theta}{\sqrt{1-\theta^2}}\right) + 2\sqrt{1-\theta^2}\,\sin^{-1}\theta + 2\theta$ 　　Simplifying

Now we can substitute all the results found above in the original equation as follows

$\dfrac{du}{d\theta} = 2\theta + \dfrac{2\sin^{-1}\theta}{\sqrt{1-\theta^2}} - \left(2\theta\sin^{-1}\theta\left(\dfrac{-\theta}{\sqrt{1-\theta^2}}\right) + 2\sqrt{1-\theta^2}\,\sin^{-1}\theta + 2\theta\right)$

$= 2\theta + \dfrac{2\sin^{-1}\theta}{\sqrt{1-\theta^2}} + \dfrac{2\theta^2\sin^{-1}\theta}{\sqrt{1-\theta^2}} - 2\sqrt{1-\theta^2}\,\sin^{-1}\theta - 2\theta$ 　　Simplifying

$= \dfrac{2\sin^{-1}\theta}{\sqrt{1-\theta^2}} + \dfrac{2\theta^2\sin^{-1}\theta}{\sqrt{1-\theta^2}} - 2\sqrt{1-\theta^2}\,\sin^{-1}\theta$ 　　Simplifying

Therefore

$$\sqrt{1-\theta^2}\,\frac{du}{dx} = \sqrt{1-\theta^2}\left(\frac{2\sin^{-1}\theta}{\sqrt{1-\theta^2}} + \frac{2\theta^2\sin^{-1}\theta}{\sqrt{1-\theta^2}} - 2\sqrt{1-\theta^2}\,\sin^{-1}\theta\right)$$

$$= 2\sin^{-1}\theta + 2\theta^2\sin^{-1}\theta - 2(1-\theta^2)\sin^{-1}\theta$$

$$= 2\sin^{-1}\theta + 2\theta^2\sin^{-1}\theta - 2\sin^{-1}\theta + 2\theta^2\sin^{-1}\theta$$

$$= 4\theta^2\sin^{-1}\theta$$

15. Find the value of $\frac{d^2y}{dx^2}$ when $y = 2\sin^{-1}x + 3\cos^{-1}x$ and $x = \frac{1}{2}$

$$\frac{dy}{dx} = \frac{2}{\sqrt{1-x^2}} - \frac{3}{\sqrt{1-x^2}} = \frac{-1}{\sqrt{1-x^2}}$$
Using standard results for $\frac{d}{dx}\sin^{-1}x$ and $\frac{d}{dx}\cos^{-1}x$

$$\frac{d^2y}{dx^2} = \frac{(1-x^2)^{\frac{1}{2}} \times 0 - (-1) \times \frac{1}{2}(1-x^2)^{-\frac{1}{2}} \times (-2x)}{1-x^2}$$
Using quotient rule

$$= \frac{-x(1-x^2)^{-\frac{1}{2}}}{1-x^2}$$
Simplifying

$$= \frac{-x}{(1-x^2)\sqrt{1-x^2}}$$

$$= \frac{x}{(x^2-1)\sqrt{1-x^2}}$$

When $x = \frac{1}{2}$, $\frac{d^2y}{dx^2} = \frac{\frac{1}{2}}{\left(\left(\frac{1}{2}\right)^2 - 1\right)\sqrt{1-\left(\frac{1}{2}\right)^2}} = \frac{\frac{1}{2}}{\left(\frac{-3}{4}\right)\sqrt{\frac{1}{4}}}$

$$= \frac{\frac{1}{2}}{-\frac{3}{4} \times \frac{1}{2}} = -\frac{4}{3}$$
Simplifying

16. Differentiate $\sin^{-1}(x)\cos^{-1}(x)$ with respect to x

The solution of this type of problem is quite straightforward and we can solve it with the product rule for differentiation.

We know from previous results that $\frac{d}{dx}\sin^{-1}x = \frac{1}{\sqrt{1-x^2}}$ and $\frac{d}{dx}\cos^{-1}x = -\frac{1}{\sqrt{1-x^2}}$

Hence we can proceed using the product rule as follows:

$$\frac{dy}{dx} = \sin^{-1}x \times -\frac{1}{\sqrt{1-x^2}} + \cos^{-1}x \times \frac{1}{\sqrt{1-x^2}}$$

$$= -\frac{1}{\sqrt{1-x^2}}\left(\sin^{-1}x - \cos^{-1}x\right)$$

$$= \frac{1}{\sqrt{1-x^2}}\left(\cos^{-1}x - \sin^{-1}x\right)$$

17. Differentiate $\sec^{-1} x$ with respect to x

Let $y = \sec^{-1} x$

Then $x = \sec y$

And $\dfrac{dx}{dy} = \sec x \tan x$ Standard result

Hence $\dfrac{dy}{dx} = \dfrac{1}{\sec y \tan y}$

$$= \dfrac{1}{\sqrt{\sec^2 y \tan^2 y}}$$

$$= \dfrac{1}{\sqrt{\sec^2 y - 1}}$$ Since $\sec^2 y = 1 - \tan^2 y$

$$= \dfrac{1}{\sqrt{x^2 - 1}}$$

18. Differentiate $\cot^{-1} x$ with respect to x

Let $y = \cot^{-1} x$

Then $x = \cot y = \dfrac{1}{\tan y}$

And $\dfrac{dx}{dy} = \dfrac{\tan y \times 0 - 1 \times \sec^2 y}{\tan^2 y} = \dfrac{-\sec^2 y}{\tan^2 y}$ Using quotient rule

$$\Rightarrow \dfrac{dy}{dx} = -\dfrac{\tan^2 y}{\sec^2 y} = -\dfrac{\tan^2 y}{1 + \tan^2 y}$$

$$= -\left(\dfrac{\frac{1}{x^2}}{1 + \frac{1}{x^2}} \right) = -\left(\dfrac{\frac{1}{x^2}}{\frac{x^2 + 1}{x^2}} \right) = -\dfrac{1}{x^2 + 1}$$ Since $x = \dfrac{1}{\tan y}$

19. Differentiate $\operatorname{cosec}^{-1} x$ with respect to x

Let $y = \operatorname{cosec} x$

Then $x = \operatorname{cosec} y = \dfrac{1}{\sin y}$

And $\dfrac{dx}{dy} = \dfrac{\sin y \times 0 - 1 \times \cos y}{\sin^2 y} = \dfrac{-\cos y}{\sin^2 y}$ Using quotient rule

Hence $\dfrac{dy}{dx} = \dfrac{\sin^2 y}{-\sqrt{1 - \sin^2 y}}$ Since $\cos^2 y = 1 - \sin^2 y$

$$= -\frac{\dfrac{1}{x^2}}{\sqrt{1-\left(\dfrac{1}{x^2}\right)}} = -\frac{1}{x^2\sqrt{\dfrac{x-1}{x^2}}}$$

Since $x = \dfrac{1}{\sin y}$

$$= \frac{1}{x\sqrt{x-1}}$$

Rearranging

MODULE IN1

MODULE IN1 GENERAL INTEGRATION

INTRODUCTION

Indefinite Integration

When we differentiate x^2 with respect to x, the derived function is $2x$. conversely, if we are given an unknown function whose derived function is $2x$ then clearly, the unknown function could be x^2. The process of finding a function from its derived function is called ***integration.***

The Constant of Integration

If we consider the functions $x^2 + 2$ and $x^2 - 5$ and differentiate them we would get:

$$\frac{d}{dx}\left(x^2 + 2\right) = 2x \text{ and } \frac{d}{dx}\left(x^2 - 5\right) = 2x$$

This show that $2x$ is the derivative of not just x^2 but also of x^2 ***plus any constant C.***

Hence the result of integrating $2x$ which is called the ***integral*** of $2x$ is not a unique function but a function of the form $2x + C$ where C is an arbitrary constant called the ***constant of integration.*** The resulting function is written as:

$\int 2x\,dx = x^2 + C$ Where $\int f(x)\,dx$ means the integral of $f(x)$ with respect to x

As integration is the reverse procedure to differentiation then for any given function $f(x)$ we have $\int \frac{d}{dx} f(x)\,dx = f(x) + C$

Similarly, when differentiating x^3 we obtain $3x^2$ we can say that $\int 3x^2\,dx = x^3 + C$. Note that is not necessary to write the constant as $\frac{C}{3}$ as C represents ***any*** constant.

In general, since differentiating the term x^{n+1} w.r.t. x (with respect to x) gives $(n+1)x^n$ we have the general result:

$$\int x^n\,dx = \frac{x^{n+1}}{n+1} + C$$

This simply means that when integrating a power of x we increase the power by 1 and divide by this new power.

This rule can be used to integrate x^n for any number n except for -1 which we will discuss later.

Integrating a constant

The result of differentiating ax where a is a constant is a. Hence:

$$\int a\,dx = ax + C$$

Integrating ax^n

Differentiating ax^{n+1} gives $(n+1)(a)x^n$. Hence: $\int (n+1)(a)x^n\,dx = ax^{n+1} + C$ or:

$$\int ax^n\,dx = \frac{a}{n+1}x^{n+1} + C$$

In view of the fact that differentiation is a distributive process with respect to addition and subtraction, it follows that since integration is a reversal of the differentiating process, integration is also distributive in this respect :

e.g $\int \left(x^3 + \frac{1}{x^2} - \sqrt{x} \right) dx = \int x^3\,dx + \int x^{-2}\,dx - \int x^{\frac{1}{2}}\,dx$

$$= \frac{x^4}{4} + \frac{x^{-1}}{-1} - \frac{x^{\frac{3}{2}}}{\frac{3}{2}} + C$$

$$= \frac{x^4}{4} - \frac{1}{x} - \frac{2}{3}x^{\frac{3}{2}} + C$$

Worked Examples of Indefinite Integrals

1.1 Evaluate $\int (x^2 + x + 2)\,dx$

$$\int (x^2 + x + 2)\,dx = \int x^2\,dx + \int x\,dx + \int 2\,dx = \frac{x^3}{3} + \frac{x^2}{2} + 2x + C$$

1.2 $\int \left(x^2 + 3x - \frac{1}{x^2} \right) dx$

$$\int \left(x^2 + 3x - \frac{1}{x^2} \right) dx = \int (x^2 + 3x - x^{-2}) = \int x^2\,dx + \int 3x\,dx - \int x^{-2}\,dx$$

$$= \frac{x^3}{3} + \frac{3}{2}x^2 - (-x^{-1}) = \frac{x^3}{3} + \frac{3}{2}x^2 + x^{-1} + C \; or \; \frac{x^3}{3} + \frac{3}{2}x^2 + \frac{1}{x} + C$$

1.3 Evaluate $\int (2x + 1)^3\,dx$

We can write $(2x + 1)^3$ as $8x^3 + 12x^2 + 6x + 1$ Expanding the bracket
Hence

$$\int (2x + 1)^3\,dx = 8\int x^3\,dx + 12\int x^2\,dx + 6\int x\,dx + \int dx$$

$$= 8\left(\frac{x^4}{4} \right) + 12\left(\frac{x^3}{3} \right) + 6\left(\frac{x^2}{2} \right) + x + C = 2x^4 + 4x^3 + 3x^2 + x + C$$

1.4 Find $\int\left(\dfrac{t^4+1}{t^2}\right)dx$

$$\int\left(\dfrac{t^4+1}{t^2}\right)dt = \int\left(t^2+\dfrac{1}{t^2}\right)dt = \int t^2\,dt + \int t^{-2}\,dt = \dfrac{t^3}{3}-\dfrac{1}{t}+C$$

Divide numerator by t^2 first

1.5 Evaluate $\int 2\sin\theta + \cos\theta\,d\theta$

$$\int 2\sin\theta + \cos\theta\,d\theta = 2\int\sin\theta\,d\theta + \int\cos\theta\,d\theta = -2\cos\theta + \sin\theta + C$$
Or $\sin\theta - 2\cos\theta + C$

1.6 Find $\int 3\theta + \cos\theta\,d\theta$

$$\int 3\theta + \cos\theta\,d\theta = 3\int\theta\,d\theta + \int\cos\theta\,d\theta = 3\left(\dfrac{\theta^2}{2}\right)+\sin\theta + C$$

$$= \dfrac{3\theta^2}{2}+\sin\theta + C$$

1.7 Find $\int\left(x^2+\dfrac{1}{x}\right)^2 dx$

Since $\left(x^2+\dfrac{1}{x}\right)^2 = x^4 + 2x^2 + \dfrac{1}{x^2}$

Expand the bracketed term

$$\int\left(x^2+\dfrac{1}{x}\right)^2 dx = \int x^4 + 2\int x\,dx + \int x^{-2} = \dfrac{x^5}{5}+x^2-x^{-1}+C = \dfrac{x^5}{5}+x^2-\dfrac{1}{x}+C$$

1.8 Evaluate $\int\left(1+x^2\right)^3$ with respect to x

Since $\left(1+x^2\right)^3 = x^6 + 3x^4 + 3x^2 + 1$

Expand the bracketed term

$$\int\left(1+x^2\right)^3 dx = \int x^6\,dx + 3\int x^4\,dx + 3\int x^2\,dx + 1\int dx = \dfrac{x^7}{7}+\dfrac{3}{5}x^5 + 3\dfrac{x^3}{3}+x+C$$

$$= \dfrac{x^7}{7}+\dfrac{3}{5}x^5 + x^3 + x + C$$

1.9 Find $\int\dfrac{\left(ax^{-2}+bx^{-1}+cx+d\right)}{x^{-3}}dx$

$$\dfrac{\left(ax^{-2}+bx^{-1}+cx+d\right)}{x^{-3}} = ax + bx^2 + cx^4 + dx^3$$

Dividing top & bottom by x^{-3}

$$\int\dfrac{\left(ax^{-2}+bx^{-1}+cx+d\right)}{x^{-3}}dx = \int ax\,dx + \int bx^2\,dx + \int cx^4\,dx + \int dx^3 = \dfrac{ax^2}{2}+\dfrac{bx^3}{3}+\dfrac{cx^5}{5}+\dfrac{dx^4}{4}+C$$

1.10 Find $\int\sin x + \cos x\,dx$

$$\int\left(\sin x + \cos x\right)dx = \int\sin x\,dx + \int\cos x = -\cos x + \sin x + +C \text{ Or } \sin x - \cos x + C$$

1.11 Evaluate $\int 2\sec^2 x\, dx$

$$\int 2\sec^2 x\, dx = 2\int \sec^2 x = 2\tan x \qquad \text{Standard result}$$

We know that $\dfrac{d}{dx}\tan x = \sec^2 x$ hence $\int \dfrac{d}{dx}\tan x = \int \sec^2 x \Rightarrow \int \sec^2 x = \tan x + C$

1.12 Determine $\int\left(\dfrac{1}{\theta^3} + \dfrac{2}{\theta^2} + \theta + 3\right)d\theta$

Since $\dfrac{1}{\theta^3} + \dfrac{2}{\theta^2} + \theta + 3 = \theta^{-3} + \theta^{-2} + \theta + 3 \qquad \text{Rearranging}$

$$\int\left(\dfrac{1}{\theta^3} + \dfrac{2}{\theta^2} + \theta + 3\right)d\theta = \int \theta^{-3}\,d\theta + 2\int \theta^{-2}\,d\theta + \int \theta\, d\theta + 3\int d\theta$$

$$= -\dfrac{\theta^{-2}}{2} - 2\theta^{-1} + \dfrac{\theta^2}{2} + 3\theta + C$$

Or $\dfrac{\theta^2}{2} - \dfrac{2}{\theta} - \dfrac{1}{2\theta^2} + 3\theta + C \qquad \text{Rearranging}$

1.13 If $\left(1 + x^2\right)\dfrac{dy}{dx} = 1$ Find the general value of y

If $\left(1 + x^2\right)\dfrac{dy}{dx} = 1$ Then $\dfrac{dy}{dx} = \dfrac{1}{\left(1 + x^2\right)} \Rightarrow dy = \dfrac{1}{\left(1 + x^2\right)}dx \Rightarrow \int dy = \int \dfrac{1}{\left(1 + x^2\right)}dx$

Let $x = \tan t \Rightarrow \dfrac{dx}{dt} = \sec^2 t \Rightarrow dx = \sec^2 t\, dt$

Hence $\dfrac{1}{\left(1 + x^2\right)} = \dfrac{1}{\sec^2 t}$

So $\dfrac{1}{1 + x^2}dx = dt$

Hence $y = \int \dfrac{1}{\left(1 + x^2\right)}dx = \int dt = t$

$$= \tan^{-1} x + C \qquad \text{since } t = \tan^{-1} x \text{ (standard result)}$$

1.14 Find $\int \dfrac{1}{\sqrt{(x + c)}}\, dx$

$$\int \dfrac{1}{\sqrt{(x + b)}}\, dx = \int \left(x + b\right)^{-\frac{1}{2}}\, dx = \dfrac{\left(x + b\right)^{\frac{1}{2}}}{\left(\dfrac{1}{2}\right)} = 2\sqrt{(x + b)} + C$$

1.15 Evaluate $\int \dfrac{1}{\sqrt{ax+b}}\,dx$

$$\int \frac{1}{\sqrt{ax+b}}\,dx = \int (ax+b)^{-\frac{1}{2}}\,dx = \frac{1}{a}\frac{(ax+b)^{\frac{1}{2}}}{\left(\frac{1}{2}\right)}$$

Standard result

$$= \frac{2}{a}\sqrt{(ax+b)} + C$$

1.16 Find $\int \left(\sqrt[3]{x} + \dfrac{1}{\sqrt{x}}\right)dx$

$$\int \left(\sqrt[3]{x} + \frac{1}{\sqrt{x}}\right)dx = \int \left(x^{\frac{1}{3}} + \int x^{-\frac{1}{2}}\right)dx = \int x^{\frac{1}{3}}\,dx + \int x^{-\frac{1}{2}}\,dx$$

Rearranging as a sum of integrals

$$= \frac{x^{\frac{4}{3}}}{\left(\frac{4}{3}\right)} + \frac{x^{\frac{1}{2}}}{\left(\frac{1}{2}\right)} + C = \frac{3}{4}x^{\frac{4}{3}} + 2\sqrt{x} + C$$

1.17 Determine $\int 5x^3\left(1+x^2\right)dx$

$$\int 5x^3\left(1+x^2\right)dx = \left(\int 5x^3 + 5x^5\right)dx = \int 5x^3\,dx + \int 5x^5\,dx$$

Expand bracket and write as sum

$$= \frac{5}{4}x^4 + \frac{5}{6}x^6 + C$$

1.18 Find $\int \dfrac{(2x+1)}{\left(x^2+2x+5\right)}\,dx$

$$\int \frac{(2x+1)}{\left(x^2+2x+5\right)}\,dx = \ln\left(x^2+2x+5\right) + C$$

The numerator is the derivative of denominator (standard result) i.e.

$$\frac{d}{dx}\left(x^2+2x+5\right) = 2x+2$$

1.19 Find $\int ag\,dt$

$$\int ag\,dt = ag\int dt = agt + C$$

Don't forget we are being asked to integrate with respect to t

1.20 Evaluate $\int \left(\dfrac{\pi}{4} - 4x^{0.5} + 4\right)dx$

$$\left(\frac{\pi}{4} - 4x^{0.5} + 4\right) = \left(\frac{\pi}{4} - 4x^{\frac{1}{2}} + 4\right)$$

Rewriting

Hence $\int \left(\dfrac{\pi}{4} - 4x^{0.5} + 4 \right) dx = \int \left(\dfrac{\pi}{4} - 4x^{\frac{1}{2}} + 4 \right) = \dfrac{\pi}{4}x - \dfrac{4x^{\frac{3}{2}}}{\left(\dfrac{3}{2}\right)} + 4x + C$

$= = \left(\dfrac{\pi}{4} + 4 \right) x - \dfrac{8}{3}x^{\frac{3}{2}} + C$ Rearranging and simplifying

1.21   Find $\int \dfrac{3}{(x+1)} + \dfrac{5}{(x-3)}\, dx$

Rearrange to express $\int \left(\dfrac{3}{(x+1)} + \dfrac{5}{(x-3)} \right)$ as the sum of two integrals each of which uses

the standard result $\int \dfrac{a}{(x \pm b)} = a \ln (x \pm b)$

$\int \left(\dfrac{3}{(x+1)} + \dfrac{5}{(x-3)} \right) dx = \int \dfrac{3}{(x+1)}\, dx + \int \dfrac{5}{(x-3)}\, dx = 3 \ln (x+1) + 5 \ln (x-3) + C$

1.22 Find $\int (x+1)(x^2 - 2)\, dx$

First multiply out the brackets

$(x+1)(x^2 - 2) = x^3 + x^2 - 2x - 2$

Hence

$\int (x+1)(x^2 - 2)\, dx = \int x^3 dx + \int x^2 dx - 2\int x\, dx - 2\int dx$

$= \dfrac{x^4}{4} + \dfrac{x^3}{3} - x^2 - 2x + C$

Note. It is more usual to evaluate the integral directly as $\int \left(x^3 + x^2 - 2x - 2 \right) dx$ to obtain
the same result.

1.23 Find $\int \sin \theta \cos \theta\, dx$

We first have to recognise that $\sin 2\theta = 2 \sin \theta \cos \theta$

Hence $\sin \theta \cos \theta = \dfrac{1}{2} \sin 2\theta$

Therefore we can say that $\int \sin \theta \cos \theta\, d\theta = \dfrac{1}{2} \int \sin 2\theta\, d\theta$

Hence $\int \sin \theta \cos \theta\, d\theta = -\dfrac{1}{2} \dfrac{\cos 2\theta}{2} + C = -\dfrac{1}{4} \cos 2\theta + C$

Definite Integration

Suppose we wish to find the area bound by the curve $y = 2x^2$, the x axis and the lines $x = 2$ and $x = 4$ in Fig 1.1 below.

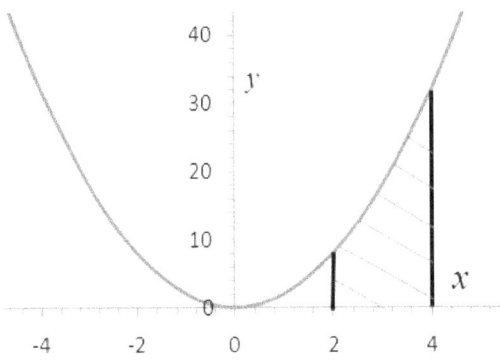

Fig 1.1 Graph of $y = 2x^2$

The total area under a graph between two limits a and b is given by $\int_a^b y\, dx$ where a and b are called the boundary limits of the integration where b is the upper limit and a the lower limit.

In this case we require the area between the limits 2 and 4 so $a = 2$ and $b = 4$ hence we require:

$$\int_2^4 2x^2\, dx = \left(\frac{2}{3}x^3\right)_{x=4} - \left(\frac{2}{3}x^3\right)_{x=2}$$

The RHS of this equation is usually denoted by $\left[\frac{2}{3}x^3\right]_2^4$ Hence it is usual to say:

$$\int_2^4 2x^2\, dx = \left[\frac{2}{3}x^3\right]_2^4 = \frac{2}{3}(4)^3 - \frac{2}{3}(2)^3 = \frac{2}{3}(64) - \frac{2}{3}(8) = \frac{128}{3} - \frac{16}{3} = \frac{112}{3}$$

Worked examples of Definite Integrals

1.24   Evaluate $\int_0^{\frac{\pi}{4}} \sin(\theta - 2\theta)\, d\theta$

First note that $\sin(\theta - 2\theta) = \sin(-\theta) = -\sin\theta$

Hence $\int_0^{\frac{\pi}{4}} \sin(\theta - 2\theta)\, d\theta = -\int_0^{\frac{\pi}{4}}(\sin\theta)\, d\theta = -\left[-\cos\theta\right]_0^{\frac{\pi}{4}} = \left[\cos\theta\right]_0^{\frac{\pi}{4}} = \frac{\sqrt{2}}{2} - 1$

1.25   Determine the value of $\int_0^{\frac{\pi}{2}} 2\cos^2\theta + 3\sin 3\theta$

First we need to do some work with the expression $2\cos^2\theta + 3\sin 3\theta$

Now $2\cos^2\theta = 1 + \cos 2\theta$ therefore we can rewrite the integrand as $1 + \cos 2\theta + 3\sin 3\theta$

Hence

$$\int_0^{\frac{\pi}{2}} 2\cos^2\theta + 3\sin 3\theta = \int_0^{\frac{\pi}{2}}(1+\cos 2\theta + 3\sin 3\theta)\,d\theta = \left[\theta + \frac{\sin 2\theta}{2} - 3\frac{\cos 3\theta}{3}\right]_0^{\frac{\pi}{2}}$$

$$= \left[\theta + \frac{\sin 2\theta}{2} - \cos 3\theta\right]_0^{\frac{\pi}{2}} = \left(\frac{\pi}{2} + 0 - 0\right) - (0 + 0 - 1) = \frac{\pi}{2} + 1$$

1.26 Evaluate $\int_0^1\left(\frac{1}{\sqrt{(x+3)}} + \sqrt{(x+3)}\right)dx$

$$\int_0^1\left(\frac{1}{\sqrt{(x+3)}} + \sqrt{(x+3)}\right)dx = \int_0^1\left((x+3)^{-\frac{1}{2}} + (x+3)^{\frac{1}{2}}\right)dx$$

$$= \left[2(x+3)^{\frac{1}{2}} + \frac{2}{3}(x+3)^{\frac{3}{2}}\right]_0^1 = \left(4 + \frac{16}{3}\right) - \left(2\sqrt{3} + 2\sqrt{3}\right)$$

$$= \frac{28}{3} - 4\sqrt{3}$$

1.27 Evaluate $\int_0^{\frac{\pi}{4}}\cos^2 x\,dx$

$$\int_0^{\frac{\pi}{4}}\cos^2 x\,dx = \frac{1}{2}\int_0^{\frac{\pi}{4}}(1+\cos 2x)\,dx \qquad\qquad \text{Since } \cos^2 x = \frac{1}{2}(1+\cos 2x)$$

$$= \frac{1}{2}\left[x + \frac{1}{2}\sin 2x\right]_0^{\frac{\pi}{4}} = \frac{1}{2}\left(\frac{\pi}{4} + \frac{1}{2}\right) - (0) = \frac{\pi}{8} + \frac{1}{4}$$

1.28 Evaluate $\int_0^{\frac{\pi}{4}}(3\cos^2 x + \sin^2 x)\,dx$

$$3\cos^2 x + \sin^2 x = \frac{3}{2}(1+\cos 2x) + \frac{1}{2}(1-\cos 2x) \quad \text{Since } \cos^2 x = \frac{1}{2}(1+\cos 2x) \text{ and } \sin^2 x = \frac{1}{2}(1-\cos 2x)$$

Hence $\int_0^{\frac{\pi}{4}}(3\cos^2 x + \sin^2 x)\,dx = \int_0^{\frac{\pi}{4}}(2+\cos 2x)\,dx$

$$= \left[2x + \frac{1}{2}\sin 2x\right]_0^{\frac{\pi}{4}} = \frac{\pi}{2} + \frac{1}{2} - 0 = \frac{\pi+1}{2}$$

1.29 Evaluate the definite integral $\int_{\frac{\pi}{4}}^{\frac{3\pi}{4}}\cos\left(x - \frac{\pi}{4}\right)dx$

$$\int_{\frac{\pi}{4}}^{\frac{3\pi}{4}}\cos\left(x - \frac{\pi}{4}\right)dx = \left[\sin\left(x - \frac{\pi}{4}\right)\right]_{\frac{\pi}{4}}^{\frac{3\pi}{4}}$$

$$= \sin\left(\frac{\pi}{2}\right) - (\sin(0)) = \sin\left(\frac{\pi}{2}\right) = 1$$

Note: $\int \cos(ax \pm b)\,dx = \dfrac{1}{a}\sin(ax \pm b) + C$ is a standard result. In this case $a = 1; b = -\dfrac{\pi}{4}$

1.30　Determine $\int_0^{\frac{\pi}{3}} 2\cos\dfrac{3x}{2}\,dx$

$$\int_0^{\frac{\pi}{3}} 2\cos\frac{3x}{2}\,dx = 2\int_0^{\frac{\pi}{3}}\cos\frac{3x}{2}\,dx = 2\left[\frac{2}{3}\sin\frac{3x}{2}\right]_0^{\frac{\pi}{3}}$$

$$= 2\left[\left(\frac{2}{3}\sin\frac{\pi}{2}\right) - \frac{2}{3}\sin 0\right] = \frac{4}{3}$$

Note. Here we have used the same standard form as above with $a = \dfrac{3}{2}; b = 0$

1.31　Find $\int \sin^2 2x\,dx$

Now $\sin^2 x = \dfrac{1}{2}(1 - \cos 2x)$　　　　　　　　　　　Standard result

Hence $\sin^2 2x = \dfrac{1}{2}(1 - \cos 4x)$　　　　　　　　　　Putting $x = 2x$ in above eq.

Hence $\int \sin^2 2x\,dx = \dfrac{1}{2}\int (1 - \cos 4x)\,dx$

$$\frac{1}{2}\left(x - \frac{\sin 4x}{4}\right) = \frac{x}{2} - \frac{1}{8}\sin 4x + C$$

MODULE IN2

INTEGRATION BY SUBSTITUTION OR BY CHANGE OF VARIABLE

WORKED SOLUTIONS

2.1 Find $\int x\left(1+x^2\right)^3 dx$

Let $t = 1 + x^2$

Then $\dfrac{dt}{dx} = 2x$ .... $dx = \dfrac{dt}{2x}$

And $x\,dx = \dfrac{dt}{2}$

$\therefore \int x\left(1+x^2\right)^3 dx = \dfrac{1}{2}\int t^3 dt = \dfrac{1}{8}t^4 = \dfrac{1}{8}\left(1+x\right)^4$ Substituting for $t = 1 + x^2$

2.2 Evaluate $\int_0^1 \dfrac{1}{\left(3x+3\right)} dx$ correct to three decimal places

Let $t = 3x + 2$

Then $\dfrac{dt}{dx} = 3$ .... $dx = \dfrac{dt}{3}$

$\therefore \int \dfrac{1}{\left(3x+2\right)} dx = \dfrac{1}{3}\int \dfrac{1}{t} dt = \dfrac{1}{3}\ln t + C = \dfrac{1}{3}\ln\left(3x+2\right) + C$ Substituting for t$= 3x + 3$

Note that this is a standard result i.e $\int \dfrac{1}{\left(ax+b\right)} dx = \dfrac{1}{a}\ln\left(ax+b\right) + C$

Hence $\int_0^1 \dfrac{1}{\left(3x+2\right)} dx = \dfrac{1}{3}\ln\left[3x+2\right]_0^1 = \dfrac{1}{3}\left(\ln 5 - \ln 2\right) = \dfrac{1}{3}\ln\dfrac{5}{2}$

$= \dfrac{1}{3}\left(0.916291\right) = 0.3054303$ Or 0.305 to three places of decimals

2.3 Integrate $\int x\sqrt{1+x^2}\,dx$ with respect to x

Let $t = x^2$

Then $\dfrac{dt}{dx} = 2x$ .... $dx = \dfrac{dt}{2x}$

Substituting for $t = x^2$ we get

$$\int x\sqrt{1+x^2}\,dx = \frac{1}{2}\int\sqrt{(1+t)}\,dt = \frac{1}{2}\int(1+t)^{\frac{1}{2}}$$

$$= \frac{1}{2}\frac{(1+t)^{\frac{3}{2}}}{\frac{3}{2}} = \frac{1}{3}(1+t)^{\frac{3}{2}}$$

$$= \frac{1}{3}(1+x^2)^{\frac{3}{2}} + C$$

2.4 Find $\int \dfrac{x}{(1+x^2)}\,dx$

Let $t = x^2$ as before

Then $\dfrac{dt}{dx} = 2x \ldots dx = \dfrac{dt}{2x}$

Substituting for $t = x^2$ we get

$$\int \frac{x}{(1+x^2)}\,dx = \frac{1}{2}\int\frac{1}{1+t^2}\,dt = \frac{1}{2}\tan^{-1}t = \frac{1}{2}\tan^{-1}x^2 + C$$

Note that $\int \dfrac{1}{1+t^2} = \tan^{-1}t$ is a standard result

2.5 Find $\int \dfrac{1}{\sqrt{a^2-x^2}}\,dx$

Let $x = a\sin t$

Then $\dfrac{dx}{dt} = a\cos t \ldots dx = a\cos t\,dt$

Hence
$$\int\frac{1}{\sqrt{a^2-x^2}}\,dx = \int\frac{a\cos t}{\sqrt{(a^2-a^2\sin^2 t)}}\,dt = \int\frac{a\cos t}{\sqrt{a^2(1-\sin^2 t)}}\,dt$$

$$= \int\frac{a\cos t}{\sqrt{a^2\cos^2 t}}\,dt = \int 1\,dt = t \qquad\qquad (2.1)$$

Since $x = a\sin t$ then $t = \sin^{-1}\dfrac{x}{a}$

Hence substituting for $t = \sin^{-1}\dfrac{x}{a}$ in (2.1) we obtain

$$\int\frac{1}{\sqrt{a^2-x^2}}\,dx = \sin^{-1}\frac{x}{a} + C \qquad\qquad \text{This is a standard result}$$

Note. If the integrand contains (a^2-x^2) the substitution of $x = a\sin t$ or $x = a\cos t$ can often work. Note also that if we had used the substitution $x = a\cos t$ we would have obtained the

result $\int\dfrac{1}{\sqrt{a^2-x^2}}\,dx = -\cos^{-1}\dfrac{x}{a} + C_1$. This apparent discrepancy between the two results is due to the relationship $\sin^{-1} x + \cos^{-1} x = \dfrac{\pi}{2}$; hence the two results will differ by $\dfrac{\pi}{2}$. If we therefore

write $C = C_1 - \dfrac{\pi}{2}$ the two results will be in agreement. This is another example of indefinite

integrals giving different answers resulting from the use of alternative methods, differing by a constant.

2.6 Find $\displaystyle\int \dfrac{dx}{\left(a^2 + x^2\right)}$

If the integrand contains $a^2 + x^2$ (as in this case), the substitution of $x = a \tan t$ is usually used.

Hence we will let $x = a \tan t$ for this problem. Differentiating, we get:

$$\dfrac{dx}{dt} = a \sec^2 t \dots dx = a \sec^2 t \, dt \qquad\qquad\qquad \dfrac{d}{dt} \tan t = \sec^2 t \ \text{(std result)}$$

Hence $\displaystyle\int \dfrac{dx}{\left(a^2 + x^2\right)} = \int \dfrac{a \sec^2 t}{\left(a^2 + a^2 \tan^2 t\right)} \, dt = \int \dfrac{a \sec^2 t}{a^2 \left(1 + \tan^2 t\right)} \, dt$ Using substitution $x = a \tan t$

$$= \int \dfrac{a \sec^2 t}{a^2 \sec^2 t} \, dt = \dfrac{1}{a} \int dt = \dfrac{t}{a} + C \qquad\qquad (2.2) \qquad \text{Since } 1 + \tan^2 t = \sec^2 t$$

Hence $\displaystyle\int \dfrac{dx}{\left(a^2 + x^2\right)} = \dfrac{1}{a} \tan^{-1} \dfrac{x}{a} + C \qquad\qquad$ Substituting in (2.2) for $t = \tan^{-1} \dfrac{x}{a}$

Note: $x = a \tan t \Leftrightarrow t = \tan^{-1} \dfrac{a}{x}$

2.7 Find $\displaystyle\int_0^{\frac{\pi}{2}} \sin^3 \theta \, d\theta$

Let $t = \cos \theta$

Then $\dfrac{dt}{d\theta} = -\sin \theta \dots d\theta = -\dfrac{dt}{\sin \theta}$

Now $\sin^3 \theta = \sin \theta \left(1 - \cos^2 \theta\right)$

So $\displaystyle\int \sin^3 \theta \, d\theta = \int \sin \theta \left(1 - \cos^2 \theta\right) d\theta \qquad\qquad (2.3)$

$$= \int \dfrac{\sin \theta \left(1 - t^2\right)}{-\sin \theta} \, dt \qquad\qquad\qquad \text{Substituting for } \cos \theta = t \text{ in (2.3)}$$

$$= -\int \left(1 - t^2\right) dt = -\left(t - \dfrac{t^3}{3}\right) + C \qquad\qquad (2.4)$$

$$= -\int \left(1 - \cos^2 \theta\right) d\theta = -\left(\cos \theta - \dfrac{\cos^3 \theta}{3}\right) + C \qquad\qquad \text{Substituting for } t = \cos \theta \text{ in (2.4)}$$

Hence $\displaystyle\int_0^{\frac{\pi}{2}} \sin^3 \theta \, d\theta = -\left[\cos \theta + \cos^3 \theta \right]_0^{\frac{\pi}{2}} = -\left\{(0) - \left(1 - \dfrac{1}{3}\right)\right\} = 1 - \dfrac{1}{3} = \dfrac{2}{3}$

Note: We could also use (2.4) directly if we change the limits of integration so that when $\theta = 0$, $t = \cos 0 = 1$ and when $\theta = \dfrac{\pi}{2}$, $t = \cos\dfrac{\pi}{2} = 0$. Hence we could say

that $\int_0^{\frac{\pi}{2}} \sin^3\theta \, d\theta = -\int_1^0 \left(t - \dfrac{t^3}{3} \right) dt = \dfrac{2}{3}$ as before. The method chosen is optional depending on

your preference but either method be used for solving any integral where you use substitution.

2.8 Determine $\int_0^{\frac{\pi}{4}} \sec^4 x \, dx$

We will choose the substitution $t = \tan x$ because we know that $\sec^2 x = 1 + \tan^2 x$ which will be helpful if we think ahead a little. First we rearrange $\sec^2 x$ as follows:

$$\int \sec^4 x \, dx = \int \sec^2 x \sec^2 x \, dx = \int \left(1 + \tan^2 x\right) \sec^2 x \, dx \qquad (2.5)$$

Now Let $t = \tan x$

Then $\quad \dfrac{dt}{dx} = \dfrac{1}{\cos^2 x} ....dx = \cos^2 x \, dt$

Hence $\int \sec^4 x \, dx = \int \left(1 + t^2\right) \sec^2 x \cos^2 x \, dt \qquad$ Substituting for $t = \tan x$ in (2.5)

$\quad = \int \left(1 + t^2\right) + C = t + \dfrac{t^3}{3} + C \qquad$ Since $\sec^2 x \cos^2 x = 1$

$\quad = \cos x + \dfrac{\cos^3 x}{3} \qquad$ Reversing the substitution

$\therefore \int_0^{\frac{\pi}{4}} \sec^4 x \, dx = \left[\tan x + \dfrac{\tan^3 x}{3} \right]_0^{\frac{\pi}{4}} = 2 - 0 = \dfrac{4}{3}$

2.9 Find $\int \dfrac{x^3}{\left(1 + x^8\right)} \, dx$

In this case, we should recognize that if we let $t = x^4$ then we can obtain a term in x^3 when we differentiate that will cancel out the x^3 in the numerator of the integrand as follows:

Let $t = x^4$

Then $\dfrac{dt}{dx} = 4x^3 .... dx = \dfrac{dt}{4x^3}$

Hence $\int \dfrac{x^3}{\left(1 + x^8\right)} \, dx = \int \dfrac{x^3 \, dt}{4x^3 \left(1 + t^2\right)} \qquad$ Substituting for $t = x^4$

$= \dfrac{1}{4} \int \dfrac{dt}{1 + t^2} = \dfrac{1}{4} \tan^{-1} t + C$

Note that $\int \dfrac{1}{\left(1 + t^2\right)} = \tan^{-1} t$ is a standard result

Hence $\int \frac{x^3}{(1+x^8)} \, dx = \tan^{-1}(x^4) + C$ Reversing the substitution

2.10 Evaluate $\int \sqrt{4-x^2} \, dx$

Let $x = 2\sin\theta$ (2.6) You will soon see why this was chosen.

Then $\frac{dx}{dt} = 2\cos\theta \dots dx = 2\cos\theta \, d\theta$

Hence $\int \sqrt{4-x^2} \, dx = \int \left(\sqrt{4-4\sin^2\theta}\right)(2\cos\theta) \, d\theta$ Substituting for $x = 2\sin t$

$= \int \left(\sqrt{4(1-\sin^2 t)}\right)(2\cos\theta) \, d\theta$

$= 2\int \left(\sqrt{\cos^2\theta}\right)2\cos\theta \, d\theta = 4\int \cos^2\theta \, d\theta$

$= \int 4\frac{1}{2}(1+\cos 2\theta) \, d\theta$

$= 2\int (1+\cos 2\theta) \, d\theta$

$= 2\left(\theta + \frac{1}{2}\sin 2\theta\right) + C$

$= 2\left(\theta + \frac{1}{2}2\sin\theta\cos\theta\right)$ Since $\sin 2\theta = 2\sin\theta\cos\theta$

$= 2\theta + \sin\theta\cos\theta$ (2.7)

Now we reverse the substitution using from (2.6) as follows:

$x = 2\sin\theta \Rightarrow \theta = \sin^{-1}\left(\frac{x}{2}\right)$

Also $\cos\theta = \sqrt{1-\sin^2\theta}$ so $2\cos\theta = 2\sqrt{1-\sin^2\theta} = \sqrt{4-4\sin^2\theta}$

Therefore $\cos\theta = \frac{1}{2}\sqrt{4-x^2}$ since $x^2 = 4\sin^2\theta$

Now we substitute for x and $\cos\theta$ in (2.7)

Hence $\int \sqrt{4-x^2} \, dx = 2\sin^{-1}\left(\frac{x}{2}\right) + \frac{1}{2}x\sqrt{4-x^2} + C$

2.11 Find $\int \frac{dx}{\sqrt{a^2-x^2}}$

Let $x = a\sin\theta$ (2.8)

Then $\frac{dx}{dt} = a\cos\theta \dots dx = a\cos\theta \, d\theta$

Hence $\int \frac{dx}{\sqrt{a^2-x^2}} \, dx = \int \frac{a\cos\theta}{\sqrt{(a^2-a^2\sin^2\theta)}} \, d\theta$

$$= \int \frac{a\cos\theta}{\sqrt{a^2 - a^2\sin^2\theta}}\, d\theta = \int \frac{a\cos\theta}{a\sqrt{1-\sin^2\theta}}\, d\theta$$

$$\int \frac{a\cos\theta}{a\cos\theta}\, d\theta = \int d\theta = \theta + C$$

$$= \sin^{-1}\left(\frac{x}{a}\right) + C \qquad \text{Reversing the substitution used in (2.8) giving } \theta = \sin^{-1}\left(\frac{x}{a}\right)$$

2.12 Find $\int \frac{dx}{a^2 + x^2}$

We will choose the substitution $x = a\tan\theta$ because we know that $\sec^2 x = 1 + \tan^2 x$ hence we will able to change the denominator to $\sec^2\theta$ and the numerator will become $a^2\sec^2\theta$ as follows:

Let $x = a\tan\theta$

Then $\dfrac{dx}{d\theta} = a\sec^2\theta\,....dx = a\sec^2\theta d\theta$

Therefore $\int \dfrac{dx}{a^2 + x^2} = \dfrac{a\sec^2\theta}{\left(a^2 + a^2\tan^2\theta\right)}\, d\theta$

$$\int \frac{a\sec^2\theta}{a^2(1 + \tan^2\theta)}\, d\theta = \int \frac{a\sec^2\theta}{a^2\sec^2\theta}\, d\theta$$

$$\int \frac{1}{a}\, d\theta = \frac{1}{a}\int d\theta = \frac{1}{a}\theta + C$$

Reversing the substitution and using $\theta = \tan^{-1}\left(\dfrac{x}{a}\right)$ in the last result we obtain:

$$\int \frac{dx}{a^2 + x^2} = \frac{1}{a}\tan^{-1}\left(\frac{x}{a}\right) + C$$

2.13 Evaluate $\int \dfrac{(x - 1)}{\sqrt{1 - 9x^2}}\, dx$

Here we should notice that $\sqrt{1 - 9x^2} = \sqrt{1 - (3x)^2}$ so we will choose the substitution $3x = \sin\theta$ which will enable us to change the denominator to $\cos\theta$ as follows:

Let $3x = \sin\theta$ \hfill (2.9)

Then $x = \dfrac{1}{3}\sin\theta\,....dx = \dfrac{1}{3}\cos\theta d\theta$

Substituting for these results we obtain:

$$\int \frac{(x-1)}{\sqrt{1-9x^2}}\,dx = \int \frac{\left(\frac{1}{3}\sin\theta-1\right)\left(\frac{1}{3}\cos\theta\right)}{\sqrt{1-9\left(\frac{1}{3}\sin\theta\right)^2}}\,d\theta$$

$$= \int \frac{\left(\frac{1}{3}\sin\theta-1\right)\frac{1}{3}\cos\theta}{\sqrt{1-\sin^2\theta}}\,d\theta = \int \frac{\left(\frac{1}{9}\sin\theta-\frac{1}{3}\right)\cos\theta}{\cos\theta}\,d\theta$$

$$= \int \left(\frac{1}{9}\sin\theta-\frac{1}{3}\right)d\theta = -\frac{1}{9}\cos\theta-\frac{1}{3}\theta+C \qquad (2.10)$$

Now $\cos\theta = \sqrt{1-\sin^2\theta}$

But $\sin^2\theta = 9x^2$ hence $\cos\theta = \sqrt{1-9x^2}$

And $\theta = \sin^{-1}(3x)$ from (2.9)

Substituting for θ and $\cos\theta$ in (2.10) we obtain:

$$\int \frac{(x-1)}{\sqrt{1-9x^2}}\,dx = -\frac{1}{9}\sqrt{1-9x^2}-\frac{1}{3}\sin^{-1}(3x)+C$$

2.14 Find $\int \operatorname{cosec} x\,dx$

Sometimes the substitution $t = \frac{1}{2}\tan x$ is useful for solving some problems where:

$$\sin x = \frac{2\tan\left(\frac{1}{2}x\right)}{1+\tan^2\left(\frac{1}{2}x\right)} \quad \text{and} \quad \cos x = \frac{1-\tan^2\left(\frac{1}{2}x\right)}{1+\tan^2\left(\frac{1}{2}x\right)}$$

Then $\sin x = \frac{2t}{1+t^2}$ and $\cos x = \frac{1-t^2}{1+t^2}$. We will now attack this problem in the usual way:

Let $t = \tan\left(\frac{1}{2}x\right)$ \qquad (2.11)

Then $\frac{dt}{dx} = \frac{1}{2}\sec^2 x \ldots dx = \frac{2\,dt}{\sec^2\left(\frac{1}{2}x\right)} = \frac{2\,dt}{1+t^2}$ \qquad (2.12)

We also know that $\operatorname{cosec} x = \frac{1}{\sin x}$

Hence $\int \operatorname{cosec} x\,dx = \int \frac{dx}{\sin x}$

We can now use the substitution $\sin x = \frac{2t}{1+t^2}$ and the result from (2.12) to give

$$\int \frac{dx}{\sin x} = \int \left(\frac{1+t^2}{2t} \right) \left(\frac{2}{1+t^2} \right) dt = \int \frac{dt}{t} = \ln t + C$$

Therefore by reversing the substitution using (2.11) we obtain:

$$\int \text{cosec} x \, dx = \ln \left| \tan \frac{1}{2} x \right| + C$$

In general, the substitution of $t = \tan \left(\frac{1}{2} x \right)$ is useful for determining integrals of the form

$$\int \frac{dx}{a + b \sin x} \quad \text{or} \int \frac{dx}{a + b \cos x}$$

Another example of the use of this substitution is shown below.

2.15 Find $\int \frac{dx}{(5 + 4\cos\theta)}$

Let $t = \tan \left(\frac{1}{2} x \right)$

....$dx = \frac{2dt}{1+t^2}$ As before

Then $\int \frac{dx}{(5 + 4\cos\theta)} = \int \frac{2\,dt}{\left(5 + 4 \left(\frac{1-t^2}{1+t^2} \right) \right)(1+t^2)}$ Since $\cos x = \frac{1-t^2}{1+t^2}$ and ....$dx = \frac{2dt}{1+t^2}$

$$= \int \frac{2dt}{5(1+t^2) + 4(1-t^2)}$$

$$= \int \frac{2dt}{(9+t^2)} = 2\int \frac{dt}{(9+t^2)} = 2\int \frac{dt}{(3^2+t^2)}$$

$$= \frac{2}{3} \tan^{-1} \left(\frac{t}{3} \right) + C$$ Since $\int \frac{dx}{a^2 + x^2} = \frac{1}{a} \tan^{-1} \left(\frac{x}{a} \right) + C$ (standard result)

$$= \frac{2}{3} \tan^{-1} \left\{ \frac{\tan(\frac{x}{2})}{3} \right\} + C$$

2.16 Integrate $x(1+x^2)^3$ (2.14)

Let $u = 1 + x^2$

Then $\frac{du}{dx} = 2x$ .... $dx = \frac{du}{2x}$ or $xdx = \frac{1}{2}du$

Hence $\int x(1+x^2)^3 = \frac{1}{2} \int u^3 \, du$ Substituting for $u = 1+x^2$ and $xdx = \frac{1}{2}du$ in (2.14)

$$= \frac{1}{2} \left(\frac{1}{4} u^4 \right) + C = \frac{1}{8} (1+x^2) + C$$ Reversing the substitution with $u = 1 + x^2$

2.17 Integrate $\dfrac{1}{\sqrt{x}\left(2+\sqrt{x}\right)}$

Let $u = \sqrt{x}$ so that $x = u^2$ .... $dx = 2u$

Hence $\displaystyle\int \dfrac{dx}{\sqrt{x}\left(2+\sqrt{x}\right)} = \int \dfrac{2u\,du}{u\left(2+u\right)}$ Substituting for $u = \sqrt{x}$ and .... $dx = 2u$

$= \displaystyle\int \dfrac{du}{2+u} = 2\ln\left(2+u\right) + C$

$= 2\ln\left(2+\sqrt{x}\right) + C$ Reversing the substitution with $u = \sqrt{x}$

2.18 Find $\displaystyle\int \dfrac{dx}{x^2 + 4x + 7}$

This type of problem can be solved by completing the square as follows:

$$x^2 + 4x + 7 = x^2 + 4x + 4 + 3 = \left(x+2\right)^2 + \left(\sqrt{3}\right)^2$$

Now let $u = 2x + 2$

Then $\dfrac{du}{dx} = 2 .... dx = \dfrac{du}{2}$

Therefore $\displaystyle\int \dfrac{dx}{x^2 + 4x + 7} = \dfrac{1}{2}\int \dfrac{du}{u^2 + \left(\sqrt{3}\right)^2}$

Notice that this result is of the form $\displaystyle\int \dfrac{dx}{a^2 + x^2} = \dfrac{1}{a}\tan^{-1}\left(\dfrac{x}{a}\right) + C$ where $a = \sqrt{3}$ and

$x = u = \left(x - 2\right)$

Hence $\displaystyle\int \dfrac{dx}{x^2 + 4x + 7} = \dfrac{1}{\sqrt{3}}\tan^{-1}\left(\dfrac{x+2}{\sqrt{3}}\right) + C$

Note that we could also have solved this problem more directly by noticing that $\dfrac{d}{dx}\left(x+2\right) = dx$.

Therefore after completing the square we could say that:

$$\int \dfrac{dx}{x^2 + 4x + 7} = \int \dfrac{dx(x+2)}{\left(x+2\right)^2 + \left(\sqrt{3}\right)^2} = \int \dfrac{dx}{\left(x+2\right)^2 + \left(\sqrt{3}\right)^2} = \dfrac{1}{\sqrt{3}}\tan^{-1}\left(\dfrac{x+2}{\sqrt{3}}\right) + C \text{ as before}$$

2.19 Find the integral of $x\left(2x^2 + 3\right)^3$

Let $u = 2x^2 + 3$

Then $\dfrac{du}{dx} = 4x .... dx = \dfrac{du}{4x}$

Therefore $\int x\left(2x^2+3\right)^3 dx = \frac{1}{4}\int u^3 du = \frac{1}{4}\left(\frac{u^4}{4}\right)+C$

Substituting for $u = 2x^2 + 3$ and $dx = \dfrac{du}{4x}$

$\dfrac{1}{16}u^4 + C = \dfrac{1}{16}\left(2x^2+3\right)^4 + C$ $\qquad$ Reversing the substitution with $u = 2x^2 + 3$

2.20 Determine $\int \dfrac{5-2x}{\left(x^2-5x+4\right)^2} dx$

Let $u = x^2 - 5x + 4$

Then $\dfrac{du}{dx} = 2x - 5 .... dx = \dfrac{du}{\left(2x-5\right)}$

Therefore $\int \dfrac{5-2x}{\left(x^2-5x+4\right)^2} dx = \int \dfrac{-du}{u^2}$

Substituting for $u = x^2 - 5x + 4$ and $\left(5-2x\right)dx = -du$

Note that $5 - 2x = -(2x-5)$ hence $\left(5-2x\right)dx = \dfrac{\left(5-2x\right)du}{\left(2x-5\right)} = -du$

Hence $\int \dfrac{5-2x}{\left(x^2-5x+4\right)^2} dx = -\int u^{-2} du = -\dfrac{u^{-1}}{(-1)} + C$

$= u^{-1} + C = \dfrac{1}{u} + C = \dfrac{1}{\left(x^2-5x+4\right)} + C$

2.21 Determine $\int \dfrac{4x-6}{\sqrt{x^2-3x+2}} dx$

Let $u = x^2 - 3x + 2$

Then $\dfrac{du}{dx} = 2x - 3 ... dx = \dfrac{du}{2x-3}$ and $\left(4-6x\right)dx = \dfrac{\left(4x-6\right)du}{\left(2x-3\right)} = 2du$

Hence $\int \dfrac{4-6x}{\sqrt{x^2-3x+2}} dx = 2\int \dfrac{du}{u^{\frac{1}{2}}} = 2\int u^{-\frac{1}{2}} du$

Substituting for $u = x^2 - 3x + 2$ and $\left(4-6x\right)dx = 2du$

Therefore $\int \dfrac{4x-6}{\sqrt{x^2-3x+2}} dx = 2\dfrac{u^{\frac{1}{2}}}{\left(\frac{1}{2}\right)} + C = 4\sqrt{x^2-3x+2} + C$

2.22 Integrate the function $y = \dfrac{e^x}{\left(2e^x+1\right)}$

Let $u = 2e^x + 1$

Then $\dfrac{du}{dx} = 2e^x \ldots\ldots dx = \dfrac{du}{2e^x}$

Hence $e^x dx = \dfrac{e^x\,du}{2e^x} = \dfrac{1}{2}du$

Therefore $\displaystyle\int y = \int \dfrac{e^x}{\left(2e^x + 1\right)}\,dx = \dfrac{1}{2}\int \dfrac{du}{u}$ Substituting for u and du

$= \dfrac{1}{2}\ln u + C = \dfrac{1}{2}\ln(2e^x + 1) + C$ Note: $\displaystyle\int \dfrac{1}{x}dx = \ln x + C$ is a standard result

2.23 Find $\displaystyle\int \dfrac{4x - 3}{2x^2 - 3x + 5}\,dx$

Let $u = 2x^2 - 3x + 5$

Then $\dfrac{du}{dx} = 4x - 3 \ldots\ldots dx = \dfrac{du}{4x - 3}$

Hence $\displaystyle\int \dfrac{4x - 3}{2x^2 - 3x + 5}\,dx = \int \dfrac{du}{u} = \ln u + c$

$= \ln\left(2x^2 - 3x + 5\right) + C$ Reversing the substitution

2.24 Find $\displaystyle\int x\cos(x^2 - 1)\,dx$

Let $t = x^2 - 1$

Then $\dfrac{dt}{dx} = 2x \ldots\ldots dx = \dfrac{dt}{2x} \Rightarrow x\,dx = \dfrac{1}{2}dt$

Therefore $\displaystyle\int x\cos(x^2 - 1)\,dx = \dfrac{1}{2}\int \cos t\,dt$ Using the above substitutions

$= \dfrac{1}{2}\sin t + C = \dfrac{1}{2}\sin(x^2 - 1) + C$

2.25 $\displaystyle\int \sin^3(2x)\,dx$

The first task is to rearrange $\sin^3(2x)$ into a more convenient form for integration as follows:

$\sin^3(2x) = \sin 2x \sin^2 2x$

$= \sin 2x\left(1 - \cos^2 2x\right) = \sin 2x - \sin 2x \cos^2 2x$

Now we know that if we differentiate $\cos 2x$ we will obtain $-\sin 2x$ and we will therefore be able to substitute for $\sin 2x$ and $\cos 2x$ as follows:

$\displaystyle\int \sin^3(2x)\,dx = \int \sin 2x(1 - \cos^2 2x)\,dx$

Now let $t = \cos 2x$

Then $\dfrac{dt}{dx} = -2\sin 2x \ldots\ldots dx = -\dfrac{dt}{2\sin 2x}$

Hence $\int \sin^3(2x)\,dx = -\frac{1}{2}\int(1-t^2)\,dt$ substituting for t and dt

$$= -\frac{1}{2}(t - t^3) + C = -\frac{1}{2}(\cos 2x - \cos^3 2x) + C$$

2.26 Evaluate $\int_0^6 \dfrac{dx}{(x^2 + 5x + 6)}$

This is another example of completing the square with a small amount of extra rearrangement before we can integrate the given function. What we are trying to achieve is to get the integral into

the form $\int \dfrac{dx}{a^2 - x^2}$ which is a standard form that we can integrate.

Now $x^2 + 5x + 6 = \boxed{x^2 + 5 + \left(\frac{5}{2}\right)^2} - \boxed{}$ Note the effect of adding and subtracting

$\left(\dfrac{5}{2}\right)^2$ leaves the original expression unchanged

$$= \left(x + \frac{5}{2}\right)^2 -$$

But where did the $\left(\dfrac{1}{2}\right)^2$ come from? Well, $-\left(\dfrac{5}{2}\right)^2 + 6 = -\dfrac{25}{4} + 6 = -\dfrac{25}{24} + \dfrac{24}{4} = -\dfrac{1}{4} = -\left(\dfrac{1}{2}\right)^2$

Using these results we can now say that:

$$\int_0^6 \frac{dx}{(x^2 + 5x + 6)} = \int_0^6 \frac{dx}{\left(\frac{1}{2}\right)^2 - \left(x + \frac{5}{2}\right)^2}$$

This is now in the required standard form $\int \dfrac{dx}{a^2 - x^2} = \dfrac{1}{2a}\ln\left(\dfrac{x - a}{a + a}\right) + C$ where $a = \dfrac{1}{2}$ and

$$x = \left(x + \frac{5}{2}\right)$$

Hence $\displaystyle\int_0^6 \frac{dx}{(x^2 + 5x + 6)} = \left[\frac{1}{2a}\ln\left(\frac{x - a}{x + a}\right)\right]_0^6 = \left[\frac{1}{2 \cdot \frac{1}{2}}\ln\left(\frac{x + \frac{5}{2} - \frac{1}{2}}{x + \frac{5}{2} + \frac{1}{2}}\right)\right]_0^6$

$= \left[\ln\left(\dfrac{x + 2}{x + 3}\right)\right]_0^6 = \ln\left(\dfrac{8}{9}\right) - \ln\left(\dfrac{2}{3}\right) = \ln\left(\dfrac{\frac{8}{9}}{\frac{2}{3}}\right) = \ln\left(\dfrac{4}{3}\right)$ Since $\ln(y) - \ln(z) = \ln\left(\dfrac{y}{z}\right)$

2.27 Find $\int_0^a \sqrt{a^2 - x^2}\,dx$

This type of problem suggests the use of $x = a\sin t$ as described in Q 2.5

Let $x = a\sin t$

Then $\dfrac{dx}{dt} = a\cos t .... dx = a\cos t\,dt$

Hence $\int_0^a \sqrt{a^2 - x^2}\, dx = \int_0^a \left(\sqrt{a^2 - a^2\sin^2 t}\right) a\cos t\, dt$

$\left(\sqrt{a^2\cos^2 t}\right) a\cos t\, dt = \int_0^a a^2\cos^2 t\, dt = a^2\int_0^a \cos^2 t\, dt$

$= a^2\int_0^a \frac{1}{2}(1+\cos 2t)\, dt = \left[\frac{a^2}{2}\left(t + \frac{1}{2}\sin 2t\right)\right]_0^{-a}$

$= \frac{a^2}{2}[t + \sin t\cos t]_0^a$ (2.15)

Now $\sin t = \dfrac{x}{a}$; $t = \sin^{-1}\dfrac{x}{a}$

And $\cos^2 t = 1 - \sin^2 t = 1 - \dfrac{x^2}{a^2} = \dfrac{a^2 - x^2}{a^2}$

So $\cos t = \dfrac{\sqrt{a^2 - x^2}}{a}$

Therefore, substituting for t, $\sin t$ and $\cos t$ in (2.15) we can say that:

$\dfrac{a^2}{2}[t + \sin t\cos t]_0^a = \dfrac{a^2}{2}\left[\sin^{-1}\dfrac{x}{a} + \dfrac{x}{a}\dfrac{\left(\sqrt{a^2 - x^2}\right)}{a}\right]_0^{-a}$

$= \dfrac{a^2}{2}\left[\sin^{-1}\dfrac{x}{a} + \dfrac{x\sqrt{a^2 - x^2}}{a^2}\right]_0^{-a} = \dfrac{a^2}{2}\left(\sin^{-1} 1 + \dfrac{a\sqrt{a^2 - a^2}}{a^2} - 0\right)$

$= \dfrac{a^2}{2}\left(\dfrac{\pi}{2}\right) = \dfrac{\pi a^2}{4}$

Hence $\int_0^a \sqrt{a^2 - x^2}\, dx = \dfrac{\pi a^2}{4}$

2.28 Find $\int \dfrac{dx}{\cos x}$

$\int \dfrac{dx}{\cos x} = \int \sec x\, dx$

We can write $\sec x$ as: $\sec x = \sec x\dfrac{(\sec x + \tan x)}{(\sec x + \tan x)} = \dfrac{(\sec^2 x + \sec x\tan x)}{(\sec x + \tan x)}$

Let $t = \sec x + \tan x$

Then $\dfrac{dt}{dx} = \sec x\tan x + \sec^2 x \dots dx = \dfrac{dt}{\sec x\tan x + \sec^2 x}$

Since $\dfrac{d}{dx}(\sec x) = \sec x\tan x$ and $\dfrac{d}{dx}(\tan) = \sec^2 x$

Then $(\sec^2 x + \sec x\tan x)\, dx = dt$

Substituting for t and dx we obtain:

$$\int \sec x \, dx = \int \frac{dt}{t} = \ln t = \ln(\sec x + \tan x) + C \quad \text{Since} \int \frac{dt}{t} = \ln t + C \text{ is a standard result}$$

Reversing the substitution

2.29 Evaluate $\int_0^{\frac{\pi}{2}} \frac{\cos \theta}{1 + \sin^2 \theta} \, d\theta$

Let $t = \sin \theta$

Then $\dfrac{dt}{d\theta} = \cos \theta \dots d\theta = \dfrac{dt}{\cos \theta}$

Hence $\displaystyle\int \frac{\cos \theta}{1 + \sin^2 \theta} \, d\theta = \int \frac{d\theta}{1 + t^2}$ 　　　　Substituting for $\sin \theta = t$ and $d\theta = \dfrac{dt}{\cos \theta}$

$= \left[\tan^{-1}(t) \right] + C$ 　　　　Since $\displaystyle\int \frac{d\theta}{1 + t^2} = \tan^{-1}(t)$ is a standard result

Therefore $\displaystyle\int_0^{\frac{\pi}{2}} \frac{\cos \theta}{1 + \sin^2 \theta} \, d\theta = \left[\tan^{-1}(\sin \theta) \right]_0^{\frac{\pi}{2}}$ 　　Reversing the substitution

$= \tan^{-1}\left(\sin\left(\tfrac{\pi}{2} \right) \right) - \tan^{-1}\left(\sin(0) \right)$

$= \tan^{-1}(1) - \tan^{-1}(0) = \dfrac{\pi}{4}$ 　　　　Since $\tan^{-1}(1) = \dfrac{\pi}{4}$ and $\tan^{-1}(0) = 0$

2.30 $\int \sin x \cos^4 x \, dx$

This problem is quite straightforward to solve since we can see by a very quick inspection that
$\dfrac{d}{dx} \cos x = -\sin x$. Hence we will use this substitution as follows:

Let $t = \cos x$

Then $\dfrac{dt}{dx} = -\sin x \dots dx = -\dfrac{dt}{\sin x}$

Therefore $\displaystyle\int \sin x \cos^4 x \, dx = -\int t^4 dt$ 　　　Substituting for $\cos x = t$ and $dx = -\dfrac{dt}{\sin x}$

$= -\dfrac{t^5}{5} + C$

Hence $\displaystyle\int \sin x \cos^4 x \, dx = -\frac{1}{5}\cos x + C$

2.31 $\int_0^{\frac{\pi}{2}} \frac{\cos \theta}{\sqrt{\sin \theta}} \, d\theta$

Again the substitution here is fairly obvious as follows:

Let $t = \sin \theta$

Then $\dfrac{dt}{d\theta} = \cos \theta \dots d\theta = \dfrac{dt}{\cos \theta}$

Hence $\displaystyle\int \frac{\cos \theta}{\sqrt{\sin \theta}} \, d\theta = \int \frac{dt}{t^{\frac{1}{2}}} = \int t^{-\frac{1}{2}} dt = 2t^{\frac{1}{2}} + C$

$$= 2\sqrt{\sin\theta} + C \qquad\qquad\qquad \text{Reversing the substitution}$$

Hence $\displaystyle\int_0^{\frac{\pi}{4}} \frac{\cos\theta}{\sqrt{\sin\theta}}\,d\theta = \left[2\sqrt{\sin\theta}\,\right]_0^{\frac{\pi}{4}}$

$$= 2\sqrt{\sin\left(\tfrac{\pi}{4}\right)} - 2\sin(0) = 2\sqrt{\frac{\sqrt{2}}{2}} = 2.\frac{(2)^{\frac{1}{4}}}{2^{\frac{1}{2}}} = 2.(2)^{\frac{1}{4}}(2)^{-\frac{1}{2}} = 2.2^{-\frac{1}{4}} = 2^{\frac{3}{4}} \quad \text{Note: } \sin\left(\frac{\pi}{4}\right) = \frac{\sqrt{2}}{2}$$

2.32   Find $\displaystyle\int x(1+x^2)^3\,dx$

Let $t = 1 + x^2$

Then $\dfrac{dt}{dx} = 2x$

Hence $x\,dx = \dfrac{dt}{2}$

$\therefore \displaystyle\int x(1+x^2)^3\,dx = \frac{1}{2}\int t^3\,dt$

$$= \frac{1}{8}t^4 = \frac{1}{8}(1+x^2)^4 \qquad\qquad\qquad \text{Reversing the substitution}$$

MODULE IN3

MODULE IN3 INTEGRATION BY PARTS

WORKED SOLUTIONS

3.1 Find $\int x \cos x \, dx$

Using the reverse of the formula $\frac{d}{dx}(uv) = v\frac{du}{dx} + u\frac{dv}{dx}$ we obtain:

$$\int u \frac{dv}{dx} \, dx = uv - \int v \frac{du}{dx} \, dx$$

Don't forget to always choose $\frac{dv}{dx}$ as the option that makes the function the easiest one to integrate.

In this case we will take $u = x$ and $\frac{dv}{dx} = \cos x$

$$\Rightarrow v = \int \frac{dv}{dx} = \sin x \text{ and } \frac{du}{dx} = 1$$

Hence $\int x \cos x \, dx = x \sin x - \int \sin x \, dx$

$= x \sin x + \cos x + C$

Note. We could have chosen $u = \cos x$ and $\frac{dv}{dx} = x \Rightarrow v = \frac{1}{2}x^2$ leading to the integral

$\int v \frac{du}{dx} \, dx = \int \frac{1}{2}x^2 \sin x \, dx$ but this result is more complex than the original integral

3.2 Find $\int x \sin x \, dx$

For the same reasoning as in the previous question we will choose:

$u = x$ and $\frac{dv}{dx} = \sin x$

$$\Rightarrow v = \int \sin x = -\cos x \text{ and } \frac{du}{dx} = 1$$

Hence $\int x \sin x \, dx = -x \cos x - \int -\cos x \, dx$

$= -x \cos x + \sin x + C$ Or $\sin x - x \cos x + C$

3.3 Find $\int_0^\pi x^2 \cos x \, dx$

Let $u = x^2$ and $\frac{dv}{dx} = \cos x$

$$\Rightarrow v = \int \cos x \, dx = \sin x + C \text{ and } \frac{du}{dx} = 2x$$

Hence $\int_0^\pi x^2 \cos x \, dx = \left[x^2 \sin x \right]_0^\pi - \int_0^\pi 2x \sin x \, dx$

$= -\int_0^\pi 2x \sin x \, dx$ Since the bracketed term is zero.

However, this result still cannot be integrated as it stands so we will have to repeat the process.

Let $u = 2x$ and $\dfrac{dv}{dx} = \sin x$

$\Rightarrow \dfrac{du}{dx} = 2$ and $v = -\cos x$

Hence $\displaystyle\int_0^\pi x^2 \cos x \, dx = \left[-2x \cos x\right]_0^\pi + \int_0^\pi 2\cos x \, dx$

$= -2\pi + \left[2\sin x\right]_0^\pi = -2\pi$ Since the bracketed term is zero

Note: In general $\displaystyle\int x^n \sin x \, dx$ and $\displaystyle\int x^n \cos x \, dx$ can be evaluated in the same way. Choose the x^n term as u and use repeated integration by parts until n reduces to either 1 or x as shown above. This method is known as integration by successive reduction. The above two examples are simple, special cases of this technique and this method is discussed in detail in section 7 – Integration using Reduction Formulae.

3.4 Integrate xe^x

Again we will choose $u = x$ as it will reduce 1 following differentiation as follows:

Let $u = x$ and $\dfrac{dv}{dx} = e^x$

$\Rightarrow \dfrac{du}{dx} = 1$ and $v = e^x$

Hence $\displaystyle\int xe^x = xe^x - \int 1.e^x \, dx = xe^x - e^x + C$

Or $e^x(x-1) + C$ Rearranging

3.5 Find $\displaystyle\int \ln x \, dx$

In this problem it appears that that we do not have a product to apply the method of integration by parts but in fact we do have a product of $1 \times \ln x$.

Let $u = \ln x$ and $\dfrac{dv}{dx} = 1$

$\Rightarrow \dfrac{du}{dx} = \dfrac{1}{x}$ and $v = x$

Hence $\displaystyle\int \ln x \, dx = x \ln x - \int x.\dfrac{1}{x} \, dx$

$= x \ln x - \int 1.dx = x \ln x - x + C$

Or $x(\ln x - 1) + C$ Rearranging

3.6 Find $\displaystyle\int e^x \cos x \, dx$

This is another example of having to carry out integration by parts twice as we will see

Let $u = e^x$ and $\dfrac{dv}{dx} = \cos x$

Then $\dfrac{du}{dx} = e^x$ and $v = \sin x$

Hence $\int e^x \cos x\, dx = e^x \sin x - \int e^x \sin x\, dx$

$e^x \sin x$ is still not suitable for integration but if we apply integration by parts again we can change $\sin x$ to $\cos x$ which will allow us to solve the problem as follows:

Let $u = e^x$ and $\dfrac{dv}{dx} = \sin x$

$\Rightarrow \dfrac{du}{dx} = e^x$ and $v = -\cos x$

Then $\int e^x \cos x\, dx = e^x \sin x - \left\{ e^x(-\cos x) - \int e^x(-\cos x)\, dx \right\}$

Hence $\int e^x \cos x\, dx = e^x \sin x + e^x \cos x - \int e^x \cos x\, dx$

Therefore $2 \int e^x \cos x\, dx = e^x \sin x + e^x \cos x + C = e^x(\sin x + \cos x) + C$

So that $\int e^x \cos x\, dx = \dfrac{1}{2} e^x(\sin x + \cos x) + C$ 　　　　　　　　Rearranging

3.7 Find $\int e^x \cos x\, dx$

We will use the same approach as in Q 3.5

Let $u = e^x$ and $\dfrac{dv}{dx} = \cos x$

$\Rightarrow \dfrac{du}{dx} = e^x$ and $v = \sin x$

Then $\int e^x \cos x\, dx = e^x \sin x - \int e^x \sin x\, dx$

Just as before we still cannot evaluate the last integral and we will have to apply integration by parts again.

Let $u = e^x$ and $\dfrac{dv}{dx} = \sin x$

$\Rightarrow \dfrac{du}{dx} = e^x$ and $v = -\sin x$

Hence $\int e^x \cos x\, dx = e^x \sin x - \left\{ -e^x \cos x - \int -e^x \cos x\, dx \right\}$

$= e^x \sin x + e^x \cos x - \int e^x \cos x\, dx$

Therefore $\int e^x \cos x\, dx = e^x \sin x + e^x \cos x - \int e^x \cos x\, dx$

So that $2 \int e^x \cos x\, dx = e^x \sin x + e^x \cos x + C$

Hence $\int e^x \cos x\, dx = \dfrac{1}{2} e^x(\sin x + \cos x) + C$ 　　　　　　　　Rearranging

3.8 $\int x e^{2x}\, dx$

In this case will choose $u = x$ as on differentiation $\dfrac{du}{dx} = 1$ as follows:

Let $u = x$ and $\dfrac{dv}{dx} = e^{2x}$

$\Rightarrow \dfrac{du}{dx} = 1$ and $v = \dfrac{1}{2}e^{2x}$

Hence $\displaystyle\int xe^{2x}\,dx = x\left(\dfrac{1}{2}e^{2x}\right) - \int\left(\dfrac{1}{2}e^{2x}\right)dx$

$= x\left(\dfrac{1}{2}e^{2x}\right) - \dfrac{1}{4}e^{2x} + C$

$= e^{2x}\left(\dfrac{x}{2} - \dfrac{1}{4}\right) + C$

3.9 Determine $\displaystyle\int \log_{10} x\,dx$

First of all we have to change the integrand into a form that we can integrate. To do this we have to change the base from base10 to base e as follows:

In general $\log_{10} x = \dfrac{\log_e x}{\log_e 10} = \dfrac{\ln x}{\ln 10}$
  since $\log_e N = \ln(N)$

We can now treat $\ln(10)$ as a constant
  as it does not contain the variable x

Hence $\displaystyle\int \log_{10} x\,dx = \dfrac{1}{\ln(10)}\int \ln x\,dx$

We can re-write this as $\displaystyle\int \log_{10} x\,dx = \dfrac{1}{\ln(10)}\int 1\times \ln x\,dx$
  Note this "trick" for future use

Now we use integration by parts as follows:

Let $u = \ln x$ and $\dfrac{dv}{dx} = 1$

$\Rightarrow \dfrac{du}{dx} = \dfrac{1}{x}$ and $v = x$

Therefore $\displaystyle\int \log_{10} x\,dx = \dfrac{1}{\ln(10)}\left(x\ln x - \int x\,\dfrac{1}{x}\,dx\right)$

$= \dfrac{1}{\ln(10)}\left(x\ln x - \int dx\right)$

$= \dfrac{1}{\ln(10)}(x\ln x - x) = \dfrac{1}{\ln(10)}x(\ln x - 1) + C$

3.10 Find $\displaystyle\int \sqrt{x^2 + a^2}\,dx$

As in the previous problem we can re-write $\displaystyle\int \sqrt{x^2 + a^2}\,dx$ as $\displaystyle\int 1\times \sqrt{x^2 + a^2}\,dx$

Let $u = \sqrt{x^2 + a^2}$ and $\dfrac{dv}{dx} = 1$

Then $\dfrac{du}{dx} = \dfrac{1}{2}\left(x^2 + a^2\right)^{-\frac{1}{2}}\times 2x = \dfrac{x}{\sqrt{x^2 + a^2}}$ and $v = x$

Hence $\displaystyle\int \sqrt{x^2 + a^2}\,dx = x\sqrt{x^2 + a^2} - \int \dfrac{x^2}{\sqrt{x^2 + a^2}}\,dx$

$$= x\sqrt{x^2+a^2} - \left\{ \int \frac{x^2+a^2}{\sqrt{x^2+a^2}} dx - \int \frac{a^2}{\sqrt{x^2+a^2}} dx \right\}$$

Note the "trick" of adding and subtracting of $\dfrac{a^2}{\sqrt{x^2+a^2}}$ here

$$= x\sqrt{x^2+a^2} - \int \sqrt{x^2+a^2}\, dx + \int \frac{a^2}{\sqrt{x^2+a^2}} dx$$

$$= x\sqrt{x^2+a^2} - \int \sqrt{x^2+a^2}\, dx + a^2 \ln\left(x+\sqrt{x^2+a^2}\right)$$

Therefore $2\int \sqrt{x^2+a^2}\, dx = x\sqrt{x^2+a^2} + a^2 \ln\left(x+\sqrt{x^2+a^2}\right)$

Note: $\int \dfrac{1}{\sqrt{x^2+a^2}} dx = \ln\left(x+\sqrt{x^2+a^2}\right) + C$ is a standard result

$$= \frac{x}{2}\sqrt{x^2+a^2} + \frac{a^2}{2}\ln\left(x+\sqrt{x^2+a^2}\right) + C$$

Note $\int \sqrt{x^2-a^2}\, dx = \dfrac{x}{2}\sqrt{x^2-a^2} + \dfrac{a^2}{2}\ln\left(x+\sqrt{x^2-a^2}\right) + C$ (replace $+a^2$ with $-a^2$)

3.11 Find $\int_0^{\frac{\pi}{2}} x\sin x\, dx$

Let $u = x$ and $\dfrac{dv}{dx} = \sin x$

$\Rightarrow \dfrac{du}{dx} = 1$ and $v = -\cos x$

Then $\int x\sin x\, dx = -x.\cos x - \int 1.(-\cos x)\, dx$

$$= -x.\cos x + \sin x + C$$

$$= \sin x - x\cos x + C \qquad\qquad \text{Rearranging}$$

Hence $\int_0^{\frac{\pi}{2}} x\sin x\, dx = \left[\sin x - x\cos x\right]_0^{\frac{\pi}{2}} = 1 - 0 = 1$

3.12 Find $\int x^2 \sin x\, dx$

This is another example of having to apply integration by parts twice because of the x^2 in the integrand. Compare this with the previous question.

Let $u = x^2$ and $\dfrac{dv}{dx} = \sin x$

Then $\dfrac{du}{dx} = 2x$ and $v = -\cos x$

Hence $\int x^2 \sin x\, dx = -x^2 \cos x + \int 2x \cos x\, dx$

We now have to evaluate the second integral:

Let $u = 2x$ and $\dfrac{dv}{dx} = \cos x$

So that $\dfrac{du}{dx} = 2$ and $v = \sin x$

Then $\int x^2 \sin x\, dx = -x^2 \cos x + \left\{ 2x \sin x - \int 2 \sin x\, dx \right\}$

$\qquad = -x^2 \cos x + 2x \sin x - 2 \cos x + C$

$\qquad = 2x \sin x - x^2 \cos x - 2 \cos x + C$

$\qquad = 2x \sin x - \left(x^2 - 2 \right) \cos x$ $\qquad\qquad$ Rearranging

3.13 Find $\int \sin^2 x\, dx$

Integration by parts is not the best method for evaluating $\int \sin^2 x\, dx$ but it is included here to illustrate a technique that may be useful in other cases. The "trick" is to re-write the integral first.

$\int \sin^2 x\, dx = \int \sin x \sin x\, dx$

Let $u = \sin x$ and $\dfrac{dv}{dx} = \sin x$

Then $\dfrac{du}{dx} = \cos x$ and $v = -\cos x$

$\int \sin^2 x\, dx = -\sin x \cos x - \int -\cos x \cos x\, dx$

$\qquad = -\sin x \cos x + \int \cos^2 x\, dx$

$\qquad = -\sin x \cos x + \int \left(1 - \sin^2 x \right) dx$

$\qquad = -\sin x \cos x + \int 1.dx - \int \sin^2 x\, dx$

$\qquad = -\sin x \cos x + x - \int \sin^2 x\, dx$

Hence $2 \int \sin^2 x\, dx = x - \sin x \cos x + C$

Thus $\int \sin^2 x\, dx = \dfrac{1}{2} \left(x - \sin x \cos x \right) + C$

$\qquad\qquad = \dfrac{1}{2} \left(x - \dfrac{1}{2} \sin 2x \right) + C$ $\qquad\qquad$ Since $\sin 2x = 2 \sin x \cos x$

3.14 $\int x \sqrt{x + 1}\, dx$

Let $u = x$ and $\dfrac{dv}{dx} = \sqrt{x + 1}$

Then $\dfrac{du}{dx} = 1$ and $v = \dfrac{2}{3}(x+1)^{\frac{3}{2}}$

Note $v = \int \sqrt{x + 1}$ is obtained by using integration by substitution as follows:

$u = x + 1 \Rightarrow \dfrac{du}{dx} = 1 \Rightarrow dx = du \Rightarrow \int \sqrt{x+1} = \int u^{\frac{1}{2}}\, du = \dfrac{2}{3} u^{\frac{3}{2}} = \dfrac{2}{3}(x+1)^{\frac{3}{2}}$

Therefore $\int x \sqrt{x+1}\, dx = \dfrac{2}{3} x (x+1)^{\frac{3}{2}} - \int 1 . \dfrac{2}{3}(x+1)^{\frac{3}{2}}\, dx$

Now we can integrate $\dfrac{2}{3}(x+1)^{\frac{3}{2}}$ by using substitution again as follows:

Let $u = x + 1$

Then $\dfrac{du}{dx} = 1$

Hence $\displaystyle\int \dfrac{2}{3}(x+1)^{\frac{3}{2}}\,dx = \int \dfrac{2}{3}u^{\frac{3}{2}}\,du = \dfrac{2}{3}u^{\frac{5}{2}} + C = \dfrac{2}{3}\cdot\dfrac{2}{5}(x+1)^{\frac{5}{2}} = \dfrac{4}{15}(x+1)^{\frac{5}{2}} + C$

Thus $\displaystyle\int x\sqrt{x+1}\,dx = \dfrac{2}{3}x(x+1)^{\frac{3}{2}} - \dfrac{4}{15}(x+1)^{\frac{5}{2}} + C$

$$= (x+1)^{\frac{3}{2}}\left(\dfrac{2}{3}x - \dfrac{4}{15}(x+1) \right) + C$$

$$= \dfrac{2}{15}(x+1)^{\frac{3}{2}}(3x-2) + C$$

3.15 Find $\displaystyle\int x\sqrt{3-2x}\,dx$

Let $u = x$ and $\dfrac{dv}{dx} = \sqrt{3-2x}$

Then $\dfrac{du}{dx} = 1$ and $v = -\dfrac{1}{2}\cdot\dfrac{2}{3}(3-2x)^{\frac{3}{2}} = \dfrac{1}{3}(3-2x)^{\frac{3}{2}}$

As before we can evaluate $v = \displaystyle\int \sqrt{3-2x}\,dx$ by using a substitution:

Let $u = 3-2x \Rightarrow \dfrac{du}{dx} = -2 \Rightarrow dx = -\dfrac{1}{2}du$.

Then $\displaystyle\int \sqrt{3-2x}\,dx = \dfrac{1}{2}u^{\frac{1}{2}}\,du = -\dfrac{1}{2}\cdot\dfrac{2}{3}u^{\frac{3}{2}} = -\dfrac{1}{3}u^{\frac{3}{2}} + C = -\dfrac{1}{3}(3-2x)^{\frac{3}{2}} + C$

Hence $\displaystyle\int x\sqrt{3-2x}\,dx = -\dfrac{1}{3}x(3-2x)^{\frac{3}{2}} + \int 1.\dfrac{1}{3}(3-2x)^{\frac{3}{2}}\,dx$

We now use a similar substitution to evaluate $\displaystyle\int \dfrac{1}{3}(3-2x)^{\frac{3}{2}}\,dx$ as follows:

Let $u = 3-2x \Rightarrow \dfrac{du}{dx} = -2 \Rightarrow dx = -\dfrac{1}{2}du$

Then $\displaystyle\int \dfrac{1}{3}(3-2x)^{\frac{3}{2}}\,dx = -\dfrac{1}{2}\cdot\dfrac{1}{3}\int u^{\frac{3}{2}}\,du = \dfrac{1}{6}\cdot\dfrac{2}{5}u^{\frac{5}{2}} = -\dfrac{1}{15}(3-2x)^{\frac{5}{2}}$

$\displaystyle\int x\sqrt{3-2x}\,dx = -\dfrac{1}{3}x(3-2x)^{\frac{3}{2}} - \dfrac{1}{15}(3-2x)^{\frac{5}{2}}$

$$= -(3-2x)^{\frac{3}{2}}\left(\dfrac{x}{3} + \dfrac{3}{15} - \dfrac{2x}{15} \right)$$

$$= -\dfrac{1}{5}(3-2x)^{\frac{3}{2}}(x+1)$$

3.16 Find $\displaystyle\int x(x+3)^5\,dx$

Let $u = x$ and $\dfrac{dv}{dx} = (x+3)^5$

Then $\dfrac{du}{dx} = 1$ and $v = \dfrac{1}{6}(x+3)^6$

You can get this result by inspection since $\dfrac{d}{dx}\left(\dfrac{1}{6}(x+3)^6\right) = (x+3)^5 \times 1 = (x+3)^5$

Hence $\displaystyle\int x(x+3)^5\,dx = \dfrac{1}{6}x(x+3)^6 - \dfrac{1}{6}\int(x+3)^6\,dx$

Now we need to integrate $(x+3)^6$

Once again, by inspection we can see that this must be $\dfrac{1}{7}(x+3)^7$ since

$$\dfrac{d}{dx}\left(\dfrac{1}{7}(x+3)^7\right) = (x+3)^6$$

$$\int x(x+3)^5\,dx = \dfrac{1}{6}x(x+3)^6 - \dfrac{1}{6}\cdot\dfrac{1}{7}(x+3)^7 + C$$

$$= (x+3)^6\left(\dfrac{x}{6} - \dfrac{(x+3)}{42}\right) + C$$

$$= \dfrac{1}{42}(x+3)^6(2x-1) + C$$

3.17 Find $\displaystyle\int x\sec^2 x\,dx$

Let $u = x$ and $\dfrac{dv}{dx} = \sec^2 x$

Then $\dfrac{du}{dx} = 1$ and $v = \tan x$ standard result

Hence $\displaystyle\int x\sec^2 x\,dx = x\tan x - \int \tan x\,dx$

$$= x\tan x + \ln(\cos x) + C$$

Since $\displaystyle\int \tan x\,dx = -\ln(\cos x)$ (standard result)

3.18 Find $\displaystyle\int \sqrt{1-x^2}\,dx$

Let $u = \sqrt{1-x^2}$ and $\dfrac{dv}{dx} = 1$

Then $\dfrac{du}{dx} = \dfrac{1}{2}(1-x^2)^{-\frac{1}{2}} \times (-2x) = -x(1-x^2)^{-\frac{1}{2}}$ and $v = x$

$$\int \sqrt{1-x^2}\,dx = x(1-x^2)^{-\frac{1}{2}} - \int\left(-x(1-x^2)^{-\frac{1}{2}} \times x\right)dx$$

$$= x\sqrt{1-x^2} + \int\dfrac{x^2}{\sqrt{1-x^2}}\,dx$$

$$= x\sqrt{1-x^2} + \int\dfrac{1^2 - (1^2 - x^2)}{\sqrt{1-x^2}}\,dx \qquad \text{Using the same "trick" as Q3.10}$$

$$= x\sqrt{1-x^2} + \int\dfrac{1^2}{\sqrt{1-x^2}}\,dx - \int\sqrt{1-x^2}\,dx$$

Now $\displaystyle\int\dfrac{a^2}{\sqrt{a^2-x^2}}\,dx = a^2\sin^{-1}\left(\dfrac{x}{a}\right) \qquad \text{See Q 2.27 in Section 2}$

In this case $a = 1$

Hence $2\int \sqrt{1-x^2}\, dx = x\sqrt{1-x^2} + 1^2 \sin^{-1} x$

$\therefore \int \sqrt{1-x^2}\, dx = \frac{1}{2}\left\{ x\sqrt{1-x^2} + \sin^{-1} x \right\} + C$

Note. We could have solved this problem more easily by direct substitution as follows:

Let $x = \sin t$

Then $\dfrac{dx}{dt} = \cos t \dots dx = \cos t\, dt$

$\int \sqrt{1-x^2}\, dx = \int \cos t \cos t\, dt = \int \cos^2 t\, dt$

$= \int \frac{1}{2}(1 + \cos 2t)\, dt = \frac{1}{2}\left(t + \frac{\sin 2t}{2} \right) + C$ Since $\cos^2 t = \frac{1}{2}(1 + \cos 2t)$

$= \dfrac{t}{2} + \dfrac{\sin t \cos t}{2} + C$ Since $\sin 2t = 2\sin t \cos t$

$= \dfrac{\sin^{-1} x}{2} + \dfrac{\left(\sin(\sin^{-1} x) \right)\left(\cos(\sin^{-1} x) \right)}{2} + C$ Reversing the substitution

$= \left(\dfrac{\sin^{-1} x}{2} + \dfrac{x\sqrt{1-x^2}}{2} \right) + C$

Since $\sin(\sin^{-1} x) = x$

Therefore $\sin^2\left(\sin^{-1} x \right) = x^2$ and $\cos^2\left(\sin^{-1} x \right) = 1 - x^2$ (since $\sin^2 x + \cos^2 x = 1$)

Hence $\cos(\sin^{-1} x) = \sqrt{1-x^2}$

3.19 Find $\int x^2 \cos x\, dx$

Let $u = x^2$ and $\dfrac{dv}{dx} = \cos x$

Then $\dfrac{du}{dx} = 2x$ and $v = \sin x$

Hence $\int x^2 \cos x\, dx = x^2 \sin x - \int 2x \sin x\, dx$

Once again we have to carry out integration by parts to evaluate the last integral

Let $u = 2x$ and $\dfrac{dv}{dx} = \sin x$

Then $\dfrac{du}{dx} = 2$ and $v = -\cos x$

Hence $\int 2x \sin x\, dx = -2x\cos x - 2\int -\cos x\, dx$

Therefore $\int x^2 \cos x\, dx = x^2 \sin x - \left(-2x\cos x - 2\int -\cos x\, dx \right)$

$= x^2 \sin x + 2x \cos x - 2\sin x + C$

$= (x^2 - 2)\sin x + 2x \cos x + C$

First we need to re-write $\int \sin^{-1} x \, dx$ as $\int 1 \times \sin^{-1} x \, dx$

Let $u = \sin^{-1} x$ and $\dfrac{dv}{dx} = 1$

Now first of all we need to find $\dfrac{d}{dx} \sin^{-1} x$. In order to do this we proceed as follows

Let $y = \sin^{-1} x \Rightarrow \sin y = x$

Hence $\cos y \dfrac{dy}{dx} = 1$ Differentiating both sides with respect to x

So that $\dfrac{dy}{dx} = \dfrac{1}{\cos y} = \dfrac{1}{\sqrt{1 - \sin^2 y}} = \dfrac{1}{\sqrt{1 - x^2}}$ Since $\sin^2 y = x^2$

Hence $\dfrac{du}{dx} = \dfrac{1}{1 - x^2}$ and $v = x$

Therefore $\int \sin^{-1} x \, dx = x \sin^{-1} x - \int \dfrac{x}{\sqrt{1 - x^2}} \, dx$

$$= x \sin^{-1} x + \sqrt{1 - x^2} + C$$

As before we re-write $\int \cos^{-1} x \, dx$ as $\int 1 \times \cos^{-1} x \, dx$

Let $u = \cos^{-1} x$ and $\dfrac{dv}{dx} = 1$

Now let $y = \cos^{-1} x \Rightarrow \cos y = x$

Hence $-\sin y \dfrac{dy}{dx} = 1$ Differentiating both sides with respect to x

So that $\dfrac{dy}{dx} = -\dfrac{1}{\sin y} = -\dfrac{1}{\sqrt{1 - \cos^2 y}} = -\dfrac{1}{\sqrt{1 - x^2}}$ Since $\cos^2 y = x^2$

Hence $\dfrac{du}{dx} = -\dfrac{1}{1 - x^2}$ and $v = x$

Therefore $\int \cos^{-1} x \, dx = x \cos^{-1} x - \int -\dfrac{x}{\sqrt{1 - x^2}} \, dx$

$$= x \cos^{-1} x - \sqrt{1 - x} + C$$

Again we need to re-write $\int \tan^{-1} x \, dx$ as $\int 1 \times \tan^{-1} x \, dx$

Let $u = \tan^{-1} x$ and $\dfrac{dv}{dx} = 1$

Now let $y = \tan^{-1} x \Rightarrow \tan y = x$

Then $\sec^2 y \dfrac{dy}{dx} = 1$ Differentiating both sides with respect to x

So that $\dfrac{dy}{dx} = \dfrac{1}{\sec^2 y} = \dfrac{1}{1+\tan^2 y} = \dfrac{1}{1+x^2}$ Since $\tan^2 y = x^2$

Hence $\dfrac{du}{dx} = \dfrac{1}{1+x^2}$ and $v = x$

Therefore $\displaystyle\int \tan^{-1} x \, dx = x \tan^{-1} x - \int \dfrac{x}{1+x^2}\, dx$

$$= x \tan^{-1} x - \dfrac{1}{2}\ln\left(1+x^2\right) + C$$

3.23 Find $\displaystyle\int \sec^3 x \, dx$

We will re-write $\displaystyle\int \sec^3 x \, dx$ as $\displaystyle\int \sec x \sec^2 x \, dx$

Let $u = \sec x$ and $\dfrac{dv}{dx} = \sec^2 x$

Then $\dfrac{du}{dx} = \sec x \tan x$ and $v = \tan x$ Since $\displaystyle\int \sec^2 x \, dx = \tan x$

Hence $\displaystyle\int \sec^3 x \, dx = \sec x \tan x - \int \sec x \tan x \tan x \, dx$

$$= \sec x \tan x - \int \sec x \tan^2 x \, dx$$

$$= \sec x \tan x - \int \sec x \left(\sec^2 x - 1\right) dx \qquad \text{Using identity } \sec^2 x = 1 + \tan^2 x$$

$$= \sec x \tan x - \int \left(\sec^3 x - \sec x\right) dx$$

$$= \sec x \tan x - \left[\int \sec^3 x \, dx - \int \sec x \, dx\right] \qquad \text{Splitting into two integrals}$$

$$= \sec x \tan x - \int \sec^3 x \, dx + \int \sec x \, dx$$

$$\therefore 2\int \sec^3 x \, dx = \sec x \tan x + \int \sec x \, dx$$

$$\therefore 2\int \sec^3 x \, dx = \sec x \tan x + \ln\left|\sec x + \tan x\right| \qquad \text{By 2.28}$$

$$\therefore \int \sec^3 x \, dx = \dfrac{1}{2}\sec x \tan x + \dfrac{1}{2}\ln\left|\sec x + \tan x\right| + C$$

3.24 Find $\displaystyle\int \cos(\ln x)\, dx$

To solve this problem, we again re-write $\displaystyle\int \cos(\ln x)\, dx$ as $\displaystyle\int 1 \times \cos(\ln x)\, dx$

Now let $u = \cos(\ln x)$ and $\dfrac{dv}{dx} = 1$

Then $\dfrac{du}{dx} = \sin\left(\ln(x)\right) \times \dfrac{1}{x} = \dfrac{1}{x}\sin\left(\ln(x)\right)$ and $v = x$

Hence $\displaystyle\int \cos(\ln x)\, dx = x\cos(\ln x) - \int x\dfrac{1}{x}\sin(\ln x)\, dx$

$$= x\cos(\ln x) - \int \sin(\ln x)\, dx$$

$$= x\cos(\ln x) - \int 1 \times \sin(\ln x)\, dx \qquad \text{Re-writing } \int \sin(\ln x)\, dx$$

Now let $u = \sin(\ln x)$ and $\dfrac{dv}{dx} = 1$ Integrating by parts again

Then $\dfrac{du}{dx} = \cos(\ln x) \times \dfrac{1}{x} = \dfrac{1}{x}\cos(\ln x)$ and $v = x$

Therefore $\displaystyle\int \cos(\ln x)\,dx = x\cos(\ln x) - \left[x\sin(\ln x) - \int x\dfrac{1}{x}\cos(\ln x) \right]$

$$= x\cos(\ln x) - \left[x\sin(\ln x) \right] - \int \cos(\ln x)\,dx$$

Hence $2\displaystyle\int \cos(\ln x)\,dx = x\cos(\ln x) - x\sin(\ln x)$

$\therefore \displaystyle\int \cos(\ln x)\,dx = \dfrac{1}{2}x\cos(\ln x) - \dfrac{1}{2}x\sin(\ln x)$

$$= \dfrac{x}{2}\left(\cos(\ln x) - \sin(\ln x) \right) + C \qquad \text{Rearranging}$$

3.25 Find $\displaystyle\int x(\ln x)^2\,dx$

Let $u = (\ln x)^2$ and $\dfrac{dv}{dx} = x$

Then $\dfrac{du}{dx} = 2\dfrac{\ln x}{x}$ and $v = \dfrac{x^2}{2}$

Hence $\displaystyle\int x(\ln x)^2\,dx = \dfrac{x^2}{2}(\ln x)^2 - \int \dfrac{x^2}{2}\left(2\dfrac{\ln x}{x} \right)dx$

$$= \dfrac{x^2}{2}(\ln x)^2 - \int x\ln x\,dx$$

Now let $u = \ln x$ and $\dfrac{dv}{dx} = x$ \qquad Integrating by parts again

Then $\dfrac{du}{dx} = \dfrac{1}{x}$ and $v = \dfrac{x^2}{2}$

Hence $\displaystyle\int x(\ln x)^2\,dx = \dfrac{x^2}{2}(\ln x)^2 - \left[\dfrac{x^2}{2}\ln x - \int \left(\dfrac{x^2}{2} \times \dfrac{1}{x} \right)dx \right]$

$$= \dfrac{x^2}{2}(\ln x)^2 - \dfrac{x^2}{2}\ln x + \int \dfrac{x}{2}\,dx$$

$$= \dfrac{x^2}{2}(\ln x)^2 - \dfrac{x^2}{2}\ln x + \dfrac{x^2}{4} + C$$

$$= \dfrac{x^2}{2}\left((\ln x)^2 - \ln x + \dfrac{1}{4} \right) + C \qquad \text{Rearranging}$$

3.26 Find $\displaystyle\int x^2 e^{x}\,dx$

This is the type of problem that needs a little experience to solve so follow through the reasoning shown below so that you will be able to resolve similar problems of this type in the future.

Fist we will use the "trick" we have seen several times before and re-write the integral as follows:

$$\int x^3 e^{x^2} dx = \int x^2 \left(x e^{x^2} \right) dx$$

Now let $u = x^2$ and $\dfrac{dv}{dx} = x e^{x^2}$

Now we can easily obtain $\dfrac{du}{dx} = 2x$ but we want to find $v = \int x e^{x^2} dx$

At first sight it looks as if we need to use integration by parts to find v but if we examine the integrand carefully we can note the following:

$$\frac{d}{dx}\left(\frac{1}{2} e^{x^2} \right) = \frac{1}{2} \times 2x e^{x^2} = x e^{x^2} \qquad\qquad \text{Since } \frac{d}{dx}\left(e^{x^n} \right) = n x e^{x^n}$$

Hence $v = \int x e^{x^2} dx = \dfrac{1}{2} e^{x^2}$

Always look to see if you can integrate a function like this by inspection first, rather than get bogged down in trying to integrate using substitutions or integration by parts.

So to recap we now have $u = x^2$ and $v = \dfrac{1}{2} e^{x^2}$

Now we can use integration by parts as normal:

$$\int x^3 e^{x^2} dx = x^2 \times \frac{1}{2} e^{x^2} - \int \left(\frac{1}{2} e^{x^2} \times 2x \right) dx$$

$$= \frac{1}{2} x^2 e^{x^2} - \int x e^{x^2} dx \qquad\qquad \text{Rearranging}$$

$$= \frac{1}{2} x^2 e^{x^2} - \frac{1}{2} e^{x^2} + C \qquad\qquad \text{Since } \int x e^{x^2} dx = \frac{1}{2} e^{x^2}$$

$$= \frac{1}{2} e^{x^2} \left(x^2 - 1 \right) + C$$

3.27 Find $\int \dfrac{1}{x} \ln(\ln x)\, dx$

Let $u = \ln(\ln x)$ and $\dfrac{dv}{dx} = \dfrac{1}{x}$

Then $\dfrac{du}{dx} = \dfrac{1}{\ln x} \times \dfrac{1}{x} = \dfrac{1}{x \ln x}$ and $v = \ln x$

Note: $\dfrac{du}{dx}$ was obtained by using the chain rule as follows:

Let $u = \ln(\ln x)$

Let $z = \ln x$

Therefore $u = \ln(z)$

Then $\dfrac{du}{dz} = \dfrac{1}{z}$ and $\dfrac{dz}{dx} = \dfrac{1}{x}$

So that $\dfrac{du}{dx} = \dfrac{du}{dz} \cdot \dfrac{dz}{dx} = \dfrac{1}{zx} = \dfrac{1}{x \ln x}$ \qquad Substituting for $z = \ln x$

Hence $\int \dfrac{1}{x} \ln(\ln x)\, dx = \ln x . \ln(\ln x) - \int \ln x \times \dfrac{1}{x \ln x}\, dx$

$$= \ln x . \ln(\ln x) - \int \frac{1}{x} dx$$

$$= \ln x . \ln(\ln x) - \ln x + C$$

3.28 Find $\int \cos(\ln x) dx$

Once again we will re-write $\int \cos(\ln x) dx$ as $\int 1 \times \cos(\ln x) dx$

Let $u = \cos(\ln x)$ and $\frac{dv}{dx} = 1$

Then $\frac{du}{dx} = -\sin(\ln x) \times \frac{1}{x} = -\frac{\sin(\ln x)}{x}$ and $v = x$

$\frac{du}{dx} = -\frac{1}{x} \sin(\ln x)$ Obtained by use of chain rule

Hence $\int \cos(\ln x) dx = x \cos(\ln x) - \int x(-\frac{\sin(\ln x)}{x})) dx$

$$= x \cos(\ln x) + \int x \sin(\ln x) dx \qquad \text{Rearranging}$$

Now we have to re-write $\int x \sin(\ln x) dx$ as $\int 1 \times x \sin(\ln x) dx$ and integrate the last integral by parts again

Let $u = \sin(\ln x)$ and $\frac{dv}{dx} = 1$

Then $\frac{du}{dx} = \cos(\ln x) \times \frac{1}{x} = \frac{1}{x} \cos(\ln x)$ and $v = x$

$\frac{du}{dx} = \frac{1}{x} \cos(\ln x)$ Obtained by use of chain rule

Hence $\int x \sin(\ln x) dx = x \sin(\ln x) - \int x \frac{1}{x} \cos(\ln x) dx$

$$= x \sin(\ln x) - \int \cos(\ln x) dx \qquad \text{Rearranging}$$

Therefore $\int \cos(\ln x) dx = x \cos(\ln x) + x \sin(\ln x) - \int \cos(\ln x) dx$

Hence $2 \int \cos(\ln x) dx = x \cos(\ln x) + x \sin(\ln x)$

Therefore $\int \cos(\ln x) dx = \frac{1}{2}(x \cos(\ln x) + x \sin(\ln x))$

3.29 Integrate $e^{-x} \sin x$

Let $u = e^{-x}$ and $\frac{dv}{dx} = \sin x$

Then $\frac{du}{dx} = -e^{-x}$ and $v = -\cos x$

Hence $\int e^{-x} \sin x dx = -e^{-x} \cos x - \int -\cos x \times (-e^{-x}) dx$

$$= -e^{-x} \cos x - \int e^{-x} \cos x dx$$

Now we have to integrate by parts again to integrate $\int e^{-x} \cos x dx$

Let $u = e^{-x}$ and $\frac{dv}{dx} = \cos x$

Then $\dfrac{du}{dx} = -e^{-x}$ and $v = \sin x$

Hence $\displaystyle\int e^{-x}\cos x\,dx = e^{-x}\sin x - \int -e^{-x}\sin x\,dx$

Therefore $\displaystyle\int e^{-x}\sin x\,dx = -e^{-x}\cos x - \left[e^{-x}\sin x + \int e^{-x}\sin x\,dx \right]$

$$= -e^{-x}\cos x - e^{-x}\sin x - \int e^{-x}\sin x\,dx$$

Hence $2\displaystyle\int e^{-x}\sin x\,dx = -e^{-x}\sin x - e^{-x}\cos x$

So that $\displaystyle\int e^{-x}\sin x\,dx = -\dfrac{1}{2}e^{-x}\left(\sin x + \cos x\right) + C$

3.30 Determine $\displaystyle\int x^2 \cos\left(\dfrac{1}{2}x\right)dx$

Let $u = x^2$ and $\dfrac{dv}{dx} = \cos\left(\dfrac{1}{2}x\right)$

Then $\dfrac{du}{dx} = 2x$ and $v = 2\sin\left(\dfrac{1}{2}x\right)$ $\qquad$ Since $\dfrac{d}{dx}\left(2\sin\left(\dfrac{1}{2}x\right)\right) = \cos\left(\dfrac{1}{2}x\right)$

Hence $\displaystyle\int x^2\cos\left(\dfrac{1}{2}x\right)dx = 2x^2\sin\left(\dfrac{1}{2}x\right) - \left(\int 2\sin\left(\dfrac{1}{2}x\right)\times 2x\,dx\right)$

$$= 2x^2\sin\left(\dfrac{1}{2}x\right) - 4\int x\sin\left(\dfrac{1}{2}x\right)dx$$

We now need to use integration by parts again on the last integral:

Let $u = x$ and $\dfrac{dv}{dx} = \sin\left(\dfrac{1}{2}x\right)$

$\dfrac{du}{dx} = 1$ and $v = -2\cos\left(\dfrac{1}{2}x\right)$ $\qquad$ Since $\dfrac{d}{dx}\left(-2\cos\left(\dfrac{1}{2}x\right)\right) = \sin\left(\dfrac{1}{2}x\right)$

Then $\displaystyle\int x\sin\left(\dfrac{1}{2}x\right)dx = -2x\cos\left(\dfrac{1}{2}x\right) - \int -2\cos\left(\dfrac{1}{2}x\right)dx$

$$= -2x\cos\left(\dfrac{1}{2}x\right) + 4\sin\left(\dfrac{1}{2}x\right)$$

Hence $\displaystyle\int x^2\cos\left(\dfrac{1}{2}x\right)dx = 2x^2\sin\left(\dfrac{1}{2}x\right) - 4\left(-2\cos\left(\dfrac{1}{2}x\right) + 4\sin\left(\dfrac{1}{2}x\right)\right) + C$

$$= 2x^2\sin\left(\dfrac{1}{2}x\right) + 8x\cos\left(\dfrac{1}{2}x\right) - 16\sin\left(\dfrac{1}{2}x\right) + C$$

$$= 8x\cos\left(\dfrac{1}{2}x\right) + \left(2x^2 - 16\right)\sin\left(\dfrac{1}{2}x\right) + C \qquad\text{Rearranging}$$

3.31 Integrate $\ln x$

Rewriting $\displaystyle\int \ln x\,dx$ as $\displaystyle\int \ln x \times 1\,dx$ we can use integration by parts as follows:

Let $u = \ln x$ and $\dfrac{dv}{dx} = 1$

Then $\dfrac{du}{dx} = \dfrac{1}{x}$ and $v = x$

Hence $\displaystyle\int \ln x \, dx = x \ln x - \int \dfrac{1}{x} \times x \, d$

$$= x \ln x - \int dx$$

$$= x \ln x - x + C$$

We can use integration by parts two way to integrate this function

(1) Let $u = x$ and $\dfrac{dv}{dx} = \ln x$

Then $\dfrac{du}{dx} = 1$ and $v = x \ln x - x$ from above

Hence $\displaystyle\int x \ln x \, dx = x(x \ln x - x) - \int x(\ln x - 1) \, dx$

$$= x(x \ln x - x) - \int x \ln x \, dx + \int x \, dx$$

$$= x(x \ln x - x) - \int x \ln x \, dx + \dfrac{x^2}{2}$$

Therefore $2\displaystyle\int x \ln x \, dx = x(x \ln x - x) + \dfrac{x^2}{2}$

Hence $\displaystyle\int x \ln x \, dx = \dfrac{1}{2}x(x \ln x - x) + \dfrac{x^2}{4} + C$

$$= \dfrac{1}{2}x^2 \ln x - \dfrac{1}{2}x^2 + \dfrac{1}{4}x^2 + C$$

$$= \dfrac{1}{4}x^2(2 \ln x - 1) + C$$

(ii) Let $u = \ln x$ and $\dfrac{dv}{dx} = x$

Then $\dfrac{du}{dx} = \dfrac{1}{x}$ and $v = x^2$

Therefore $\displaystyle\int x \ln x \, dx = \dfrac{1}{2}x^2 \ln x - \int \left(\dfrac{1}{2}x^2 \times \dfrac{1}{x} \right) dx$

$$= \dfrac{1}{2}x^2 \ln x - \dfrac{1}{2}\int x \, dx$$

$$= \dfrac{1}{2}x^2 \ln x - \dfrac{1}{4}x^2 + C$$

$$= \dfrac{1}{4}x^2(2 \ln x - 1) + C \text{ as before}$$

Using the previous technique, we will express $\displaystyle\int \sqrt{a^2 - x^2} \, dx$ as $\displaystyle\int 1.\sqrt{a^2 - x^2} \, dx$

Let $u = \sqrt{a^2 - x^2}$ and $\dfrac{dv}{dx} = 1$

Then $\dfrac{du}{dx} = \dfrac{1}{2}\left(a^2 - x^2\right)^{-\frac{1}{2}} \times (-2x) = \dfrac{-x}{\sqrt{a^2 - x^2}}$ and $v = x$

Then $\displaystyle\int \sqrt{a^2 - x^2}\, dx = \sqrt{a^2 - x^2} \cdot x - \int \dfrac{-x}{\sqrt{a^2 - x^2}} \cdot x\, dx$

$= x\sqrt{a^2 - x^2} + \displaystyle\int \dfrac{x^2}{\sqrt{a^2 - x^2}}\, dx$

$= x\sqrt{a^2 - x^2} + \displaystyle\int \dfrac{a^2 - \left(a^2 - x^2\right)}{\sqrt{a^2 - x^2}}\, dx$

Note the "trick" in the last integral of splitting it up into two integrals by re-writing it as shown. This type of "trick" is frequently used in integrating by parts and should be remembered.

$= x\sqrt{a^2 - x^2} + \displaystyle\int \dfrac{a^2}{\sqrt{a^2 - x^2}}\, dx - \int \dfrac{\left(a^2 - x^2\right)}{\sqrt{a^2 - x^2}}\, dx$

$= x\sqrt{a^2 - x^2} + \displaystyle\int \dfrac{a^2}{\sqrt{a^2 - x^2}}\, dx - \int \left(a^2 - x^2\right)\left(a^2 - x^2\right)^{-\frac{1}{2}}\, dx$

$= x\sqrt{a^2 - x^2} + \displaystyle\int \dfrac{a^2}{\sqrt{a^2 - x^2}}\, dx - \int \left(a^2 - x^2\right)^{\frac{1}{2}}\, dx$

$= x\sqrt{a^2 - x^2} + \displaystyle\int \dfrac{a^2}{\sqrt{a^2 - x^2}}\, dx - \int \sqrt{a^2 - x^2}\, dx$

Now $\displaystyle\int \dfrac{a^2}{\sqrt{a^2 - x^2}}\, dx = a^2 \sin^{-1}\dfrac{x}{a}$ \hspace{1cm} Standard result

Hence $\displaystyle\int \sqrt{a^2 - x^2}\, dx = x\sqrt{a^2 - x^2} + \sin^{-1}\dfrac{x}{a} - \int \sqrt{a^2 - x^2}\, dx$

Therefore $2\displaystyle\int \sqrt{a^2 - x^2} = x\sqrt{a^2 - x^2} + \sin^{-1}\dfrac{x}{a} + C$ \hspace{1cm} Rearranging

So that $\displaystyle\int \sqrt{a^2 - x^2} = \dfrac{1}{2} x\sqrt{a^2 - x^2} + \dfrac{1}{2}\sin^{-1}\dfrac{x}{a} + C$!

MODULE IN4

INTEGRATION USING PARTIAL FRACTIONS

INTRODUCTION

The use of partial fractions is a very useful technique to re-arrange certain type of functions so that they may be integrated more easily. The procedure is the reverse of that used to combine separate fraction into a single fraction. For example the expression:

$$\frac{1}{x-2} + \frac{1}{x+2} - \frac{2x}{\left(x^2+4\right)} = \frac{\left(x+2\right)\left(x^2+4\right)+\left(x-2\right)\left(x^2+4\right)-2x\left(x-2\right)\left(x+2\right)}{\left(x-2\right)\left(x+2\right)\left(x^2+4\right)}$$

$$= \frac{x^3+2x^2+4x+8+x^3-2x^2+4x-8-2x^3+8x}{\left(x^2-1\right)\left(x+2\right)\left(x^2+4\right)}$$

$$= \frac{16x}{\left(x^2-4\right)\left(x^2+4\right)}$$

$$= \frac{16x}{x^4-16}$$

Has been simplified into a single fraction where the denominator is the lowest common denominator of all the separate fractions. The reverse method is to split a single fraction whose denominator contains factors into two or more **partial fractions.**

This process depends on the following rules

(i) If the degree of the numerator of the original fraction is equal to or greater than the degree of the denominator – divide the numerator by the denominator first until a remainder is achieved of degree less than that of the denominator.

(ii) For every linear factor like $\left(x-a\right)$ in the denominator there is a corresponding partial fraction that has the form $\frac{A}{x-a}$

(iii) For every repeated factor like $\left(x-a\right)^2$ in the denominator there are two corresponding partial fractions that have the form $\frac{A}{x-a}$ and $\frac{B}{\left(x-a\right)^2}$. For every repeated factor like $\left(x-a\right)^3$ in the denominator there are three corresponding partial fractions that have the form $\frac{A}{x-a}$ and

$\frac{B}{\left(x-a\right)^2}$ and $\frac{C}{\left(x-a\right)^3}$ and so on.

(iv) For every quadratic factor in the denominator like $ax^2 + bx + c$ there is a corresponding partial fraction $\dfrac{Cx + D}{ax^2 + bx + c}$. Additional repeated factors are treated in the same way as in (iii).

For example the factor $\left(ax^2 + bx + c\right)^2$ has corresponding partial fractions $\dfrac{Cx + D}{ax^2 + bx + c}$ and

$$\frac{Ex + F}{\left(ax^2 + bx + c\right)^2}$$

The following examples illustrate the use of the above rules:

1.0 Resolve the expression $\dfrac{3}{(x+1)(x-1)}$ into partial fractions.

$$\frac{3}{(x+1)(x-1)} \equiv \frac{A}{x+1} + \frac{B}{x-1} \equiv \frac{A(x-1) + B(x+1)}{(x+1)(x-1)} \qquad \text{Rule (ii)}$$

Then $3 \equiv A(x-1) + B(x+1)$

Hence for $x = 1$, $2B = 3 \Rightarrow B = \dfrac{3}{2}$

And for $x = -1$; $-2A = 3 \Rightarrow A = -\dfrac{3}{2}$

$$\therefore \frac{3}{(x+1)(x-1)} \equiv \frac{3}{2(x-1)} - \frac{3}{2(x+1)}$$

2.0 Separate $\dfrac{7}{(x-1)(x+2)^2}$ into partial fractions $\qquad$ (1)

$$\frac{7}{(x-1)(x+2)^2} \equiv \frac{A}{x-1} + \frac{B}{x+2} + \frac{C}{(x+2)^2} \qquad \text{Rule (iii)}$$

$$\frac{7}{(x-1)(x+2)^2} \equiv \frac{A}{x-1} + \frac{B}{x+2} + \frac{C}{(x+2)^2}$$

$$\equiv \frac{A(x+2)(x+2)^2 + B(x-1)(x+2)^2 + C(x-1)(x+2)}{(x-1)(x+2)(x+2)^2}$$

$$\equiv \frac{A(x+2)^2 + B(x-1)(x+2) + C(x-1)}{(x-1)(x+2)^2} \qquad \text{Dividing top \& bottom by } (x+2)$$

Then $7 \equiv A(x+2)^2 + B(x-1)(x+2) + C(x-1)$ (2)

When $x = 1$; $A = \dfrac{7}{9}$

When $x = -2$; $-3C = 7 \Rightarrow C = -\dfrac{7}{3}$

To find B we can equate the coefficients of x^2 in (2) to zero since there are no terms involving x^2 in the numerator of (1)

Hence $A + B = 0$

Since we have already found that $A = \dfrac{7}{9}$ then $B = -\dfrac{7}{9}$

Therefore $\dfrac{7}{(x-1)(x+2)^2} \equiv \dfrac{7}{9(x-1)} - \dfrac{7}{9(x+2)} - \dfrac{7}{3(x+2)^2}$

3.0 Resolve $\dfrac{x^3}{x^2 - 2x - 3}$ into partial fractions

Now in this case the numerator is of a higher degree than the denominator so we need to divide the numerator by the denominator until we are left with a remainder of degree less than the numerator (Rule (i)) as follows:

$$
\begin{array}{r}
x + 2 \\
x^2 - 2x - 3 \overline{\smash{\big)}\, x^3} \\
\underline{x^3 - 2x^2 - 3x} \\
2x^2 + 3x \\
\underline{2x^2 - 4x - 6x} \\
7x + 6
\end{array}
$$

Hence $\dfrac{x^3}{x^2 - 2x - 3} \equiv x + 2 + \dfrac{7x+6}{x^2 - 2x - 3}$

$\equiv x + 2 + \dfrac{7x+6}{(x+1)(x-3)}$

We now need to split the remaining fraction into partial fractions

Let $\dfrac{7x+6}{(x+1)(x-3)} \equiv \dfrac{A}{x+1} + \dfrac{B}{x-3} \equiv \dfrac{A(x-3) + B(x+1)}{(x+1)(x-3)}$

Hence $7x+6 \equiv A(x-3)+B(x+1)$

When $x=3$; $4B=27 \Rightarrow B=\dfrac{27}{4}$

When $x=-1$; $-4A=-1 \Rightarrow A=\dfrac{1}{4}$

Hence $\dfrac{x^3}{x^2-2x-3} \equiv x+2+\dfrac{1}{4(x+1)}+\dfrac{27}{4(x-3)}$

4.0 Resolve the expression $\dfrac{x^2+2x+4}{2x(x-3)(x+1)}$ into partial fractions

Let $\dfrac{x^2-2x+4}{2x(x-3)(x+1)} \equiv \dfrac{A}{2x}+\dfrac{B}{x-3}+\dfrac{C}{x+1}$

$$\equiv \dfrac{A(x-3)(x+1)+2Bx(x+1)+2Cx(x-3)}{2x(x-3)(x+1)}$$

Then $x^2-2x+4 \equiv A(x+1)(x-3)+B(2x)(x+1)+C(2x)(x-3)$

When $x=3$; $24B=7 \Rightarrow B=\dfrac{7}{24}$

When $x=-1$; $8C=7 \Rightarrow C=\dfrac{7}{8}$

When $x=0$; $-3A=4 \Rightarrow A=\left(-\dfrac{4}{3}\right)$

Hence $\dfrac{x^2+2x+4}{2x(x-3)(x+1)} \equiv \dfrac{7}{24(x-3)}-\dfrac{4}{3(2x)}+\dfrac{7}{8(x+1)}$

$$\equiv \dfrac{7}{24(x-3)}-\dfrac{2}{3x}+\dfrac{7}{8(x+1)}$$

5.0 Separate $\dfrac{x^3+3}{x^2+3x+4}$ into partial fractions

Note that there are two problems here.

(i)   The numerator is of higher degree than the denominator and

The denominator does not factorise

Therefore we will have to first divide the numerator by the denominator until it is of lower degree than the numerator and then invoke rule (iv) to deal with the denominator.

$$x^2 + 3x + 4 \overline{\smash{\big)}\begin{array}{r} x - 3 \\ x^3 + 3 \end{array}}$$

$$\begin{array}{r} x^3 + 3x^2 + 4x \\ \hline -3x^2 - 4x + 3 \\ -3x^2 - 9x - 12 \\ \hline 5x + 15 \end{array}$$

Hence $\dfrac{x^3 + 3}{x^2 + 3x + 4} \equiv x - 3 + \dfrac{5(x+3)}{x^2 + 3x + 4}$

We cannot split the remaining fraction as the denominator does not factorise.

WORKED INTEGRATION SOLUTIONS USING PARTIAL FRACTIONS

4.1 Find $\displaystyle\int \frac{dx}{x^2 - a^2}$

Use partial fractions to change the denominator as follows:

$$\frac{1}{x^2 - a^2} \equiv \frac{1}{(x-a)(x+a)} \equiv \frac{p}{x-a} + \frac{q}{x+a}$$

Where p and q are constants

$$\therefore 1 \equiv p(x+a) + q(x-a) \tag{1}$$

So that:

$$p \times 2a + q \times 0 = 1 \Rightarrow p = \frac{1}{2a}$$ Putting $x = a$ in (1)

And $1 = p(0) + q(-2a) \Rightarrow q = -\dfrac{1}{2a}$ Putting $x = -a$ in (1)

Hence $\dfrac{1}{x^2 - a^2} = \left(\dfrac{1}{2a}\right)\left(\dfrac{1}{x-a}\right) - \left(\dfrac{1}{2a}\right)\left(\dfrac{1}{x+a}\right)$

Therefore $\int \dfrac{dx}{(x^2 - a^2)} = \dfrac{1}{2a}\int \dfrac{dx}{(x-a)} - \dfrac{1}{2a}\int \dfrac{dx}{(x+a)}$

$= \dfrac{1}{2a}\ln(x-a) - \dfrac{1}{2a}\ln(x+a) + C$ $\qquad$ Since $\int \dfrac{1}{x}dx = \ln x + c$

$= \dfrac{1}{2a}\ln\left(\dfrac{x-a}{x+a}\right) + C$ $\qquad$ Since $\ln a - \ln b = \ln\dfrac{a}{b}$

4.2 Integrate $\dfrac{x+3}{(x-1)(x-2)^2}$

Let $\dfrac{x+3}{(x-1)(x-2)^2} \equiv \dfrac{A}{(x-1)} + \dfrac{B}{(x-2)} + \dfrac{C}{(x-2)^2}$

$\equiv \dfrac{A(x-2)^2 + B(x-1)(x-2) + C(x-1)}{(x-1)(x-2)^2}$

Then $x+3 \equiv A(x-2)^2 + B(x-1)(x-2) + C(x-1)$

Putting $x = 1$ we obtain $A = 4$

Putting $x = 2$ we obtain $C = 5$

Hence $x+3 = 4(x-2)^2 + B(x-1)(x-2) + 5(x-1)$

$= 4(x^2 - 4x + 4) + B(x^2 - 3x + 2) + 5(x-1)$

Now the numerator, $x+3$ contains no terms in x^2

Therefore we can equate the coefficients of x^2 to zero. I.e.

$4x^2 + Bx^2 = 0$

$\Rightarrow B = -4$

$\therefore \int \dfrac{x+3}{(x-1)(x-2)^2}dx = \int\left(\dfrac{4}{(x-1)} - \dfrac{4}{(x-2)} + \dfrac{5}{(x-2)^2}\right)dx$

$$= \int \frac{4}{(x-1)} \, dx - \int \frac{4}{(x-2)} \, dx + \int \frac{5}{(x-2)^2} \, dx$$

$$= 4 \ln (x-1) - 4 \ln (x-2) + \int \frac{5}{(x-2)^2} \, dx$$

To determine the remaining integral let $t = x - 2 \Rightarrow \dfrac{dt}{dx} = 1 \Rightarrow dx = dt$

Hence $\displaystyle \int \frac{5}{(x-2)^2} \, dx = \int \frac{5}{t^2} \, dt = 5 \int t^{-2} \, dt = -5t^{-1} = -\frac{5}{t} = -\frac{5}{x-2}$

$$\therefore \int \frac{x+3}{(x-1)(x-2)^2} \, dx = 4 \ln (x-1) - 4 \ln (x-2) - \frac{5}{x-2} + C$$

4.3 Find $\displaystyle \int \frac{dx}{(x-2)(x-3)}$

Let $\dfrac{1}{(x-2)(x-3)} \equiv \dfrac{A}{x-2} + \dfrac{B}{x-3}$

$$\equiv \frac{A(x-3) + B(x-2)}{(x-2)(x-3)}$$

Hence $A(x-3) + B(x-2) \equiv 1$

When $x = 2, A = -1$

When $x = 3, B = 1$

$$\therefore 1 = \frac{-1}{x-2} + \frac{1}{x-3}$$

So that:

$$\int \frac{dx}{(x-2)(x-3)} = \int \left(\frac{1}{x-3} - \frac{1}{x-2} \right) dx$$

$$= \int \frac{dx}{x-3} - \int \frac{dx}{x-2}$$

$$= \ln(x-3) - \ln(x-2) + C$$

$$= \ln\left(\frac{x-3}{x-2} \right) + C$$

4.4 Find $\int \dfrac{dx}{x(x+5)}$

Let $\dfrac{1}{x(x+5)} \equiv \dfrac{A}{x} + \dfrac{B}{(x+5)}$

$$\equiv \frac{A(x+5) + Bx}{x(x+5)}$$

When $x = -5, B = -\dfrac{1}{5}$

When $x = 0, A = \dfrac{1}{5}$

$\therefore 1 = \dfrac{1}{5x} - \dfrac{1}{5(x+5)}$

So that

$$\int \frac{dx}{x(x+5)} = \frac{1}{5} \int \frac{dx}{x} - \frac{1}{5} \int \frac{dx}{x+5}$$

$$= \frac{1}{5} \ln x - \frac{1}{5} \ln(x+5)$$

$$= \frac{1}{5} \left(\ln x - \ln(x+5) \right)$$

$$= \frac{1}{5} \ln\left(\frac{x}{x+5} \right) \qquad \text{Since } \ln a - \ln b = \ln \frac{a}{b}$$

4.5 Find $\int \dfrac{x^3}{x^2 + x - 20} dx$

In this case the numerator is of a higher degree than the denominator, therefore we have to divide the numerator by the denominator until we achieve a remainder that has a lower degree.

$$\frac{x^2}{x^2+x-20} = x-1 + \frac{21x-20}{x^2+x-20}$$ Using long division

Now we can resolve the remainder into partial fractions:

$$\frac{21x-20}{x^2+x-20} = \frac{21x-20}{(x+5)(x-4)} \equiv \frac{A}{x+5} + \frac{B}{x-4} \equiv \frac{A(x-4)+B(x+5)}{(x+5)(x-4)}$$

Hence $21x - 20 \equiv A(x-4) + B(x+5)$

When $x = -5$

$$-105 - 20 = -9A \Rightarrow A = \frac{125}{9}$$

When $x = 4$

$$84 - 20 = 9B \Rightarrow B = \frac{64}{9}$$

Hence

$$\int \frac{x^3}{x^2+x-20}\, dx = \int \left(x-1 + \frac{125}{9(x+5)} x + \frac{64}{9(x-4)} \right) dx$$

$$= \int x - \int dx + \int \frac{125}{9(x+5)} x\, dx + \int \frac{64}{9(x-4)}$$

$$= \frac{x^2}{2} - x + \frac{125}{9} \ln(x+5) + \frac{64}{9} \ln(x-4) + C$$

4.6   Find $\int \frac{dx}{a^2 - x^2}$

Let $\dfrac{1}{a^2-x^2} \equiv \dfrac{P}{(a+x)} + \dfrac{Q}{(a-x)} \equiv \dfrac{P(a-x)+Q(a+x)}{(a+x)(a-x)}$

Then $P(a-x) + Q(a+x) \equiv 1$

When $x = a$

$$1 = 2aQ \Rightarrow Q = \frac{1}{2a}$$

When $x = -a$

$$1 = 2aP \Rightarrow P = \frac{1}{2a}$$

Hence

$$\int \frac{dx}{a^2 - x^2} = \frac{1}{2a} \int \frac{dx}{a-x} + \frac{1}{2a} \int \frac{dx}{a+x} \qquad \text{note that } \frac{1}{2a} \int \frac{dx}{a-x} = -\frac{1}{2a} \ln(a-x)$$

$$= -\frac{1}{2a} \ln(a-x) + \frac{1}{2a} \ln(a+x) + C$$

$$= \frac{1}{2a} \ln\left(\frac{a+x}{a-x}\right) + C \qquad \text{Since } \ln a - \ln b = \ln \frac{a}{b}$$

4.7 Integrate $\dfrac{3x}{(x-1)(x-2)(x-3)}$

Let $\dfrac{3x}{(x-1)(x-2)(x-3)} \equiv \dfrac{A}{x-1} + \dfrac{B}{x-2} + \dfrac{C}{x-3}$

$$\equiv \frac{A(x-2)(x-3) + B(x-1)(x-3) + C(x-1)(x-2)}{(x-1)(x-2)(x-3)}$$

Hence $A(x-2)(x-3) + B(x-1)(x-3) + C(x-1)(x-2) \equiv 3x$

When $x = 1$; $2A = 3 \Rightarrow A = \dfrac{3}{2}$

When $x = 2$; $-B = 6 \Rightarrow B = -6$

When $x = 3$; $2C = 9 \Rightarrow C = \dfrac{9}{2}$

Hence $\dfrac{3x}{(x-1)(x-2)(x-3)} \equiv \dfrac{3}{2(x-1)} - \dfrac{6}{x-2} + \dfrac{9}{2(x-3)}$

Therefore $\displaystyle\int \frac{3x}{(x-1)(x-2)(x-3)} \, dx = \int \left(\frac{3}{2(x-1)} - \frac{6}{x-2} + \frac{9}{2(x-3)} \right) dx$

$$= \frac{3}{2} \ln(x-1) - 6\ln(x-2) + \frac{9}{2} \ln(x-3) + C$$

4.8 Find $\displaystyle\int \frac{dx}{(x-1)^2(x^2+1)}$

Let $\displaystyle\frac{dx}{(x-1)^2(x^2+1)} \equiv \frac{A}{x-1} + \frac{B}{(x-1)^2} + \frac{Cx+D}{x^2+1}$

$$\equiv \frac{A(x-1)(x^2+1) + B(x^2+1) + (Cx+D)(x-1)^2}{(x-1)^2(x^2+1)}$$

Hence $A(x-1)(x^2+1) + B(x^2+1) + (Cx+D)(x-1)^2 \equiv 1$

When $x=1$; $2B = 1 \Rightarrow B = \dfrac{1}{2}$

Equating the coefficients of x^2 we obtain $-A + B - 2C + D = 0$ $\qquad$ (8.1)

Equating the coefficients of terms independent of x, we obtain $-A + B + D = 1$ $\quad$ (8.2)

Subtracting (8.2) from (8.1) yields

$-2C = -1 \Rightarrow C = \dfrac{1}{2}$

Equating the coefficients of x^3 we obtain:

$A + C = 0 \Rightarrow A = -C = -\dfrac{1}{2}$

Substituting for A and C in (8.2) yields:

$\dfrac{1}{2} + \dfrac{1}{2} + D = 1 \Rightarrow D = 0$

Hence:

$$\int \frac{dx}{(x-1)^2(x^2+1)} = \int \left(-\frac{1}{2(x-1)} + \frac{1}{2(x-1)^2} + \frac{\frac{1}{2}x}{x^2+1} \right) dx$$

$$= \int \left(-\frac{1}{2(x-1)} + \frac{1}{2(x-1)^2} + \frac{x}{2(x^2+1)} \right) dx$$

$$= -\frac{1}{2}\int \frac{dx}{(x-1)} + \frac{1}{2}\int \frac{dx}{(x-1)^2} + \frac{1}{2}\int \frac{x\,dx}{(x^2+1)}$$

$$= -\frac{1}{2}\ln(x-1) - \frac{1}{2(x-1)} + \frac{1}{4}\ln(x^2+1) + C$$

Note: $\dfrac{1}{2}\int \dfrac{x\,dx}{(x^2+1)} = \dfrac{1}{4}\int \dfrac{2x}{x^2+1} = \dfrac{1}{4}\int \dfrac{\frac{d}{dx}(x^2+1)}{x^2+1}\,dx = \dfrac{1}{4}\ln(x^2+1)$

4.9 Find $\displaystyle\int \frac{x}{(x-4)(x-1)}\,dx$

Let $\dfrac{x}{(x-4)(x-1)} \equiv \dfrac{A}{x-4} + \dfrac{B}{x-1}$

$$\equiv \frac{A(x-1)+B(x-4)}{(x-4)(x-1)}$$

Hence $A(x-1) + B(x-4) \equiv x$

When $x = 4;\ \ 3A = 4 \Rightarrow A = \dfrac{4}{3}$

When $x = 1;\ \ -3B = 1 \Rightarrow B = -\dfrac{1}{3}$

$$\int \frac{x}{(x-4)(x-1)}\,dx = \int \left(\frac{4}{3(x-4)} - \frac{1}{3(x-1)} \right) dx$$

$$= \frac{4}{3}\int \frac{dx}{x-4} - \frac{1}{3}\int \frac{dx}{(x-1)}$$

$$= \frac{4}{3}\ln(x-4) - \frac{1}{3}\ln(x-1)$$

4.10 Find $\displaystyle\int \frac{x-1}{(x+2)(x-2)}\,dx$

Let $\dfrac{x-1}{(x+2)(x-2)} \equiv \dfrac{A}{x+2} + \dfrac{B}{x-2}$

Then $x-1 \equiv A(x-2) + B(x+2)$

When $x = 2$; $4B = 1 \Rightarrow B\dfrac{1}{4}$

When $x = -2$; $-4A = -3 \Rightarrow A = \dfrac{3}{4}$

Therefore $\displaystyle\int \dfrac{x-1}{(x+2)(x-2)}\,dx = \int \dfrac{3}{4(x+2)}\,dx + \int \dfrac{1}{4(x-2)}\,dx$

$$= \dfrac{3}{4}\int \dfrac{dx}{x+2} + \dfrac{1}{4}\int \dfrac{dx}{x-2}$$

$$= \dfrac{3}{4}\ln(x+2) + \dfrac{1}{4}\ln(x-2)$$

4.11 Determine $\displaystyle\int \dfrac{5x - x^2}{2x(x+3)(x+4)}\,dx$

Let $\dfrac{5x - x^2}{2x(x+3)(x+4)} \equiv \dfrac{A}{2x} + \dfrac{B}{x+3} + \dfrac{C}{x+4}$

$$\equiv \dfrac{A(x+3)(x+4) + 2Bx(x+4)\,2Cx(x+3)}{2x(x+3)(x+4)}$$

Then $5x - x^2 \equiv A(x+3)(x+4) + 2Bx(x+4) + 2Cx(x+3)$

When $x = -3$; $-6B = -24 \Rightarrow B = 4$

When $x = -4$; $8C = -36 \Rightarrow C = -\dfrac{9}{2}$

Equating the coefficients of x^2 we obtain:

$A + 2B + 2C = -1$

$\therefore A + 8 - 9 = -1 \Rightarrow A = 0$

Hence $\dfrac{5x-x^2}{2x(x+3)(x+4)} \equiv \dfrac{4}{x+3} - \dfrac{9}{2(x+4)}$

Therefore $\displaystyle\int \dfrac{5x-x^2}{2x(x+3)(x+4)}\,dx = 4\int \dfrac{dx}{x+3} - \dfrac{9}{2}\int \dfrac{dx}{x+4}$

$$= 4\ln(x+3) - \dfrac{9}{2}\ln(x+4)$$

4.12 $\displaystyle\int \dfrac{x(x+1)}{(x-1)(x-2)}\,dx$

If we expand the numerator, we can see that the numerator is of equal degree to the numerator:

$$\dfrac{x(x+1)}{(x-1)(x-2)} = \dfrac{x^2+x}{x^2-3x+2}$$

Therefore we have to divide the numerator by the denominator first to ensure the numerator is of lower degree than the denominator:

$$\begin{array}{r} 1 \hspace{2.5cm} \\ x^2-3x+2\overline{)x^2 + x\hspace{1.2cm}} \\ \underline{x^2-3x+2} \\ 4x-2 \end{array}$$

Hence $\dfrac{x(x+1)}{(x-1)(x-2)} \equiv 1 + \dfrac{4x-2}{(x-1)(x-2)}$

Now $\dfrac{4x-2}{(x-1)(x-2)} \equiv \dfrac{A}{x-1} + \dfrac{B}{x-2}$

$$\equiv \dfrac{A(x-2)+B(x-1)}{(x-1)(x-2)}$$

Hence $4x-2 \equiv A(x-2)+B(x-1)$

When $x=1$; $-A=2 \Rightarrow A=-2$

When $x = 2$; $B = 6$

Hence $\dfrac{4x-2}{(x-1)(x-2)} \equiv 1 - \dfrac{2}{x-1} + \dfrac{6}{x-2}$

Hence $\displaystyle\int \dfrac{x(x+1)}{(x-1)(x-2)}\,dx = \int dx - 2\int \dfrac{dx}{x-1} + 6\int \dfrac{dx}{x-2}$

$$= x - 2\ln(x-1) + 6\ln(x-2) + C$$

4.13 Find $\displaystyle\int \dfrac{x^2 - 14x + 1}{(x^2-1)(x+3)}\,dx$

$\dfrac{x^2 - 14x + 1}{(x^2-1)(x-3)} \equiv \dfrac{x^2 - 14x + 1}{(x+1)(x-1)(x^2-1)}$

$\equiv \dfrac{A}{x+1} + \dfrac{B}{x-1} + \dfrac{C}{x-3}$

$\equiv \dfrac{A(x-1)(x-3) + B(x+1)(x-3) + C(x+1)(x-1)}{(x+1)(x-1)(x^2-1)}$

Hence $x^2 - 14x + 1 \equiv A(x-1)(x-3) + B(x+1)(x-3) + C(x+1)(x-1)$

When $x = 1$; $8A = 16 \Rightarrow A = 2$

When $x = 3$; $8C = -32 \Rightarrow C = -4$

Equating the coefficients of x^2 we obtain

$A + B + C = 1$

$\therefore 2 + B - 4 = 1 \Rightarrow B = 3$ Substituting for A and C

Therefore:

$\displaystyle\int \dfrac{x^2 - 14x + 1}{(x^2-1)(x+3)}\,dx \equiv \int \left[\dfrac{2}{x+1} + \dfrac{3}{x-1} - \dfrac{4}{x-3} \right] dx$

$\equiv 2\int \dfrac{dx}{x+1} + 3\int \dfrac{dx}{x-1} - 4\int \dfrac{dx}{x-3}$

$= 2\ln(x+1) + 3\ln(x-1) - 4\ln(x-3) + C$

4.14 Find $\displaystyle\int \frac{2x^2-10x-24}{(2x+5)(x^2-4)}\, dx$

Using rules (ii) and (iv) we could say that

$\displaystyle\frac{2x^2-10x-24}{(2x+5)(x^2-4)} \equiv \frac{A}{2x+5} + \frac{Bx+C}{x^2-4}$. However, x^2-4 will factorise into $(x+2)(x-2)$ which yields an easier solution. So we can say:

$$\frac{2x^2-10x-24}{(2x+5)(x^2-4)} \equiv \frac{2x^2-10x-24}{(2x+5)(x+2)(x-2)}$$

Hence

$$\frac{2x^2-10x-24}{(2x+5)(x^2-4)} \equiv \frac{A}{2x+5} + \frac{B}{x+2} + \frac{C}{x-2}$$

$$\equiv \frac{A(x+2)(x-2)+B(2x+5)(x-2)+C(2x+5)(x+2)}{(2x+5)(x+2)(x-2)}$$

Therefore $2x^2-10x-24 \equiv A(x+2)(x-2)+B(2x+5)(x-2)+C(2x+5)(x+2)$

When $x = 2$; $36C = -36 \Rightarrow C = -1$

When $x = -2$; $-4B = 4 \Rightarrow B = -1$

Equating coefficients of x^2

$A + 2B + 2C = 2$

Hence $A - 2 - 2 = 2 \Rightarrow A = 6$ Substituting for B and C

Therefore $\displaystyle\frac{2x^2-10x-24}{(2x+5)(x^2-4)} \equiv \frac{6}{2x+5} - \frac{1}{x+2} - \frac{1}{x-2}$

Hence $\displaystyle\int \frac{2x^2-10x-24}{(2x+5)(x^2-4)}\, dx \equiv 6\int \frac{dx}{2x+5} - \int \frac{dx}{x+2} - \int \frac{dx}{x-2}$

$$= 3\ln(2x+5) - \ln(x+2) - \ln(x-2) + C$$

Note: $6\int \dfrac{dx}{2x+5} = 3\ln(2x+5)$ since $\dfrac{d}{dx}\left(3\ln(2x+5)\right) = \dfrac{6}{2x+5}$ using the chain rule for differentiation or we can use the method of substitution as follows:

Let $u = 2x+5$

Then $\dfrac{du}{dx} = 2 \Rightarrow dx = \dfrac{du}{2}$

Hence $\quad 6\int \dfrac{dx}{2x+5} = 6\int \dfrac{du}{2u} = 3\int \dfrac{du}{u} = 3\ln u = 3\ln(2x+5)$ (reversing the substitution)

Note: also that we can write $3\ln(2x+5) - \ln(x+2) - \ln(x-2)$ as

$$3\ln(2x+5) - \ln(x+2) - \ln(x-2) = 3\ln(2x+5) - \left[\ln(x+2) + \ln(x-2)\right]$$

$$= \ln(2x+5)^3 - \ln\left[(x+2)(x-2)\right] \qquad \text{Since } n\ln a = \ln(a)^n$$

$$= \ln\left[\dfrac{(2x+5)^3}{(x+2)(x-2)}\right]$$

$$= \ln\left[\dfrac{(2x+5)^3}{x^2 - 4}\right]$$

4.15 Determine $\int \dfrac{x(x+1)}{(x-1)(x-2)(x+3)}\, dx$

$$\dfrac{x(x+1)}{(x-1)(x-2)(x+3)} \equiv \dfrac{x^2+x}{(x-1)(x-2)(x+3)} \equiv \dfrac{A}{x-1} + \dfrac{B}{x-2} + \dfrac{C}{x+3}$$

$$\equiv \dfrac{A(x-2)(x+3) + B(x-1)(x+3) + C(x-1)(x-2)}{(x-1)(x-2)(x+3)}$$

Hence $x^2 + x \equiv A(x-2)(x+3) + B(x-1)(x+3) + C(x-1)(x-2)$

When $x = 1$; $-4A = 2 \Rightarrow A = -\dfrac{1}{2}$

When $x = 2$; $5B = 6 \Rightarrow B = \dfrac{6}{5}$

When $x = -3$; $20C = 6 \Rightarrow C = \dfrac{6}{20} = \dfrac{3}{10}$

Therefore $\dfrac{x(x+1)}{(x-1)(x-2)(x+3)} \equiv -\dfrac{1}{2(x-1)} + \dfrac{6}{5(x-2)} + \dfrac{3}{10(x+3)}$

Hence $\displaystyle\int \dfrac{x(x+1)}{(x-1)(x-2)(x+3)}\,dx = -\dfrac{1}{2}\int\dfrac{dx}{(x-1)} + \dfrac{6}{5}\int\dfrac{dx}{(x-2)} + \dfrac{3}{10}\int\dfrac{dx}{(x+3)}$

$$= -\dfrac{1}{2}\ln(x-1) + \dfrac{6}{5}\ln(x-2) + \dfrac{3}{10}\ln(x+3) + C$$

4.16 Find $\displaystyle\int \dfrac{6}{(x+1)(x-1)(x-4)^2}\,dx$

Let $\dfrac{6}{(x+1)(x-1)(x-4)^2} \equiv \dfrac{A}{x+1} + \dfrac{B}{x-1} + \dfrac{C}{x-4} + \dfrac{D}{(x-4)^2}$

$\equiv \dfrac{A(x-1)(x-4)(x-4)^2 + B(x+1)(x-4)(x-4)^2 + C(x+1)(x-1)(x-4)^2 + D(x+1)(x-1)(x-4)}{(x+1)(x-4)(x-4)^2}$

Hence $6 \equiv A(x-1)(x-4)^2 + B(x+1)(x-4)^2 + C(x+1)(x-1)(x-4) + D(x+1)(x-1)$

When $x = 1$; $B = \dfrac{1}{3}$

When $x = -1$; $A = -\dfrac{3}{25}$

When $x = 4$; $D = \dfrac{2}{5}$

When $x = 0$; $= -\dfrac{16}{75}$

Hence $\dfrac{6}{(x+1)(x-1)(x-4)^2} \equiv -\dfrac{3}{25(x+1)} + \dfrac{1}{3(x-1)} - \dfrac{16}{75(x-4)} + \dfrac{2}{5(x-4)^2}$

Therefore

$\displaystyle\int \dfrac{6}{(x+1)(x-1)(x-4)^2}\,dx = -\dfrac{3}{25}\int\dfrac{dx}{(x+1)} + \dfrac{1}{3}\int\dfrac{dx}{(x-1)} - \dfrac{16}{25}\int\dfrac{dx}{(x-4)} + \dfrac{2}{5}\int\dfrac{dx}{(x-4)^2}$

$$= -\dfrac{3}{25}\ln(x+1) + \dfrac{1}{3}\ln(x-1) - \dfrac{16}{75}\ln(x-4) + \dfrac{2}{5}(x-4) + C$$

Note 1: $\dfrac{2}{5}\displaystyle\int \dfrac{dx}{(x-4)^2}$ is obtained by using the method of substitution as follows:

Let $u = x - 4$

Then $\dfrac{du}{dx} = 1 .... dx = du$

Hence $\dfrac{2}{5}\displaystyle\int \dfrac{dx}{(x-4)^2} = \dfrac{2}{5}\displaystyle\int \dfrac{du}{u^2} = \dfrac{2}{5}\displaystyle\int u^{-2}\,du = -\dfrac{2}{5}u^{-1} = -\dfrac{2}{5u} = -\dfrac{2}{5(x-4)}$ Reversing the

substitution

Note 2: We could have written $\dfrac{6}{(x+1)(x-1)(x-4)^2}$ as $\dfrac{6}{(x+1)(x-1)(x+2)(x-2)}$ in the

same way as Q 14.

4.17 Find $\displaystyle\int \dfrac{3x-2}{(x^2+9)(x^2-1)}$

We could split the integrand into partial fractions as it stands but we can see that we can re-write the denominator as follows:

$$\dfrac{3x-2}{(x^2-9)(x^2-1)} \equiv \dfrac{3x-2}{(x+3)(x-3)(x+1)(x-1)}$$

$$\equiv \dfrac{A}{x+3} + \dfrac{B}{x-3} + \dfrac{C}{x+1} + \dfrac{D}{x-1}$$

$$\equiv \dfrac{A(x-3)(x+1)(x-1) + B(x+3)(x+1)(x-1) + C(x+3)(x-3)(x-1) + D(x+3)(x-3)(x+1)}{(x+3)(x-3)(x+1)(x-1)}$$

When $x = 1$; $-16D = 1 \Rightarrow D = -\dfrac{1}{16}$

When $x = -1$; $16C = -5 \Rightarrow C = -\dfrac{5}{16}$

When $x = 3$; $48B = 7 \Rightarrow B = \dfrac{7}{48}$

When $x = 0$; $3A - 3B + 9C - 9D = -2$

Therefore $3A - \dfrac{21}{48} - \dfrac{45}{16} + \dfrac{9}{16} = -2 \Rightarrow 3A = \dfrac{129}{48} - 2 \Rightarrow A = \dfrac{33}{144} = \dfrac{11}{48}$ Substituting for B, C & D

Hence $\dfrac{3x-2}{\left(x^2-9\right)\left(x^2-1\right)} \equiv \dfrac{11}{48(x+3)} + \dfrac{7}{48(x-3)} - \dfrac{5}{16(x+1)} - \dfrac{1}{16(x-1)}$

Therefore:

$$\int \dfrac{3x-2}{\left(x^2+9\right)\left(x^2-1\right)} = \dfrac{11}{48}\int \dfrac{dx}{x+3} + \dfrac{7}{48}\int \dfrac{dx}{x-3} - \dfrac{5}{16}\int \dfrac{dx}{x+1} - \dfrac{1}{16}\int \dfrac{dx}{x-1}$$

$$= \dfrac{11}{48}\ln\left(x+3\right) + \dfrac{7}{48}\ln\left(x-3\right) - \dfrac{5}{16}\ln\left(x+1\right) - \dfrac{1}{16}\ln\left(x-1\right) + C$$

4.18 Determine $\displaystyle\int \dfrac{x^2+5}{x\left(x^2+2\right)}\,dx$

Let $\dfrac{x^2+5}{x\left(x^2+2\right)} \equiv \dfrac{A}{x} + \dfrac{Bx+C}{\left(x^2+2\right)}$

$$\equiv \dfrac{A\left(x^2+2\right) + x\left(Bx+C\right)}{x\left(x^2+2\right)}$$

Therefore $x^2+5 \equiv A\left(x^2+2\right) + x\left(Bx+C\right)$

When $x=0$; $2A=5 \Rightarrow A=\dfrac{5}{2}$

When $x=1$; $3A+B+C=6$ (1)

When $x=-1$; $3A+B-C=6$ (2)

Subtracting (2) from (1) we obtain:

$2C=0$, Giving $C=0$

Substituting for A and C in (1) yields:

$\dfrac{15}{2} + B = 6 \Rightarrow B = -\dfrac{3}{2}$

Hence $\dfrac{x^2+5}{x(x^2+2)} \equiv \dfrac{5}{2x} - \dfrac{\tfrac{3}{2}x}{(x^2+2)}$

$$\equiv \dfrac{5}{2x} + \dfrac{3x}{2(x^2+2)}$$

Therefore $\displaystyle\int \dfrac{x^2+5}{x(x^2+2)}\,dx = \dfrac{5}{2}\int \dfrac{dx}{x} - \dfrac{3}{2}\int \dfrac{x}{(x^2+2)}\,dx$

$$= \dfrac{5}{2}\ln x - \dfrac{3}{4}\ln(x^2+2) + C$$

The last integral was obtained by recognising that the numerator is $\dfrac{1}{2}$ the differential of the

denominator $= \dfrac{1}{2}\ln(\text{denominator})$

4.19 Find $\displaystyle\int \dfrac{x^3-2}{(x+2)(2x+1)(x^2+1)}\,dx$

Let $\dfrac{x^3-2}{(x+2)(2x+1)(x^2+1)} \equiv \dfrac{A}{x+2} + \dfrac{B}{2x+1} + \dfrac{Cx+D}{x^2+1}$

$$\equiv \dfrac{A(2x+1)(x^2+1) + B(x+2)(x^2+1) + (Cx+D)(x+2)(2x+1)}{(x+2)(2x+1)(x^2+1)}$$

When $x = -\dfrac{1}{2}$; $\dfrac{15}{8}B = -\dfrac{17}{8} \Rightarrow B = -\dfrac{17}{15}$

Putting $x = -2$; $-15A = -10 \Rightarrow A = \dfrac{2}{3}$

When $x = 0$; $A + 2B + 2D = -2$ (1)

Substituting for A and B in (1); we get:

$\dfrac{2}{3} - \dfrac{34}{15} + 2D = -2 \Rightarrow D = \dfrac{24}{30} - 2 = -\dfrac{1}{5}$

Equating coefficients of x^3 we obtain

$2A + B + 2C = 1$ (2)

Substituting for A and B in (2) we get:

$$\frac{4}{3} - \frac{17}{15} + 2C = 1 \Rightarrow C = \frac{2}{5}$$

Therefore

$$\frac{x^3 - 2}{(x+2)(2x+1)(x^2+1)} \equiv \frac{2}{3(x+2)} - \frac{17}{15(2x+1)} + \frac{\frac{2}{5}x - \frac{1}{5}}{(x^2+1)}$$

$$\equiv \frac{2}{3(x+2)} - \frac{17}{15(2x+1)} + \frac{2x-1}{5(x^2+1)}$$

Hence:

$$\int \frac{x^3 - 2}{(x+2)(2x+1)(x^2+1)} dx = \frac{2}{3}\int \frac{dx}{(x+2)} - \frac{17}{15}\int \frac{dx}{(2x+1)} + \frac{1}{5}\int \frac{2x-1}{(x^2+1)} dx$$

$$= \frac{2}{3}\ln(x+2) - \frac{17}{30}\ln(2x+1) + \frac{1}{5}\left(\ln(x^2+1) - \tan^{-1}(x)\right) + C$$

$$\frac{1}{30}\left[20\ln(x+2) - 17\ln(2x+1) + 6\ln(x^2+1)\right] - \frac{1}{5}\tan^{-1}(x) + C$$

$$= \frac{1}{30}\ln\left[\frac{(x+2)^{20}(x^2+1)^6}{(2x+1)^{17}}\right] - \frac{1}{5}\tan^{-1}(x) + C$$

Note: $\frac{1}{5}\int \frac{2x-1}{(x^2+1)} dx$ has been split into two fractions which can be seen easily by inspection:

$$\frac{1}{5}\int \frac{2x-1}{(x^2+1)} dx = \frac{1}{5}\left[\int \frac{2x}{x^2+1} dx - \int \frac{dx}{x^2+1}\right]$$

$$= \frac{1}{5}\ln(x^2+1) - \frac{1}{5}\tan^{-1}(x) \qquad \text{These are both standard results}$$

4.20 Find $\int \frac{3x^3 - 2x^2 + 5x - 1}{(x-2)^5} dx$

Let $\frac{3x^3 - 2x^2 + 5x - 1}{(x-2)^5} \equiv \frac{A}{(x-2)} + \frac{B}{(x-2)^2} + \frac{C}{(x-2)^3} + \frac{D}{(x-2)^4} + \frac{E}{(x-2)^5}$

$$\equiv \frac{A(x-2)^4 + B(x-2)^3 + C(x-2)^2 + D(x-2) + E}{(x-2)^5}$$

Hence $3x^3 - 2x^2 + 5x - 1 \equiv A(x-2)^4 + B(x-2)^3 + C(x-2)^2 + D(x-2) + E$

When $x = 2$, $E = 24 - 8 + 10 - 1 = 25$

Equating the coefficients of x^4 we obtain

$A = 0$

Equating the coefficients of x^3 we obtain

$8A + B = 3 \Rightarrow B = 3$

Equating the coefficients of x^2 we obtain

$20A - 6B + C = -2 \Rightarrow -18 + C = -2 \Rightarrow C = 16$

Equating the coefficients independent of x, we obtain

$16A - 8B + 4C - 2D + E = -1$
$\Rightarrow 0 - 24 + 64 - 2D + 25 = -1$
$\Rightarrow 2D = 66$
$\Rightarrow D = 33$

Hence $\dfrac{3x^3 - 2x^2 + 5x - 1}{(x-2)^5} = \dfrac{3}{(x-2)^2} + \dfrac{16}{(x-2)^3} + \dfrac{33}{(x-2)^4} + \dfrac{25}{(x-2)^5}$

Therefore $\displaystyle\int \frac{3x^3 - 2x^2 + 5x - 1}{(x-2)^5}\,dx = 3\int \frac{dx}{(x-2)^2} + 16\int \frac{dx}{(x-2)^3} + 33\int \frac{dx}{(x-2)^4} + 25\int \frac{dx}{(x-2)^5}$

Let $u = x - 2$

Then $\dfrac{du}{dx} = 1 .... dx = du$

Hence $3\displaystyle\int \frac{dx}{(x-2)^2} = 3\int \frac{du}{u^2} = 3\int u^{-2}\,du = -3u^{-1} = -\frac{3}{x-2}$

$16\displaystyle\int \frac{dx}{(x-2)^3} = 16\int \frac{du}{u^3} = 16\int u^{-3}\,du = -\frac{16}{2}u^{-2} = -\frac{8}{(x-2)^2}$

$$33\int \frac{dx}{(x-2)^4} = 33\int \frac{du}{u^4} = 33\int u^{-4}\,du = -\frac{33}{3}u^{-3} = -\frac{11}{(x-2)^3}$$

$$25\int \frac{dx}{(x-2)^5} = 25\int \frac{du}{u^5} = 25\int u^{-5}\,du = -\frac{25}{4}u^{-4} = -\frac{25}{4(x-2)^4}$$

$$\int \frac{3x^3 - 2x^2 + 5x - 1}{(x-2)^5}\,dx = -\frac{3}{x-2} - \frac{8}{(x-2)^2} - \frac{11}{(x-2)^3} - \frac{25}{4(x-2)^4} + C$$

$$= \frac{-12(x-2)^3 - 32(x-2)^2 - 44(x-2) - 25}{4(x-2)^4} + C$$

$$= \frac{-12x^3 + 40x^2 - 60x + 31}{4(x-2)^4} + C \qquad \text{Simplifying}$$

MODULE IN5

INTEGRATION OF PRODUCTS OF SINES AND COSINES

WORKED SOLUTIONS

5.1 Find $\int \cos 3x \cos 2x \, dx$

$$\int \cos 3x \cos 2x \, dx = \frac{1}{2} \int (\cos x + \cos 5x) \, dx$$

Using $\cos A \cos B = \frac{1}{2}(\cos(A-B) + \cos(A+B))$

$$= \frac{1}{2}\left(\sin x + \frac{1}{5}\sin x\right) + C$$

$$= \frac{1}{2}\sin x + \frac{1}{10}\sin x + C$$

5.2 Find $\int \sin^2 x \, dx$

$$\int \sin^2 x \, dx = \frac{1}{2}\int (1 - \cos 2x) \, dx$$

$$= \frac{1}{2}\left(x - \frac{1}{2}\sin 2x\right) + C \qquad \text{Using } \cos 2x = 1 - 2\sin^2 x$$

$$= \frac{1}{2}x - \frac{1}{4}\sin 2x + C$$

5.3 Find $\int \cos^2 x \, dx$

$$\int \cos^2 x \, dx = \frac{1}{2}\int (1 + \cos 2x) \, dx$$

$$= \frac{1}{2}\left(x + \frac{1}{2}\sin 2x\right) + C \qquad \text{Using } \cos 2x = 1 + 2\cos^2 x$$

$$= \frac{1}{2}x + \frac{1}{4}\sin 2x + C$$

5.4 Evaluate $\int \sin^2 2x \, dx$

$$\int \sin^2 2x \, dx = \frac{1}{2}\int (1 - \cos 4x)$$

$$= \frac{1}{2}\left(x - \frac{1}{4}\sin 4x\right) + C \qquad \text{Using } \cos 2x = 1 - 2\sin^2 x$$

$$= \frac{1}{2}x - \frac{1}{8}\sin 4x + C$$

5.5 Evaluate $\int_0^{\frac{\pi}{4}} \sin 2x \cos x \, dx$

$$\int_0^{\frac{\pi}{4}} \sin 2x \cos x \, dx = \frac{1}{2}\int_0^{\frac{\pi}{4}} (\sin x + \sin 3x) \, dx$$

$$= \frac{1}{2}\left[-\cos x - \frac{1}{3}\cos 3x \right]_0^{\frac{\pi}{4}}$$

$$= \frac{1}{2}\left[-\frac{\sqrt{2}}{2} - \frac{1}{3}\frac{\sqrt{2}}{2} \right] - \frac{1}{2}\left[-1 - \frac{1}{3} \right]$$

$$= \frac{1}{2}\left[-\frac{\sqrt{2}}{3} \right] - \frac{1}{2}\left[-\frac{4}{3} \right]$$

$$= -= \frac{\sqrt{2}}{6} + \frac{2}{3}$$

$$= \frac{-\sqrt{2}+4}{6}$$

$$= \frac{4-\sqrt{2}}{6}$$

5.6 Find $\int_0^{\frac{\pi}{2}} 3\sin^2 x \, dx$

$$\int_0^{\frac{\pi}{2}} 3\sin^2 x \, dx = 3\int_0^{\frac{\pi}{2}} \frac{1}{2}(1-\cos 2x) \, dx \qquad\qquad \text{Using } \cos 2x = 1 - 2\sin^2 x$$

$$= \frac{3}{2}\int_0^{\frac{\pi}{2}} (1-\cos 2x) \, dx$$

$$= \frac{3}{2}\left[x - \frac{1}{2}\sin 2x \right]_0^{\frac{\pi}{2}}$$

$$= \frac{3}{2}\left[\frac{\pi}{2} - 0 \right] = \frac{3\pi}{4}$$

5.7 Evaluate $\int_0^{\frac{\pi}{2}} 2\sin^2 x \, dx$

From Q 2 $\int \sin^2 x \, dx = \frac{1}{2}x - \frac{1}{4}\sin 2x + C$

Hence $\int 2\sin^2 x \, dx = 2\int \sin^2 x \, dx = x - \frac{1}{2}\sin 2x + C$

Therefore $\int_0^{\frac{\pi}{2}} 2\sin^2 x \, dx = \left[x - \frac{1}{2}\sin 2x \right]_0^{\frac{\pi}{2}} = \frac{\pi}{2}$

5.8 Find $\int_0^{\frac{\pi}{2}} \sin x \cos x \, dx$

$$\int_0^{\frac{\pi}{2}} \sin x \cos x \, dx = \frac{1}{2}\int_0^{\frac{\pi}{2}} \sin 2x \, dx \qquad\qquad \text{Since } \sin 2x = 2\sin x \cos x$$

$$= \left[-\frac{1}{2} \times \frac{1}{2} \cos 2x \right]_0^{\frac{\pi}{2}} = \left[-\frac{1}{4} \cos 2x \right]_0^{\frac{\pi}{2}} = \frac{1}{4} + \frac{1}{4} = \frac{1}{2}$$

5.9. Determine $\int \cos^2 (x - \theta) dx$.

Let $u = 2x + 2\theta$

Then $\dfrac{du}{dx} = 2 .... dx = \dfrac{du}{2}$

Hence $\int \cos^2 (2x - 2\theta) = \dfrac{1}{2} \int \cos^2 u \, du$

$$= \frac{1}{2} \int \left(\frac{1}{2} (1 + \cos 2u) \right) du$$

$$= \frac{1}{4} \left(u + \frac{1}{2} \sin 2u \right)$$

$$= \frac{1}{4} \left[2x - 2\theta + \frac{1}{2} \sin (4x - 4\theta) \right] \quad \text{Reversing substitution}$$

$$\therefore \int_0^{\frac{\pi}{2}} \cos^2 (x - \theta) \, dx = \frac{1}{4} \left[2x - 2\theta + \frac{\sin (4x - 4\theta)}{2} \right]_0^{\frac{\pi}{2}}$$

$$= \frac{1}{4} \left(\pi - 2\theta + \frac{\sin (2\pi - 4\theta)}{2} \right) - \frac{1}{4} \left(-2\theta + \frac{\sin (-4\theta)}{2} \right)$$

$$= \frac{1}{4} \left(\pi - 2\theta - \frac{\sin 4\theta}{2} \right) - \frac{1}{4} \left(-2\theta - \frac{\sin (4\theta)}{2} \right)$$

$$= \frac{\pi}{4}$$

5.10. Find $\int_0^{\frac{\pi}{2}} (2 \cos^2 x + 4 \sin 3x) dx$.

$$\int (2 \cos^2 x + 4 \sin 3x) dx = \int 2 \left(\frac{1}{2} (1 + \cos 2x) \right) dx + \int 4 \sin 3x \, dx$$

Using $\cos 2x = 1 + 2 \cos^2 x$

$$= x + \frac{\sin 2x}{2} - \frac{4 \cos 3x}{3}$$

Hence $\int_0^{\frac{\pi}{2}} (2 \cos^2 x + 4 \sin 3x) dx = \left[x + \frac{\sin 2x}{2} - \frac{4 \cos 3x}{3} \right]_0^{\frac{\pi}{2}}$

$$= \frac{\pi}{2} + \frac{4}{3} \, 10.$$

5.11 Find $\int_0^{\frac{\pi}{2}} 2 \sin 3\theta \cos 2\theta \, d\theta$.

We cannot integrate this function as it stands but we can use the addition formulae to change $\sin 3\theta \cos 2\theta$ into a form that we can integrate as follows:

$2 \sin A \cos B = \sin(A - B) + \sin(A + B)$

Hence $2\sin 3\theta \cos 2\theta = \sin(3\theta - 2\theta) + \sin(3\theta + 2\theta)$

$$= \sin\theta + \sin 5\theta$$

Therefore $\displaystyle\int_0^{\frac{\pi}{4}} 2\sin 3\theta \cos 2\theta\, d\theta = \int_0^{\frac{\pi}{4}} (\sin\theta + \sin 5\theta)\, d\theta$

$$= \left[-\cos\theta - \frac{\cos\theta}{5} \right]_0^{\frac{\pi}{4}}$$

$$= \left(-\frac{\sqrt{2}}{2} - \frac{\sqrt{2}}{10} \right) - \left(-1 - \frac{1}{5} \right)$$

$$= \left(-\frac{2\sqrt{2}}{5} \right) - \left(-\frac{6}{5} \right)$$

$$= \frac{2}{5}\left(3 - \sqrt{2} \right)$$

5.12 Evaluate $\displaystyle\int_0^{\frac{\pi}{2}} 3\cos^2\theta + 2\sin^2\theta\, d\theta$

$3\cos^2\theta + 2\sin^2\theta = 3\cos^2\theta + 2(1 - \cos^2\theta)$

$$= \cos^2\theta + 2$$

$$= \frac{1}{2}(1 + \cos 2\theta) + 2$$

$$= \frac{1}{2} + \frac{1}{2}\cos 2\theta + 2$$

$$= \frac{1}{2}\cos 2\theta + \frac{5}{2}$$

$$= \frac{1}{2}(\cos 2\theta + 5)$$

Hence $\displaystyle\int_0^{\frac{\pi}{2}} 3\cos^2\theta + 2\sin^2\theta\, d\theta = \frac{1}{2}\int_0^{\frac{\pi}{2}} (\cos 2\theta + 5)\, d\theta$

$$= \frac{1}{2}\left[\frac{\sin 2\theta}{2} + 5\theta \right]_0^{\frac{\pi}{2}}$$

$$= \frac{1}{2}\left(\frac{5\pi}{2} \right) = \frac{5\pi}{4}$$

5.13 Find $\displaystyle\int_0^{\frac{3\pi}{2}} \sin 3x \cos x\, dx$

$2\sin 3x \cos x\, dx = \sin(3x - x) + \sin(3x + x)$

$$= \sin 2x + \sin 4x$$

$$\therefore \sin 3x \cos x = \frac{1}{2}(\sin 2x + \sin 4x)$$

$$\therefore \int_0^{\frac{3\pi}{2}} \sin 3x \cos x \, dx = \frac{1}{2} \int_0^{\frac{3\pi}{2}} \left(\sin 2x + \sin 4x \right) dx$$

$$= \frac{1}{2} \left[-\frac{\cos 2x}{2} - \frac{\cos 4x}{4} \right]_0^{\frac{3\pi}{2}} = \frac{1}{2} \left[\left(\frac{1}{2} - \frac{1}{4} \right) - \left(-\frac{1}{2} - \frac{1}{4} \right) \right] = \frac{1}{2} \left(\frac{1}{4} + \frac{3}{4} \right) = \frac{1}{2}$$

5.14 Evaluate $\int_0^{\frac{\pi}{2}} \left(\sin 3x + \sin x \cos x \right) dx$

$$\int_0^{\frac{\pi}{2}} \left(\sin 3x + \sin x \cos x \right) dx = \int_0^{\frac{\pi}{2}} \left(\sin 3x + \frac{1}{2} \sin 2x \right) dx$$

$$= \left[-\frac{1}{3} \cos 3x - \frac{1}{4} \cos 2x \right]_0^{\frac{\pi}{2}}$$

$$= \left(0 + \frac{1}{4} \right) - \left(-\frac{1}{3} - \frac{1}{4} \right) = +\frac{1}{4} + \frac{7}{12} = \frac{5}{6}$$

5.15 Evaluate $\int \sin m\theta \sin n\theta \, d\theta$

$$\int \sin m\theta \sin n\theta \, d\theta = \frac{1}{2} \int \left(\cos(m - n) - \cos(m + n) \right) d\theta$$

Note. This is a standard result obtained from the relationship

$$2 \sin A \cos B = \cos(A - B) - \cos(A + B)$$

$$\therefore \sin A \cos B = \frac{1}{2} \left(\cos(A - B) - \cos(A + B) \right)$$

$$\int \sin m\theta \sin n\theta \, d\theta = \frac{1}{2} \int \left(\cos(m\theta - n\theta) - \cos(m\theta + n\theta) \right)$$

Hence

$$= \frac{1}{2} \left(\frac{\sin(m - n)\theta}{(m - n)\theta} - \frac{\sin(m + n)\theta}{(m + n)\theta} \right) + C$$

15. Find $\int \cos^2 x \sin 2x \, dx$

$\sin 2x = 2 \sin x \cos x$

Hence $\int \cos^2 x \sin 2x = \int \cos^2 x (2 \sin x \cos x)$

$$= \int 2 \sin x \cos^3 x \, dx$$

Now let $t = \cos x$

Then $\dfrac{dt}{dx} = -\sin x \, ... \, dx = -\dfrac{dt}{\sin x}$

Hence $\int \cos^2 x \sin 2x \, dx = -\int t^3 \, dt = \dfrac{t^4}{4}$

$$= -\frac{1}{4} \cos^4 x + C \qquad \text{Reversing substitution}$$

5.16 Find $\int \cos^2 x \cos 2x \, dx$

$$\cos^2 x \cos 2x = \left(1 - \sin^2 x\right)\left(1 - 2\sin^2 x\right)$$
$$= 1 - 2\sin^2 x - \sin^2 x + 2\sin^4 x$$
$$= 1 - 3\sin^2 x + 2\sin^4 x$$

5.17 Find $\int \sin^2 x \cos^2 x \, dx$

Now $\sin^2 x \cos^2 x = \left(\sin x \cos x\right)^2$

And $\sin 2x = 2\sin x \cos x \Rightarrow \sin x \cos x = \dfrac{1}{2}\sin 2x$

Hence $\sin^2 x \cos^2 x = \left(\sin x \cos x\right)^2 = \left(\dfrac{1}{2}\sin 2x\right)^2 = \dfrac{1}{4}\sin^2 2x$

But $\sin^2 2x = \dfrac{1}{2}\left(1 - \cos 4x\right)$

Hence $\dfrac{1}{4}\sin^2 2x = \dfrac{1}{8}\left(1 - \cos 4x\right)$

$$\int \sin^2 x \cos^2 x \, dx = \dfrac{1}{8}\int \left(1 - \cos 4x\right)$$

Therefore
$$= \dfrac{1}{8}\left(x - \dfrac{\sin 4x}{4}\right) + C = \dfrac{x}{8} - \dfrac{\sin 4x}{32} + C$$

5.18 Find $\int \sin^4 x \cos^2 x \, dx$

$\sin 2x = 2\sin x \cos x$

And $2\sin^2 x = 1 - \cos 2x$

$$\sin^2 x \left(\dfrac{1}{4}\right)\left(\sin 2x\right)^2 dx = \dfrac{1}{8}\left(1 - \cos 2x\right)\left(\sin 2x\right)^2 \quad \text{Since } \sin^2 x = \dfrac{1}{2}\left(1 - \cos 2x\right)$$

$$= \dfrac{1}{8}\left(\sin 2x\right)^2 - \dfrac{1}{8}\left(\sin 2x\right)^2 \cos 2x \, dx \qquad \text{Rearranging}$$

$$= \dfrac{1}{16}\left(1 - \cos 4x\right) - \dfrac{1}{16}\left(\sin 2x\right)^2 d\left(\sin 2x\right) \qquad \text{Rearranging}$$

$$= \dfrac{1}{16}x - \dfrac{1}{64}\sin 4x - \dfrac{1}{48}\sin^3 x$$

5.19 Find $\int \sin^3 x \, dx$

$$\int \sin^3 x \, dx = \int \sin^2 x \sin x = \int \left(1 - \cos^2 x\right)\sin x \, dx \qquad \text{Since } \sin^2 x = 1 - \cos^2 x$$

Now let $z = \cos x$

Then $-\sin\dfrac{dx}{dz} = -\sin x$

And $\Rightarrow \dfrac{dx}{dz} = -\dfrac{1}{\sin x} = -\cos ecx \dots dz = -\cos ecx \, dx$

Hence

$$\int \sin^3 x = \int \sin x (1 - \cos^2 x)(-\cos ecx) dz$$

$$= -\int \sin x (1 - z^2)(-\cos ecx) d$$

$$= -\int \frac{\sin x}{\cos ecx} (1 - z^2) dz$$

$$= -\int (1 - z^2) dz$$

$$= -\left(z - \frac{z^3}{3} \right)$$

$$= -\left(\cos x - \frac{\cos^3}{3} \right) + C$$

$$= -\cos x + \frac{\cos^3}{3} + C$$

$$= -\cos - \int \cos^2 x \, d(\cos x)$$

$$= -\cos x + \frac{\cos^3 x}{3} + C \qquad \text{Reversing substitution}$$

5 20 Find $\int \cos^3 x \, dx$

$$\int \cos^3 x \, dx = \int \cos^2 x \cos x \, dx$$

$$= \int (1 - \sin^2 x) \cos x \, dx$$

$$= \int \cos x \, dx - \int \sin^2 x \cos x \, dx$$

$$= \sin x - \frac{\sin^3 x}{3} + C$$

5 21 Find $\int \cos^4 x \, dx$

$$\int \cos^4 x \, dx = \int (\sin^2 x)^2 \, dx$$

$$= \int \frac{(1 - \cos 2x)^2}{4} \, dx$$

$$= \frac{\int 1 - \cos 2x + \cos^2 2x}{4} \, dx \qquad \text{Since } \cos^2 x = \frac{1}{2}(1 + \cos 2x)$$

$$= \frac{1}{4} \int \left(1 - 2\cos 2x + \frac{1}{2} + \frac{1}{2}\cos 4x \right) dx \qquad \text{Since } \cos^2 2x = \frac{1}{2}(1 + \cos 4x)$$

$$= \frac{1}{4} \int \left(\frac{3}{2} - 2\cos 2x + \frac{1}{2}\cos 4x \right) dx \qquad \text{Rearranging}$$

$$= \frac{1}{4} \left(\frac{3x}{2} - \frac{2\sin 2x}{2} + \frac{1}{2}\frac{\sin 4x}{4} \right) + C$$

$$= \frac{3x}{8} - \frac{\sin 2x}{4} + \frac{\sin 4x}{32} + C \qquad \text{Rearranging}$$

5.22 Find $\int \sin^5 x\ dx$

$\int \sin^5 x\ dx$ and $\int \cos^5 x\ dx$ can be determined in a similar manner to that used above to solve $\int \sin^3 x\ dx$ and $\int \cos^5 x\ dx$.

$$\int \sin^5 x\ dx = \int \sin^4 x \sin x\ dx$$

$$= \int \left(1 - \cos^2 x\right)^2 \sin x\ dx$$

$$= \int \left(1 - 2\cos^2 x + \cos^4 x\right) \sin x\ dx$$

$$= \int \sin x\ dx - 2\int \cos^2 x \sin x\ dx + \int \cos^4 x \sin x\ dx$$

$$= -\cos x + \frac{2\cos^3 x}{3} + \frac{\cos^5 x}{5} + C$$

20. Find $\int \cos^5 x\ dx$

$$\int \cos^5 x\ dx = \int \cos^4 x \cos x\ dx$$

$$= \int \left(1 - \sin x\right)^2 \cos x\ dx$$

$$= \int \left(1 - 2\sin^2 x + \sin^4 x\right) \cos x\ dx$$

$$= \int \cos x\ dx - 2\int \sin^2 x \cos x\ dx + \int \sin^4 x \cos x\ dx$$

$$= \sin x - \frac{2\sin^3}{3} + \frac{\sin^5 x}{5} + C$$

MODULE IN6

INTEGRATION USING REDUCTION FORMULAE

INTRODUCTION

We have already dealt with integration by parts in Section Module IN3 where we used the result:

$$\int u \, dv = uv - \int v \, du$$

To refresh our memory, we will use this technique to determine $\int x^2 e^x \, dx$. In this case we will

let: $u = x^2 \dots du = 2x$

and $dv = e^x \Rightarrow v = e^x$ Since $\int e^x \, dx = e^x$

Hence $\int x^2 e^x \, dx = x^2 e^x - \int 2x e^x \, dx$

Since we cannot determine the last integral we have to carry out integration by parts again:
Let $u = 2x \Rightarrow du = 2$

And $dv = e^x \Rightarrow v = e^x$ Since $\int e^x \, dx = e^x$

Hence $\int x^2 e^x \, dx = x^2 e^x - \left[2x e^x - \int 2 e^x \right] dx$ Repeating integration by parts

$$= x^2 e^x - \left[2x e^x - 2 \int e^x \right]$$

$$= x^2 e^x - 2x e^x + 2 e^x + C$$

$$= e^x \left(x^2 - 2x + 2 \right) + C$$

If we now replace the x^2 term in $\int x^2 e^x \, dx$ with x^n we would then have $\int x^n e^x \, dx$ which we will evaluate in the same way as the example above.

In this case we will let:
$u = x^n \Rightarrow du = n x^{n-1}$
$dv = e^x \Rightarrow v = e^x$ Since $\int e^x \, dx = e^x$

Hence $\int x^n e^x \, dx = x^n e^x - \int e^x n x^{n-1} \, dx$

$$= x^n e^x - n \int e^x x^{n-1} \, dx$$

Notice that $\int e^x x^{n-1} \, dx$ is of exactly the same form as the integral we started with - $\int x^n e^x \, dx$ - except that x^n has been replaced by x^{n-1}

If we now denote $\int x^n e^x \, dx$ as I_n we can denote $\int e^x x^{n-1} \, dx$ as I_{n-1}

Hence $\int x^n e^x \, dx = x^n e^x - n \int e^x x^{n-1} \, dx$ can now be written as:

$$I_n = x^n e^x - n I_{n-1} \tag{1}$$

This result allows us to write down the answer to any integral of the form $\int x^n e^x \, dx$ relatively

easily. So that our original integral $\int x^2 e^x \, dx$ we can now put $n = 2$ into the above result (2) to obtain:

$$I_2 = x^2 e^x - 2 I_1$$

Also if we put $n = 1$ we get:

$$I_1 = xe^x - 2I_0$$

Where $I_0 = \int x^0 e^x\, dx = \int 1 \times x\, dx = \int e^x\, dx = e^x + C$ (2)

Hence $I_2 = x^2 e^x - 2I_1$

And $I_2 = x^2 e^x - 2xe^x + 2e^x + C$

$$= e^x\left(x^2 - 2x + 2\right) + C$$

The use of this technique is a powerful method of solving integrals of the form $\int x^n e^x\, dx$ as we can simply substitute different values for n to evaluate all integrals of this form.

Note that we start with n equal to the power of x in the original expression and then progressively reduce the value of n until we reach $n = 1$ where the x term will disappear altogether.

The idea being that we progressively reduce the power of x hence the title Reduction Formula for this technique.

For example we can now evaluate $\int x^3 e^x\, dx$ using the result obtained from (1) above:

$$I_n = x^n e^x - nI_{n-1}.$$ (3)

In this case we start by putting $n = 3$ (since the x term in $\int x^3 e^x\, dx$ is x^3)

Hence we have $I_3 = x^3 e^x - 3I_2$ (4)

Putting $n = 2$ in (3) we have

$$I_2 = x^2 e^x - 2I_1$$ (5)

Putting $n = 1$ in (3) we have

$$I_1 = xe^x - I_0$$ (6)

Hence:

$$I_3 = x^3 e^x - 3\left(x^2 e^x - 2I_1\right)$$ Substituting for I_2 from (5) in (4)

$$= x^3 e^x - 3\left(x^2 e^x - 2\left(xe^x - I_0\right)\right)$$ Substituting for I_1 from (6) in (5)

$$= x^3 e^x - 3\left(x^2 e^x - 2xe^x + 2I_0\right)$$ Expanding

$$= x^3 e^x - 3x^2 e^x + 6xe^x - 6e^x + C$$ Substituting for $I_0 = e^x$ from (2)

$$= e^x\left(x^3 - 3x^2 + 6x - 6\right) + C$$

Note that we can use Reduction Formulae to solve a wide variety of other integrals as shown in the following examples. This technique is not restricted to the examples discussed above.

WORKED SOLUTIONS

Let $I_n = \int x^n \sin x\, dx$ (6.1)

Let $u = x^n \Rightarrow du = nx^{n-1}$

Let $dv = \sin x \Rightarrow v = -\cos x$

Then $I_n = -x^n \cos x - \int (-\cos x)\, nx^{n-1}\, dx$

$\qquad = -x^n \cos x + n\int (\cos x) x^{n-1}\, dx$

Since we cannot integrate the last integral directly, we must use integration by parts again.

Let $u = x^{n-1} \Rightarrow du = (n-1)\left(x^{n-2}\right)$

Let $dv = \cos x \Rightarrow v = \sin x$

Then $I_n = -x^n \cos x + n\left[x^{n-1} \sin x - (n-1)\int (\sin x)\left(x^{n-2}\right) dx \right]$

$\qquad = -x^n \cos x + nx^{n-1} \sin x - n(n-1)I_{n-2}$ (6.2)

To find $\int x^2 \sin x\, dx$ put $n = 2$ in (6.2)

$I_2 = -x^2 \cos x + 2x \sin x - 2I_0$ (6.3)

Now $2I_0 = 2\int \sin x\, dx = -2\cos x + C$ Putting $n = 0$ in (6.1)

Hence $I_2 = I_2 = -x^2 \cos x + 2x \sin x - (-2\cos x) + C$ Substituting for $2I_0$ in (6.3)

$-x^2 \cos x + 2x \sin x + 2\cos x + C$

$= \left(2 - x^2\right)\cos x + 2x \sin x + C$

This is solved in a similar manner to the previous question.

Let $I_n = \int x^n \cos x\, dx$ (6.4)

Let $u = x^n \Rightarrow du = nx^{n-1}$

Let $dv = \cos x \Rightarrow x = \sin x$

Then $I_n = x^n \sin x - \int \sin x (nx^{n-1})\, dx$

To determine the last integral we must integrate by parts again.

Let $u = nx^{n-1} \Rightarrow du = n(n-1)x^{n-2}$

Let $dv = \sin x \Rightarrow v = -\cos x$

Then $I_n = x^n \sin x - \left[-nx^{n-1} \cos x - \int (-\cos x) n(n-1)x^{n-2}\, dx \right]$

$\qquad = x^n \sin x + nx^{n-1} \cos x - n(n-1)\int (\cos x)\left(x^{n-1}\right) dx$

$\qquad = x^n \sin x + nx^{n-1} \cos x - n(n-1)I_{n-2}$ (6.5)

To find $\int x^2 \cos x\, dx$ put $n = 2$

$I_2 = x^2 \sin x + 2x \cos x - 2I_0$ (6.6)

Now $2I_0 = 2\int x^0 \cos x\, dx = 2\int \cos dx = 2\sin x$ Putting $n = 0$ in (6.4)

Hence $I_2 = x^2 \sin x + 2x \cos x - 2\sin x + C$ Substituting for $2I_0$ in (6.6)

$$= 2x\cos x + \left(x^2 - 2\right)\sin x + C$$

6.3 Find a reduction formula for $\int \sin^n x \, dx$ and use the result to determine $\int \sin^5 x \, dx$

$$\int \sin^n x \, dx = \int \sin^{n-1} x \sin x \, dx$$

Let $I_n = \int \sin^{n-1} x \sin x \, dx$

Let $u = \sin^{n-1} x \Rightarrow du = (n-1)\sin^{n-2} x \sin x \cos x$

Let $dv = \sin x \Rightarrow v = -\cos x$

Hence $I_n = -\sin x \cos x - \int (-\cos x)(n-1)\sin^{n-1} x \cos x \, dx$

$$= -\sin^{n-1} \cos x + (n-1)\int \cos^2 x \sin^{n-2} x \, dx$$

$$= -\sin^{n-1} x \cos x + (n-1)\int \left(1 - \sin^2 x\right)\sin^{n-2} \, dx$$

$$= -\sin^{n-1} x \cos x + (n-1)\int \left(\sin^{n-2} x - \sin^2 x \sin^{n-2} x\right) dx$$

$$= -\sin^{n-1} x \cos x + (n-1)\int \left(\sin^{n-2} x - \sin^n x\right) dx$$

$$= -\sin^{n-1} x \cos x + (n-1)\int \sin^{n-2} \, dx - (n-1)\int \sin^n x \, dx$$

Hence $I_n = -\sin^{n-1} x \cos x + (n-1)I_{n-2} - (n-1)I_n$

And $I_n + (n-1)I_n = -\sin^{n-1} x \cos x + (n-1)I_{n-2}$

$$\Rightarrow \left(1 + (n-1)\right)I_n = -\sin^{n-1} x \cos x + (n-1)I_{n-2}$$

$$nI_n = -\sin^{n-1} x \cos x + (n-1)I_{n-2}$$

$$I_n = -\frac{1}{n}\sin^{n-1} x \cos x + \frac{1}{n}(n-1)I_{n-2} \qquad\qquad (6.7)$$

To find $\int \sin^5 x \, dx$

$$I_5 = -\frac{1}{5}\sin^4 x \cos x + \frac{4}{5}I_3 \qquad\qquad\qquad \text{Putting } n = 5 \text{ in (6.7)}$$

$$I_3 = -\frac{1}{3}\sin^2 x \cos x + \frac{2}{3}I_1 \qquad\qquad\qquad \text{Putting } n = 3 \text{ in (6.7)}$$

$$I_1 = -\frac{1}{5}\sin^0 \cos x = -\frac{1}{5}\cos x \qquad\qquad\qquad \text{Putting } n = 1 \text{ in (6.7)}$$

$$\text{Hence } I_5 = -\frac{1}{5}\sin^4 x \cos x + \frac{4}{5}\left(-\frac{1}{3}\sin^2 x \cos x + \frac{2}{3}\left(-\frac{1}{5}\cos x\right)\right) \qquad \begin{array}{l}\text{Substituting for}\\ I_3 \text{ and } I_1 \text{ in } I_5\end{array}$$

$$= -\frac{1}{5}\sin^4 x \cos x - \frac{4}{14}\sin^2 x \cos x - \frac{8}{15}\cos x$$

$$= \left(-\frac{1}{5}\sin^4 x - \frac{4}{15}\sin^2 x - \frac{8}{15}\right)\cos x + C$$

6.4 Find a reduction formula for $\int \cos^n x \, dx$ and hence find $\int \cos^3 x \, dx$

Just as before, we re-write the first integral as:

$$\int \cos^n x \, dx = \int \cos^{n-1} x \cos x \, dx \qquad\qquad (6.8)$$

Let $I_n = \int \cos^{n-1} x \cos x \, dx$

Let $u = \cos^{n-1} x \Rightarrow du = -(n-1)\cos^{n-2} x \sin x$

Let $dv = \cos x \Rightarrow v = \sin x$

Then $I_n = \cos^{n-1}\sin x + \int \left(\sin x (n-1)\cos^{n-2}x\sin x\right)dx$

$$= \cos^{n-1}\sin x + (n-1)\int \sin^2 \cos^{n-2}dx$$

$$= \cos^{n-1}\sin x + (n-1)\int \left((1-\cos^2 x)\cos^{n-2}x\right)dx$$

$$= \cos^{n-1}\sin x + (n-1)\int \left(\cos^{n-2} - \cos^n x\right)dx$$

$$= \cos^{n-1}\sin x + (n-1)\int \cos^{n-2}x\, dx - (n-1)\int \cos^n dx$$

$$= \cos^{n-1}\sin x + (n-1)I_{n-2} - (n-1)I_n$$

Hence $I_n + (n-1)I_n = \cos^{n-1}\sin x - (n-1)I_{n-2}$

$$\Rightarrow (1+(n-1))I_n = \cos^{n-1}\sin x - (n-1)I_{n-2}$$

$$\Rightarrow nI_n = \cos^{n-1}\sin x - (n-1)I_{n-2}$$

$$\Rightarrow I_n = \frac{1}{n}\cos^{n-1}\sin x - \frac{1}{n}(n-1)I_{n-2} \qquad (6.9)$$

To find $\int \cos^4 x\, dx$

$$I_4 = \frac{1}{4}\cos^3 \sin x - \frac{3}{4}I_2 \qquad\qquad \text{Putting } n = 4 \text{ in (6.9)}$$

$$I_2 = \frac{1}{2}\cos x\sin x - \frac{1}{2}I_0 \qquad\qquad \text{Putting } n = 2 \text{ in (6.9)}$$

$$I_0 = \int dx = x \qquad\qquad\qquad\qquad \text{Putting } n = 0 \text{ in (6.8)}$$

$$I_4 = \frac{1}{4}\cos^3 \sin x - \frac{3}{4}\left(\frac{1}{2}\cos x\sin x - \frac{1}{2}x\right)$$

$$= \frac{1}{4}\cos^3 \sin x - \frac{3}{8}\cos x\sin x - \frac{3}{8}x$$

Some useful results are obtained with $\int \sin^n x\, dx$ and $\int \cos^n x\, dx$ with limits $x = 0$ and $x = \dfrac{\pi}{2}$ which we will now demonstrate in questions 6.5 and 6.6

6.5 Find $\int_0^{\frac{\pi}{2}} \sin^n x\, dx$ using a reduction formula

We already know from Q 6.3 that the reduction formula for $\int \sin^n x\, dx$ is:

$$I_n = -\frac{1}{n}\sin^{n-1}x\cos x + \frac{1}{n}(n-1)I_{n-2} \qquad (6.10)$$

Hence $\int_0^{\frac{\pi}{2}} \sin^n x\, dx = \left[-\frac{1}{n}\sin^{n-1}x\cos x \Big|_0^{\frac{\pi}{2}} + \frac{1}{n}(n-1)I_{n-2}\right]$

$$= [0-0] + \frac{1}{n}(n-1)I_{n-2}$$

Therefore $I_n = \frac{1}{n}(n-1)I_{n-2}$

Additionally if n is even, the formula eventually reduces down to I_0 so that

$$\int_0^{\frac{\pi}{2}} 1 \, dx = [x]_0^{\frac{\pi}{2}} = \frac{\pi}{2} \Rightarrow I_0 = \frac{\pi}{2}$$

and to I_1 if n is odd so that

$$\int_0^{\frac{\pi}{2}} \sin x \, dx = [-\cos x]_0^{\frac{\pi}{2}} = -(-1) = 1 \qquad (6.11)$$

Hence to evaluate $\int_0^{\frac{\pi}{2}} \sin^5 x \, dx$, we proceed as follows:

$I_5 = \frac{4}{5} I_3$ Substituting for $n = 5$ in (6.10)

$I_3 = \frac{2}{3} I_1$ Substituting for $n = 3$ in (6.10)

$I_1 = 1$ From (6.11)

Hence: $I_5 = \frac{4}{5} \cdot \frac{2}{3} \cdot 1 = \frac{8}{15}$

6.6 Find $\int_0^{\frac{\pi}{2}} \cos^n x \, dx$ using a reduction formula

Just as with question 6.5 we already know that the reduction formula for $\int \cos^n x \, dx$ is:

$$I_n = \frac{1}{n} \cos^{n-1} x \sin x - \frac{1}{n}(n-1) I_{n-2}$$

Hence: $\int_0^{\frac{\pi}{2}} \cos^n x \, dx = \left[\frac{1}{n} \cos^{n-1} x \sin x \right]_0^{\frac{\pi}{2}} - \frac{1}{n}(n-1) I_{n-2}$

$$= [0 - 0] - \left(-\frac{1}{n}(n-1) I_{n-2} \right)$$

$$= \frac{1}{n}(n-1) I_{n-2}$$

Note that this is exactly the same result as before for $\int_0^{\frac{\pi}{2}} \sin^n x \, dx$

Hence to evaluate $\int_0^{\frac{\pi}{2}} \cos^5 x \, dx$, the result is exactly the same. I.e. $\frac{8}{15}$

6.7 Evaluate $\int_0^{\frac{\pi}{2}} \cos^6 x \, dx$ (6.12)

We already know the reduction formula for $\int \cos^n x \, dx$ is:

$$I_n = \frac{1}{n}(n-1) I_{n-2} \qquad (6.13)$$

Hence $\int_0^{\frac{\pi}{2}} \cos^6 x \, dx = I_6 = \frac{5}{6} I_4$ Substituting for $n = 6$ in (6.13)

$I_4 = \frac{3}{4} I_2$ Substituting for $n = 4$ in (6.13)

$I_2 = \frac{1}{2} I_0$ Substituting for $n = 2$ in (6.13)

$$I_0 = \int_0^{\frac{\pi}{2}} \cos^0 x \, dx = \int_0^{\frac{\pi}{2}} 1 \cdot dx = \left[x \right]_0^{\frac{\pi}{2}} = \frac{\pi}{2}$$

Hence $I_6 = \frac{5}{6} \cdot \frac{3}{4} \cdot \frac{1}{2} \cdot \frac{\pi}{2} = \frac{5\pi}{32}$

Note the pattern in this result, the numbers from n down to 1 appear alternately in the order bottom – top, bottom-top .... enabling us to write down the answer straightaway

$$\frac{(n-1)(n-3)(n-5)}{n(n-4)(n-6)} \ldots etc$$

If n is even, the sequence finishes with 1 on top and the factor $\frac{\pi}{2}$ is added at the end

E.g. $\frac{5.3.1}{6.4.2} \cdot \frac{\pi}{2}$

If n is odd, the sequence finishes with 1 on the bottom

E.g. $\frac{6}{5} \cdot \frac{4}{3} \cdot \frac{2}{1}$

6.8 Determine $\int_0^{\frac{\pi}{2}} \sin^4 x \cos^2 x \, dx$

$$\int_0^{\frac{\pi}{2}} \sin^4 x \cos^2 x \, dx = \int_0^{\frac{\pi}{2}} \sin^4 x \left(1 - \cos^2 x \right) dx$$

$$= \int_0^{\frac{\pi}{2}} \left(\sin^4 x - \sin^6 x \right) dx$$

$$= \int_0^{\frac{\pi}{2}} \sin^4 x \, dx - \int_0^{\frac{\pi}{2}} \sin^6 x \, dx$$

$$= I_4 - I_6 = \left(\frac{3.1}{4.2} \cdot \frac{\pi}{2} \right) - \left(\frac{5.3.1}{6.4.2} \cdot \frac{\pi}{2} \right)$$

$$= \frac{3\pi}{16} - \frac{15}{96} = \frac{3\pi}{95} = \frac{\pi}{32}$$

6.9 Determine $\int_0^{\frac{\pi}{2}} \sin^3 x \cos^2 x \, dx$

$$\int_0^{\frac{\pi}{2}} \sin^3 x \cos^2 x \, dx = \int_0^{\frac{\pi}{2}} \sin^3 x \left(1 - \sin^2 x \right) dx$$

$$= \int_0^{\frac{\pi}{2}} \sin^3 x - \sin x^5 \, dx$$

$$= \int_0^{\frac{\pi}{2}} \sin^3 x - \int_0^{\frac{\pi}{2}} \sin^5 x$$

$$= I_3 - I_5 = \left(\frac{2}{3.1} \right) - \left(\frac{4.2}{5.3.1} \right) = \frac{2}{3} - \frac{8}{15} = \frac{2}{15}$$

6.10 If $I_n = \int x^n e^{2x} \, dx$ show that $I_n = \frac{1}{2} x^n e^{2x} - \frac{n}{2} I_{n-1}$ Hence find $\int x^2 e^{2x} \, dx$

Let $u = x^n \Rightarrow du = nx^{n-1}$

Let $dv = e^{2x} \Rightarrow v = \frac{1}{2} e^{2x}$

Note: to find v, let $t = 2x$ then $\int e^{2x} dx = \frac{1}{2} \int e^t dt = \frac{1}{2} e^t = \frac{1}{2} e^{2x}$

Hence $I_n = \frac{1}{2} x^n e^{2x} - \frac{n}{2} \int e^{2x} x^{n-1} dx$

$= \frac{1}{2} x^n e^{2x} - \frac{n}{2} I_{n-1}$ (6.14)

Now let $I_n = \int x^4 e^{2x} dx$ (6.15)

Then $I_4 = \frac{1}{2} x^4 e^{2x} - 2I_3$ Substituting for $n = 4$ in (6.14)

And $I_3 = \frac{1}{2} x^3 e^{2x} - \frac{3}{2} I_2$ Substituting for $n = 3$ in (6.14)

$I_2 = \frac{1}{2} x^2 e^{2x} - I_1$ Substituting for $n = 2$ in (6.14)

$I_1 = \frac{1}{2} x e^{2x} - \frac{1}{2} I_0$ Substituting for $n = 1$ in (6.14)

$I_0 = \int e^{2x} dx = \frac{1}{2} e^{2x}$ Substituting for $n = 0$ in (6.15)

Hence $I_4 = \frac{1}{2} x^4 e^{2x} - x^3 e^{2x} - 3I_1$

$= \frac{1}{2} x^4 e^{2x} - x^3 e^{2x} - \frac{3}{2} x^2 e^{2x} - 3I_1$

$= \frac{1}{2} x^4 e^{2x} - x^3 e^{2x} - \frac{3}{2} x^2 e^{2x} - \frac{3}{2} \cdot \frac{1}{2} e^{2x} + C$

$= \frac{1}{4} e^{2x} \left(2x^4 - 4x^3 - 6x^2 - 6x - 3 \right) + C$

6.11 Find a reduction formula for $\int \tan^n x \, dx$ and hence show that $I_n = \int_0^{\frac{\pi}{4}} \tan^n n = \frac{1}{n-1} - I_{n-2}$

Let $I_n = \int \tan^n x \, dx$ (6.16)

$\int \tan^n x \, dx = \int \tan^{n-2} x \tan^2 x \, dx$

$= \int \tan^{n-2} x \left(\sec^2 - 1 \right) dx$

$= \int \left(\tan^{n-2} x \sec^2 x - \tan^{n-2} x \right) dx$

$= \int \tan^{n-2} x \sec^2 x \, dx - \int \tan^{n-2} x \, dx$

$= \int \tan^{n-2} x \sec^2 x \, dx - I_{n-2}$ (6.17)

Remember that $\int f(x)^m . f' \, dx = \frac{\left[f(x) \right]^{m+1}}{m+1}$ (6.18)

Hence (6.17) becomes:

$\int \tan^n x \, dx = \frac{\tan^{n-1}}{n-1} - I_{n-2} + C$ Putting $m = n - 2$ in (6.18)

So that $I_n = \int_0^{\frac{\pi}{4}} \tan^n n = \left[\frac{\tan^{n-1} x}{n-1}\right]_0^{\frac{\pi}{4}} - I_{n-2} = \frac{1}{n-1} - I_{n-2}$ Since $\tan^{n-1}\left(\frac{\pi}{4}\right) = 1$

6.12 Show that $\int \cos^n x \, dx = \frac{\cos^{n-1} x}{n} + \frac{n-1}{n}\int \cos^{n-2} x \, dx$

Let $I_n = \int \cos^n dx$

$= \int \cos^{n-1} x \cos x \, dx$

$= \cos^{n-1} x \sin x - \int (n-1)\cos^{n-2} x \sin x (-\sin x) \, dx$

$= \cos^{n-1} x \sin x + \int (n-1)\cos^{n-2} x \sin^2 x \, dx$

$= \cos^{n-1} x \sin x$

$= \cos^{n-1} x \sin x + (n-1)\int \cos^{n-2} x (1 - \cos^2 x) \, dx$

$= \cos^{n-1} x \sin x + (n-1)\int (\cos^{n-2} x - \cos^n x) \, dx$

$= \cos^{n-1} x \sin x + (n-1)\int \cos^{n-2} x \, dx - (n-1)\int \cos^n x \, dx$

$= \cos^{n-1} x \sin x + (n-1)\int \cos^{n-2} x \, dx - (n-1)I_n$

Hence $I_n + (n-1)I_n = \cos^{n-1} x \sin x + (n-1)\int \cos^{n-2} x \, dx$

$nI_n = \cos^{n-1} x \sin x + (n-1)\int \cos^{n-2} x \, dx$

$I_n = \frac{\cos^{n-1} x \sin x}{n} + \frac{(n-1)}{n}\int \cos^{n-2} x \, dx$

Note: this question is similar to Q6.6

6.13 Show that $\int_0^{\frac{\pi}{2}} \sin^m x \cos x \, dx$ and $\int_0^{\frac{\pi}{2}} \cos^m x \sin x \, dx = \frac{1}{m+1}$

$\int_0^{\frac{\pi}{2}} \sin^m x \cos x \, dx = \int_0^{\frac{\pi}{2}} \sin^m x \, d(\sin x) x \, dx$

$= \left[\frac{\sin^{m+1} x}{m+1}\right]_0^{\frac{\pi}{2}} = \frac{1}{m+1}$ Since $\int f(x)^m . f \, dx = \frac{[f(x)]^{m+1}}{m+1}$

$\int_0^{\frac{\pi}{2}} \cos^m x \sin x \, dx = \int_0^{\frac{\pi}{2}} \cos^m x \, d(\cos x) \, dx$

$= \left[\frac{-\cos^{m+1} x}{m+1}\right]_0^{\frac{\pi}{2}} = \frac{1}{m+1}$ Since $\int f(x)^m . f \, dx = \frac{[f(x)]^{m+1}}{m+1}$

Where m and n are both positive integers, it can be shown as previously (Q 6.7) that when m and m are both even that:

$\int_0^{\frac{\pi}{2}} \cos^n x \sin^m x \, dx = \frac{(n-1)(n-3)\ldots \times (m-1)(m-3)\ldots}{(m+n)(m+n-2)(m+n-4)\ldots} \times \frac{\pi}{2}$

But when either m or n or both is an odd positive integer then

$\int_0^{\frac{\pi}{2}} \cos^n x \sin^m x \, dx = \frac{(n-1)(n-3)\ldots \times (m-1)(m-3)\ldots}{(m+n)(m+n-2)(m+n-4)\ldots} \times 1$

Where:

If the first term in the formula is even the formula will continue down to 2
If the first term in formula is odd the formula will continue down to 1
As shown in the following examples:

6.14 Find $\int_0^{\frac{\pi}{2}} \sin^5 x \cos^4 x \, dx$

Using the above results

$$\int_0^{\frac{\pi}{2}} \sin^5 x \cos^4 x \, dx = \frac{(5-1)(5-3) \times (4-1)(4-3)}{(4+5)(4+5-2)(4+5-4)(4+5-6)} \times 1$$

$$= \frac{4.2.3}{9.7.5.3} = \frac{24}{945} = \frac{8}{315}$$

6.15 Find $\int_0^{\frac{\pi}{2}} \sin^4 x \cos^4 x \, dx$

$$\int_0^{\frac{\pi}{2}} \sin^4 x \cos^4 x \, dx = \frac{(4-1)(4-3) \times (4-1)(4-3)}{(4+4)(4+4-2)(4+4-4)(4+4-6)} \times \frac{\pi}{2}$$

$$= \frac{3.1.3.1}{8.6.4.2} \times \frac{\pi}{2} = \frac{9\pi}{768} = \frac{3\pi}{256}$$

6.16 Find $\int_0^{\frac{\pi}{2}} \cos^7 x \tan^4 x \, dx$

$$\int_0^{\frac{\pi}{2}} \cos^7 x \tan^4 x \, dx = \int_0^{\frac{\pi}{2}} \cos^7 x \left(\frac{\sin x}{\cos x} \right)^4$$

$$= \int_0^{\frac{\pi}{2}} \cos^7 x \frac{\sin x^4}{\cos x^4} \, dx$$

$$= \int_0^{\frac{\pi}{2}} \cos^3 x \sin x^4 \, dx$$

$$= \frac{(3-1) \times (4-1)(4-3)}{(4+3)(4+3-2)(4+3-4)(4+3-6)} \times 1$$

$$= \frac{2.1.3.1}{7.5.3.1} \times 1 = \frac{6}{105} = \frac{2}{35}$$

6.17 Find a reduction formula for $\int \frac{dx}{\left(x^2 + a^2 \right)^n}$

Let $I_n = \int \frac{dx}{\left(x^2 + a^2 \right)^n} = \int 1 \times \frac{1}{\left(x^2 + a^2 \right)^n}$

Let $u = \frac{1}{\left(x^2 + a^2 \right)^n}$

Then $du = \frac{0 - 1.2nx \left(x^2 + a^2 \right)^{nn-1}}{\left(x^2 + a^2 \right)^{2n}}$ Using the result $d(uv) = \frac{vdu - udv}{v^2}$

$$= \frac{-2nx\left(x^2+a^2\right)^{n-1}}{\left(x^2+a^2\right)^{2n}}$$

$$= \frac{-2nx}{\left(x^2+a^2\right)^{n+1}}$$

Let $dv = 1 \Rightarrow v = \int 1dx = x$

Then $I_n = \dfrac{x}{\left(x^2+a^2\right)^n} - \int \dfrac{x.\left(-2nx\right)}{\left(x^2+a^2\right)^{n+1}} dx$

$$= \frac{x}{\left(x^2+a^2\right)^n} + 2n\int \frac{x^2}{\left(x^2+a^2\right)^{n+1}} dx$$

$$= \frac{x}{\left(x^2+a^2\right)^n} + 2n\int \frac{\left(x^2+a^2\right)-a^2}{\left(x^2+a^2\right)^{n+1}} dx \qquad \text{Note that } x^2+a^2-a^2 = x^2$$

Note that the numerator of this integral has been changed to allow it to be split into two integrals that we can integrate as shown below:

$$= \frac{x}{\left(x^2+a^2\right)^n} + 2n\int \frac{\left(x^2+a^2\right)}{\left(x^2+a^2\right)^{n+1}} dx - 2na^2\int \frac{dx}{\left(x^2+a^2\right)^{n+1}}$$

$$= \frac{x}{\left(x^2+a^2\right)^n} + 2n\int \frac{dx}{\left(x^2+a^2\right)^n} - 2na^2\int \frac{dx}{\left(x^2+a^2\right)^{n+1}}$$

Dividing the numerator and denominator of the middle integral by $\left(x^2+a^2\right)$

$$= \frac{x}{\left(x^2+a^2\right)^n} + 2nI_n - 2na^2I_{n+1}$$

Hence $I_n - 2nI_n = \dfrac{x}{\left(x^2-a^2\right)^n} - 2na^2I_{n+1}$

And $I_n = \dfrac{x}{\left(x^2-a^2\right)^n} - \dfrac{2na^2I_{n+1}}{1-2n}$

Note that if we put $n = n-1$ into the above result the same result could also be obtained by using the substitution $x = a\tan x$ and then using the reduction formula for $\int \cos^n x \, dx$ as in Q 6.6.

6.18 Show that if $I_n = \int_0^2\left(4-x^2\right)^n dx$ then $I_n = \dfrac{8nI_{n-1}}{2n-1}$

We can re-write $I_n = \int_0^2\left(4-x^2\right)^n dx$ as $I_n = \int 1\times\left(4-x^2\right)^n dx$

Let $u = \left(4-x^2\right)^n \Rightarrow du = 2nx\left(4-x^2\right)^{n-1}$

Let $dv = 1 \Rightarrow v = x$

Then $I_n = \left[x\left(4-x^2\right)^n \right]_0^2 - \int_0^2 2nx\left(4-x^2\right)^{n-1} x\,dx$

$= 0 - 2n\int_0^2 x^2\left(4-x^2\right)^{n-1} dx$

$= 2n\int_0^2 \left(4-\left(4-x^2\right)\right)\left(4-x^2\right)^{n-1} dx$

$= 2n\int_0^2 \left(4\left(4-x^2\right)^{n-1} - \left(4-x^2\right)^n\right) dx$

$= 8n\int_0^2 \left(4-x^2\right)^{n-1} dx - 2n\int_0^2 \left(4-x^2\right)^n dx$

$= 8nI_{n-1} - 2nI_n$

Hence $I_n + 2nI_n = 8I_{n-1}$

And $I_n\left(2n+1\right) = 8nI_{n-1}$

So $I_n = \dfrac{8nI_{n-1}}{2n+1}$

6.19 If $I_n = \int \left(\ln x\right)^n dx$ l prove that $I_n = x\left(\ln x\right)^n + I_{n-1}$ and hence find $\int \left(\ln x\right)^4 dx$

We may re-write $\int \left(\ln x\right)^n dx$ as $\int 1 \times \left(\ln x\right) dx$

Let $u = \left(\ln x\right)^n \Rightarrow du = n\left(\ln x\right)^{n-1}\dfrac{1}{x}$

Let $dv = 1 \Rightarrow v = x$

Hence $I_n = x\left(\ln x\right)^n - \int xn\left(\ln x\right)^{n-1}\left(\dfrac{1}{x}\right) dx$

$= x\left(\ln x\right)^n - n\int \left(\ln x\right)^{n-1} dx$

$= x\left(\ln x\right)^n - nI_{n-1}$

Hence $\int \left(\ln x\right)^4 dx = I_4 = x\left(\ln x\right)^4 - 4I_3$

$\qquad\qquad I_3 = x\left(\ln x\right)^3 - 3I_2$

$\qquad\qquad I_2 = x\left(\ln x\right)^2 - 2I_1$

$\qquad\qquad I_1 = x\ln x - I_0$

$\qquad\qquad I_0 = 1$

Therefore $I_4 = I4 = x\left(\ln x\right)^4 - 4I_3$

$\qquad\qquad = x\left(\ln x\right)^4 - 4\left(x\left(\ln x\right)^3 - 3I_2\right)$

$\qquad\qquad = x\left(\ln x\right)^4 - 4x\left(\ln x\right)^3 + 12\left(x\left(\ln x\right)^2 - 2I_1\right)$

$= x\left(\ln x\right)^4 - 4x\left(\ln x\right)^3 + 12x\left(\ln x\right)^2 - 24I_1$

$= x\left(\ln x\right)^4 - 4x\left(\ln x\right)^3 + 12x\left(\ln x\right)^2 - 24\left(x\ln x - I_0\right)$

$= x\left(\ln x\right)^4 - 4x\left(\ln x\right)^3 + 12x\left(\ln x\right)^2 - 24x\ln x + 24I_0$

$= x\left(\ln x\right)^4 - 4x\left(\ln x\right)^3 + 12x\left(\ln x\right)^2 - 24x\ln x + 24$

We will re-write $\int \sec^n x \, dx$ as $\int \sec^{n-1} x \sec x \, dx$

Let $u = \sec^{n-2} x \Rightarrow du = (n-2)\sec^{n-3} \sec x \tan x$

Let $dv = \sec^2 x \Rightarrow v = \tan x$

Then $I_n = \sec^{n-2} x \tan x - \int \tan x (n-2)\sec^{n-3} \sec x \tan x \, dx$

$\qquad = \sec^{n-2} x \tan x - (n-2)\int \tan^2 x \sec^{n-2} x \, dx$

$\qquad = \sec^{n-2} x \tan x - (n-2)\int (\sec^2 x - 1)(\sec^{n-2} x) \, dx$

$\qquad = \sec^{n-2} x \tan x - (n-2)\int \sec^n x \, dx + (n-2)\int \sec^{n-2} \, dx$

$\qquad = \sec^{n-2} x \tan x - (n-2)I_n + (n-2)I_{n-2}$

And $I_n + (n-2)I_n = \sec^{n-2} x \tan x + (n-2)I_{n-2}$

Therefore $(n-1)I_n = \sec^{n-2} x \tan x + (n-2)I_{n-2}$

And $I_n = \left(\dfrac{1}{n-1}\right)\sec^{n-2} x \tan x + \dfrac{(n-2)}{n-1}I_{n-2}$

To find $\int_0^{\frac{\pi}{4}} \sec^6 x \, dx$ Let $I_6 = \int \sec^6 x \, dx$

Then $I_6 = \dfrac{1}{5}\sec^4 x \tan x + \dfrac{4}{5}I_4$

$\qquad I_4 = \dfrac{1}{3}\sec x \tan x + \dfrac{2}{3}I_2$

$\qquad I_2 = \tan x + 0 = \tan x$

Hence $I_6 = \dfrac{1}{5}\sec^4 x \tan x + \dfrac{4}{5}\left(\dfrac{\sec^2 x \tan x}{3} + \dfrac{2}{3}I_2\right)$

$\qquad = \dfrac{1}{5}\sec^4 x \tan x + \dfrac{4}{15}\sec^2 \tan x + \dfrac{8}{15}I_2$

$\qquad = \dfrac{1}{5}\sec^4 x \tan x \dfrac{4}{15}\sec^2 \tan x + \dfrac{8}{15}\tan x$

So $\int_0^{\frac{\pi}{4}} \sec^6 x \, dx = \left[\dfrac{1}{5}\sec^4 x \tan x \dfrac{4}{15}\sec^2 \tan x + \dfrac{8}{15}\tan x\right]_0^{\frac{\pi}{4}}$

$\qquad = \dfrac{4}{5} + \dfrac{8}{15} + \dfrac{8}{15} = \dfrac{28}{15}$

Let $u = \cosh^{n-1} x \Rightarrow du = (n-1)\cosh^{n-2} x \sinh x \, dx$

Let $dv = \cosh x \Rightarrow v = \sinh x$

Then $I_n = \cosh^{n-1} \sinh x - \int \sinh x (n-1)\cosh^{n-2} x \sinh x \, dx$

$$= \cosh^{n-1} x \sinh x - (n-1) \int \sinh^2 x \cosh^{n-2} x\, dx$$

$$= \cosh^{n-1} x \sinh x - (n-1) \int \left(\cosh^2 x - 1 \right) \cosh^{n-2} x\, dx \qquad (\cosh^2 x - \sinh^2 x = 1)$$

$$= \cosh^{n-1} x \sinh x - (n-1) \int \left(\cosh^n x - \cosh^{n-2} x \right) dx$$

$$= \cosh^{n-1} x \sinh x - (n-1) \int \cosh^{n-1} x\, dx + (n-1) \int \cosh^{n-2} x\, dx$$

$$= \cosh^{n-1} x \sinh x - (n-1) I_n + (n-1) I_{n-2}$$

Hence $I_n + (n-1) I_n = \cosh^{n-1} x \sinh x + (n-1) I_{n-2}$

$$n I_n = \frac{1}{n} \cosh^{n-1} x \sinh x + \frac{(n-1)}{n} I_{n-2}$$

$$I_n = \frac{1}{n} \cosh^{n-1} x \sinh x + \frac{(n-1)}{n} I_{n-2}$$

To find $\int_0^t \cosh^4 x\, dx$

Let $I_4 = \int_0^t \cosh^4 x\, dx$ 　　　　　　　　　　　　　　　(1)

Then $I_4 = \frac{1}{4} \cosh^3 x \sinh x + \frac{3}{4} I_2$

$$I_2 = \frac{1}{2} \cosh^3 x \sinh x + \frac{1}{2} I_0$$

$$I_0 = x \qquad\qquad\qquad\qquad\qquad\qquad \text{Substituting for } n = 0 \text{ in (1)}$$

Hence $I_4 = \frac{1}{4} \cosh^3 x \sinh x + \frac{3}{4} \left(\frac{1}{2} \cosh x \sinh x + \frac{1}{2} x \right) \qquad \text{Substituting for } I_2 \text{ and } I_0$

$$= \frac{1}{4} \cosh^3 x \sinh x + \frac{3}{8} \cosh x \sinh x + \frac{3}{8} x$$

$$= \frac{1}{4} \left(\left(\sqrt{3} \right)^3 \times \sqrt{2} \right) + \frac{3}{8} \left(\sqrt{3} \times \sqrt{2} \right) + \frac{3}{8} \ln \left(\sqrt{3} + \sqrt{ \left(\sqrt{3} \right)^2 - 1} \right) \qquad \text{Putting } t = \cosh^{-1} \left(\sqrt{3} \right)$$

$$= \frac{1}{4}.3\sqrt{6} + \frac{3}{8} \sqrt{6} + \frac{3}{8} \ln \left(\sqrt{3} + \sqrt{2} \right)$$

$$= 1.837 + 0.919 + 0.430 = 3.186$$

Note: $\cosh \left(\cosh^{-1} x \right) = x$ and $\cosh^{-1} x = \ln \left(x - \sqrt{x^2 - 1} \right)$

MODULE T1

TRIGONOMETRY

TRIGONOMETRIC IDENTITIES AND EQUATIONS

1. Prove that $(1 - \cos A)(1 + \sec A) = \sin A \tan A$

Since this identity has not yet been proved, we cannot assume its truth by using the complete identity in our proof.

We must either

(1) Start with the LHS and prove it is identical to the RHS
(2) Start with the RHS and prove it is identical to the LHS
(3) Work on each side *independently* (see example 2)

Solution

We will start with the LHS and try to prove it is identical to the RHS as follows:

$(1 - \cos A)(1 + \sec A) \equiv 1 + \sec A - \cos A - \sec A \cos A$ Multiply out brackets

$\equiv 1 + \sec A - \cos A - \dfrac{1}{\cos A}.\cos A$ Since $\sec A = \dfrac{1}{\cos A}$

$\equiv 1 + \sec A - \cos A - 1$

$\equiv \sec A - \cos A$

$\equiv \dfrac{1}{\cos A} - \cos A$ Since $\sec A = \dfrac{1}{\cos A}$

$\equiv \dfrac{1 - \cos^2 A}{\cos A}$

$\equiv \dfrac{\sin^2 A}{\cos A}$ Since $\sin^2 A + \cos^2 A = 1 \Rightarrow 1 - \cos^2 A = \sin^2 A$

$\equiv \sin A(\dfrac{\sin A}{\cos A})$

$\equiv \sin A \tan A$ Since $\dfrac{\sin A}{\cos A} = \tan A$

We have therefore proved that the LHS is identical to the RHS hence the identity is true.

2. Prove that $(\cosec A - \sin A)(\sec A - \cos A) = \dfrac{1}{\tan A + \cot A}$

Again we will we will start with the LHS and try to prove it is identical to the RHS as follows:

$$(\cos ecA - \sin A)(\sec A - \cos A) \equiv \left(\frac{1}{\sin A} - \sin A\right)\left(\frac{1}{\cos A} - \cos A\right) \qquad \cos ecA = \frac{1}{\sin A}, \ \sec A = \frac{1}{\cos A}$$

$$\equiv \left(\frac{1 - \sin^2 A}{\sin A}\right)\left(\frac{1 - \cos^2 A}{\cos A}\right)$$

$$\equiv \left(\frac{\cos^2 A}{\sin A}\right)\left(\frac{\sin^2 A}{\cos A}\right) \qquad \text{from } \sin^2 A + \cos^2 A = 1$$

$$\equiv \sin A \cos A$$

We have finished up with a simple expression but we have not yet proved this is identical to the RHS. It is not immediately obvious.

To do this we will now work on the RHS (independently) to see if we can reduce it to this result. If we can, we have proved the identity is true.

$$\frac{1}{\tan A + \cot A} \equiv 1 \div \left(\frac{\sin A}{\cos A} + \frac{\cos A}{\sin A}\right) \qquad \tan A = \frac{\sin A}{\cos A}, \ \cot A = \frac{\cos A}{\sin A}$$

$$\equiv 1 \div \left(\frac{\sin^2 A + \cos^2 A}{\cos A \sin A}\right) \quad \text{rearranging with common denominator}$$

$$\equiv \frac{\cos A \sin A}{\sin^2 A + \cos^2 A}$$

$$\equiv \cos A \sin A \qquad \text{from } \sin^2 A + \cos^2 A = 1$$

Since both the LHS and RHS reduce to $\sin A \cos A$, they must be identical.

This illustrates a problem often encountered when trying to prove trigonometric identities where both sides of the identity must be worked on *independently* to obtain the required result.

3. Prove $\dfrac{\cos A}{1 - \sin A} = \tan A + \sec A$

In this example, the difficulty here is knowing how to start – this is often a common problem in solving trig identities.

It is always a good idea to pause for a moment and resist the temptation to dive right in without a plan (unless you can see how to solve it at first sight).

First of all look at the RHS. We know that $\tan A = \dfrac{\sin A}{\cos A}$ *and* $\sec A = \dfrac{1}{\cos A}$

This suggests that if we could somehow make the denominator of the LHS, $\cos A$, then we should be able to proceed but how do we do this?

Well, there does not appear to be an easy way. So what now? The answer is to apply a little lateral thinking – like this:

We can't see an obvious way of expressing the denominator in terms of $\cos A$ so how about $\cos^2 A$? But how does this help? We want $\cos A$ in the denominator **not** $\cos^2 A$. Well, we know that although we haven't proved it yet, it is obvious that this identity is true – otherwise the question wouldn't ask us to prove it!

Hence if we can change the denominator of the LHS to $\cos^2 A$, it should not be too difficult to change it to $\cos A$ with a little further manipulation. We start with the fact that $\sin^2 A + \cos^2 A = 1 \Rightarrow \cos^2 A = 1 - \sin^2 A$.

If we now multiply the denominator $1 - \sin A$ by its conjugate $1 + \sin A$ we get $1 - \sin^2 A = \cos^2 A$.

This is exactly what we want but we must remember that to maintain the integrity of the expression we must multiply **both** the numerator and the denominator by $1 + \sin A$.

Hence we can now write:

$$\frac{\cos A}{1 - \sin A} = \frac{\cos A (1 + \sin A)}{(1 - \sin A)(1 + \sin A)}$$ Multiply top and bottom by $1 + \sin A$

$$= \frac{\cos A + \cos A \sin A}{1 - \sin^2 A}$$ Multiply out brackets

$$= \frac{\cos A + \cos A \sin A}{\cos^2 A}$$ Since $\cos^2 A = 1 - \sin^2 A$

$$= \frac{1 + \sin A}{\cos A}$$ Simplifying

$$= \frac{1}{\cos A} + \frac{\sin A}{\cos A}$$ Split into two fractions

$$= \sec A + \tan A$$ Since $\frac{1}{\cos A} = \sec A$ and $\frac{\sin A}{\cos A} = \tan A$

4. Prove $\dfrac{\sin 2A}{1 + \cos 2A} = \tan A$

Again in this case we notice that $\tan A = \dfrac{\sin A}{\cos A}$ so we try to express the denominator of the LHS in terms of $\cos A$ as before.
This time we employ similar reasoning:

$$\frac{\sin 2A}{1+\cos 2A} = \frac{2\sin A\cos A}{1+(2\cos^2 A -1)} \qquad \text{Since } \sin 2A = 2\sin A\cos A \ \& \ \cos 2A = 2\cos^2 A -1$$

$$= \frac{2\sin A\cos A}{2\cos^2 A} \qquad \text{Simplifying}$$

$$= \frac{\sin A}{\cos A} = \tan A$$

Remember that all exam questions on trig identities are designed to be solved by use of all the standard trig identity formulae.

5. Solve $\dfrac{\sin 2A}{1-\cos 2A}$

This is very similar to the previous question and we can treat it with the same basic method of attack.

$$\frac{\sin 2A}{1-\cos 2A} = \frac{2\sin A\cos A}{1-(2\cos^2 A -1)}$$

$$= \frac{2\sin A\cos A}{2+2\cos^2 A} \qquad \text{Simplifying denominator}$$

$$= \frac{\sin A\cos A}{1+\cos^2 A} \qquad \text{Simplifying}$$

$$= \frac{\sin A\cos A}{1-(1-\sin^2 A)} \qquad \text{Since } \cos^2 A = 1-\sin^2 A$$

$$= \frac{\sin A\cos A}{\sin^2 A} \qquad \text{Simplifying}$$

$$= \frac{\cos A}{\sin A} \qquad \text{Simplifying}$$

$$= \cot A$$

6. Show that $\sin 3A = 3\sin A - 4\sin^3 A$

The method here is to expand the LHS and convert everything you obtain in terms of $\sin A$.

$$\sin 3A = \sin 2A\cos A + \cos 2A\sin A \qquad \sin 3A = \sin(2A+A) \ \text{(factor formulae)}$$

$$=2\sin A\cos A\cos A+(1-2\sin^2 A)\sin A \qquad \cos 2A = 1-2\sin^2 A$$

$$= 2\sin A\cos^2 A + \sin A -2\sin^3 A \qquad \text{Multiply out bracket}$$

$$=2\sin A(1-\sin^2 A)+\sin A -2\sin^3 A \qquad \cos^2 A = 1-\sin^2 A$$

$$= 2\sin A - \sin^3 A +\sin A - 2\sin^3 A \qquad \text{Multiply out bracket}$$

$$= 3\sin A - 4\sin^3 A \qquad \text{Simplifying}$$

This is a standard result and should be committed to memory

7. Prove that $\cos 3A = 4\cos^3 A - 3\cos A$

Hopefully you will now see that this question can be attacked in the same way. This time we start with the LHS and express everything in terms of $\cos A$. Hence

$$\cos 3A = \cos(2A + A) = \cos 2A \cos A - \sin 2A \sin A$$
$$= (2\cos^2 A - 1)\cos A - 2\sin A \cos A \sin A$$
$$= 2\cos^3 A - \cos A - 2\sin^2 A \cos A$$
$$= 2\cos^3 A - \cos A - 2(1 - \cos^2 A)\cos A$$
$$= 2\cos^3 A - \cos A - 2\cos A + 2\cos^3 A$$
$$= 4\cos^3 A - 3\cos A$$

This is also a standard result and should be committed to memory

8. Prove that $\dfrac{1 - \cos 2A}{1 + \cos 2A} = \tan^2 A$

The method of attack here is to recognise that the RHS $\tan^2 A = \dfrac{\sin^2 A}{\cos^2 A}$ so we need to work on the LHS to try express the numerator in terms of $\sin A$ and the denominator in terms of $\cos A$.

Just keep going until you achieve this and the result will follow.

$$\dfrac{1 - \cos 2A}{1 + \cos 2A} = \dfrac{1 - (2\cos^2 A - 1)}{1 + (2\cos^2 A - 1)} \qquad \cos 2A = 2\cos^2 A - 1$$
$$= \dfrac{2 - 2\cos^2 A}{2\cos^2 A} \qquad \text{Simplifying}$$
$$= \dfrac{1 - \cos^2 A}{\cos^2 A} \qquad \text{Simplifying}$$
$$= \dfrac{\sin^2 A}{\cos^2 A} \qquad 1 - \cos^2 A = \sin^2 A$$
$$= \tan^2 A$$

9. Show that $\tan \theta + \cot \theta = \sec \theta \cos ec\theta$

Once again we need to pause and analyse the question to determine our method of attack.

We know that $\sec \theta = \dfrac{1}{\cos \theta}$ and $\cos ec\theta = \dfrac{1}{\sin \theta} \Rightarrow \sec \theta \cos ec\theta = \dfrac{1}{\sin \theta \cos \theta}$

Hence we start with the LHS and try to obtain the value 1 for the numerator and $\sin \theta$ and $\cos \theta$ as a product in the denominator.

$$\tan \theta + \cot \theta = \tan \theta + \frac{1}{\tan \theta}$$

$$= \frac{\tan^2 \theta + 1}{\tan \theta} \qquad \text{Expressing as a single fraction}$$

$$= \frac{\sec^2 \theta}{\tan \theta} \qquad 1 + \tan^2 \theta = \sec^2 \theta$$

$$\frac{1}{\cos^2 \theta} \times \frac{\cos \theta}{\sin \theta} \qquad \text{rearranging}$$

$$\frac{1}{\cos \theta \sin \theta} \qquad \text{Simplifying}$$

$$= \sec \theta \cos ec\theta$$

10. Show that $(\sin \theta + \cos \theta)(\cot \theta + \tan \theta) = \sec \theta + \cos ec\theta$

The RHS can be written as $\dfrac{1}{\cos \theta} + \dfrac{1}{\sin \theta}$ and we also know that

$$\cot \theta + \tan \theta = \frac{1}{\tan \theta} + \tan \theta = \frac{1 + \tan^2 \theta}{\tan \theta} = \sec^2 \theta$$

This now gives us a method of attack:

$$(\sin \theta + \cos \theta)(\cot \theta + \tan \theta)$$

$$= (\sin \theta + \cos \theta)\left(\frac{1}{\tan \theta} + \tan \theta \right) \qquad \cot \theta = \frac{1}{\tan \theta}$$

$$= (\sin \theta + \cos \theta)\left(\frac{1 + \tan^2 \theta}{\tan \theta} \right) \qquad \text{Expressing RH bracket as a single fraction}$$

$$= (\sin \theta + \cos \theta)\left(\frac{\sec^2 \theta}{\tan \theta} \right) \qquad 1 + \tan^2 \theta = \sec^2 \theta$$

$$= (\sin \theta + \cos \theta)\left(\frac{1}{\cos^2 \theta} \times \frac{\cos \theta}{\sin \theta} \right) \qquad \sec^2 \theta = \frac{1}{\cos^2 \theta}$$

$$= (\sin \theta + \cos \theta)\left(\frac{1}{\sin \theta \cos \theta} \right) \qquad \text{Simplifying}$$

$$= \frac{1}{\cos \theta} + \frac{1}{\sin \theta} \qquad \text{Multiplying out brackets}$$

$$= \sec \theta + \cos ec\theta$$

11. Factorise $\sin 3A + \sin A$

When you are asked to factorise an expression it means you have to change it so that it is expressed as a product. This expression just begs for the use of the factor formula as follows:

$$\sin 3A = \sin\left(2A + A\right) \qquad \text{Factor formula}$$
$$= \sin 2A \cos A + \cos 2A \sin A$$
$$\sin A = \sin\left(2A - A\right) \qquad \text{Factor formula}$$
$$= \sin 2A \cos A - \cos 2A \sin A$$
$$\therefore \sin 3A + \sin A = \sin 2A \cos A + \cos 2A \sin A + \sin 2A \cos A - \cos 2A \sin A$$
$$= 2 \sin 2A \cos A$$

12. Express $\sqrt{\dfrac{1 - \sin 2\theta}{1 + \sin 2\theta}}$ in terms of $\tan \theta$

This immediately suggests the use of the half angle identity $\sin \theta = \dfrac{2t}{1 - t^2}$ where

$t = \tan \dfrac{\theta}{2}$ hence $\sin 2\theta = \dfrac{2t}{1 + t^2}$ where $t = \tan \theta$. So we can proceed as follows:

$$1 - \sin 2\theta \equiv 1 - \frac{2t}{1 + t^2} \qquad\qquad \sin 2\theta = \frac{2t}{1 + t^2}$$
$$\equiv \frac{1 + t^2 - 2t}{1 + t^2}$$
$$\equiv \frac{(1 - t)^2}{1 + t^2} \qquad\qquad \text{Factorising numerator}$$

$$1 + \sin 2\theta \equiv 1 - \frac{2t}{1 + t^2} \qquad\qquad \sin 2\theta = \frac{2t}{1 + t^2}$$
$$\equiv \frac{1 + t^2 + 2t}{1 + t^2}$$
$$\equiv \frac{(1 + t)^2}{1 + t^2} \qquad\qquad \text{Factorising numerator}$$

Hence $\dfrac{1 - \sin 2A}{1 + \sin 2A} \equiv \dfrac{(1 - t)^2}{1 + t^2} \div \dfrac{(1 + t)^2}{1 + t^2}$

$$\equiv \frac{(1 - t)^2}{(1 + t)^2}$$
$$\equiv \frac{(1 - \tan \theta)^2}{(1 + \tan \theta)^2}$$
$$\equiv \left(\frac{1 - \tan \theta}{1 + \tan \theta}\right)^2$$

Hence $\sqrt{\dfrac{1 - \sin 2\theta}{1 + \sin 2\theta}} \equiv \dfrac{1 - \tan \theta}{1 + \tan \theta}$ \qquad Taking the square root of both sides

TRIGONOMETRIC EQUATIONS

To successfully solve trigonometric equations we need to have a good knowledge of the trig identities, which is why we have linked these two topics together.

13. Solve the equation $\cos 3\theta \sin \theta + \sin 3\theta \cos \theta = \cos 4\theta$ in the interval $0 \le \theta \le \dfrac{\pi}{2}$

$\cos 3\theta \sin \theta + \sin 3\theta \cos \theta = \cos 4\theta$

$\Rightarrow \sin 4\theta = \cos 4\theta$ Since LHS $= \sin(3\theta + \theta) = \sin 4\theta$ (Factor formula)

$\Rightarrow \dfrac{\sin 4\theta}{\sin 4\theta} = 1$

$\Rightarrow \tan 4\theta = 1$

Hence

$\quad 4\theta = \tan^{-1} 1$

$\Rightarrow 4\theta = \dfrac{\pi}{4}$

$\Rightarrow \theta = (4n+1)\dfrac{\pi}{16} \quad (n = 0,1,2...)$

$\Rightarrow \theta = \dfrac{\pi}{16}, \dfrac{5\pi}{16} \quad$ on the interval $0 \le \theta \le \dfrac{\pi}{2}$

14. Express $5 - 5\cos\theta - 3\sin^2\theta = 0$ as a quadratic equation and obtain the solution on the interval $0 \le \theta \le 360^\circ$

$$5 - 5\cos\theta - 3\sin^2\theta = 5 - 5\cos\theta - 3\left(1 - \cos^2\right) \qquad \sin^2\theta = 1 - \cos^2\theta$$
$$= 5 - 5\cos\theta - 3 + 3\cos^2\theta \qquad \text{Expanding bracket}$$
$$= 2 - 5\cos\theta + 3\cos^2\theta \qquad \text{Simplifying}$$
$$= (2 - 3\cos\theta)(1 - \cos\theta) \qquad \text{Factorising}$$

Hence we can write the original equation as

$(2 - 3\cos\theta)(1 - \cos\theta) = 0$

Case1

$2 - 3\cos\theta = 0$

$\Rightarrow 3\cos\theta = 2$

$\Rightarrow \cos\theta = \dfrac{2}{3}$

$\Rightarrow \quad \theta = \cos^{-1}\left(\dfrac{2}{3}\right)$ on interval $0^\circ \le \theta \le 90^\circ$

$\quad\quad = 48.19^\circ$ on interval $0^\circ \le \theta \le 90^\circ$

$\quad\quad$ and $360^\circ - 48.19^\circ = 311.41^\circ$ on interval $270^\circ \le \theta \le 90^\circ \qquad \cos\theta = \cos(360 - \theta)$

Hence $\theta = 48.19^\circ$ and 311.41° on interval $0^\circ \le \theta \le 360^\circ$

Case 2

$1 - \cos\theta = 0$

$\Rightarrow \cos\theta = 1$

$\Rightarrow \quad \theta = \cos^{-1} 1$

$\Rightarrow \theta = 0°$ on interval $0° \le \theta \le 90°$

and $360° - 0° = 360°$ on interval $0° \le \theta \le 270°$

Hence $\theta = 0°$ and $360°$ on interval $0° \le \theta \le 360°$

The complete solution is therefore given by $\theta = 0°$, $48.19°$, $311.41°$ and $360°$

15. Find all the solutions of the equation $\cos 2x + \cos x = 0$ on the interval $0 \le x \le 2\pi$

$$\cos 2x + \cos x = 2\cos^2 x - 1 + \cos x \qquad \text{Since } \cos 2x = 2\cos^2 x - 1$$

Hence we can write the original equation as

$2\cos^2 x + \cos x - 1 = 0$

$\Rightarrow (2\cos x - 1)(\cos x + 1) = 0$

Case1

$2\cos x - 1 = 0$

$\Rightarrow \cos x = \dfrac{1}{2}$

$\Rightarrow x = \cos^{-1}\left(\dfrac{1}{2}\right) = \dfrac{\pi}{3}$ on the interval $0 \le x \le \dfrac{\pi}{2}$

and $x = 2\pi - \dfrac{\pi}{3}$ on the interval $\dfrac{3\pi}{2} \le x \le 2\pi$ $\qquad \cos x = \cos(2\pi - x)$

Hence $x = \dfrac{\pi}{3}$ and $\dfrac{5\pi}{3}$ in the interval $0 \le x \le 2\pi$

Case 2

$\cos x + 1 = 0$

$\Rightarrow \cos x = -1$

$\Rightarrow \quad x = \cos^{-1}(-1)$

$\quad = \pi$ in the interval $0 \le x \le \dfrac{\pi}{2}$

and $(2\pi - \pi) = \pi$ in the interval $\dfrac{\pi}{2} \le x \le \pi$ $\qquad \cos x = \cos(2\pi - x)$

Hence $x = \dfrac{\pi}{3}, \pi, \dfrac{5\pi}{3}$ on the interval $0 \le x \le 2\pi$

16. Find the solutions of the equation $\cos(30° + x) = 2\sin(40° + x)$ that lie between 0° and 360°.

$\cos(30° + x) = \cos 30° \cos x - \sin 30° \sin x$ (1)   Using factor formula

$\sin(40° + x) = \sin 40° \cos x + \cos 40° \sin x$ (2)   Using factor formula

Hence $\cos(30° + x) = 2\sin(40° + x)$ can be written as:

$\cos 30° \cos x - \sin 30° \sin x = 2(\sin 40° \cos x + \cos 40° \sin x)$ Using (1) and (2)

$\Rightarrow 0.8660 \cos x - 0.5 \sin x = 2(0.6428 \cos x + 0.7660 \sin x)$

$\Rightarrow 0.8660 \cos x - 0.5 \sin x = 1.2856 \cos x + 1.5320 \sin x$

$\Rightarrow 0.8660 \cos x - 1.2856 \cos x = 1.5320 \sin x + 0.5 \sin x$ Rearranging

$\Rightarrow -0.4196 \cos x = 2.0320 \sin x$ Simplifying

Hence $\tan x = \dfrac{-0.4196}{2.0320} = -0.2065$

$\Rightarrow x = \tan^{-1}(-0.2065) = -11.67°$

tan is negative in the second and fourth quadrants hence

$x = (180° - 11.67°) = 168.33°$ in the second quadrant $(90° \leq x \leq 180°)$

$x = (360° - 11.67°) = 348.33°$ in the fourth quadrant $(270° \leq x \leq 360°)$

Hence the complete solution is: $168.33°$ and $348.33°$

17. Find all values of x satisfying the equation $\cos^2 2x + 3\sin 2x - 3 = 0$

$\cos^2 2x + 3\sin 2x - 3 = 0$

$\Rightarrow 1 - \sin^2 2x + 3\sin 2x - 3 = 0$ Since $\cos^2 2x = 1 - \sin^2 2x$

$\Rightarrow -2 + 3\sin 2x - \sin^2 2x = 0$ Simplifying

$\Rightarrow \sin^2 2x - 3\sin 2x + 2 = 0$ Multiplying throughout by -1

This is clearly a quadratic equation which we can solve either by factorising or by use of the standard formula. Always check to see if an expression factorises first – and in this case it factorises into:

$(\sin 2x - 1)(\sin 2x - 2) = 0$

Case 1

$\sin 2x - 1 = 0$

$\Rightarrow \sin 2x = 1$

$\Rightarrow \quad 2x = \sin^{-1} 1$

$\qquad\qquad = \dfrac{\pi}{2} + 2n\pi$

$\therefore \qquad x = \dfrac{\pi}{4} + n\pi$

Case 2

$\sin 2x - 2 = 0$ Which has no solution   Because the maximum value for $\sin(2x)$ is 1

Note that this question does not state an interval within which the solution is required. It asks for all values of x, therefore we have to provide the general solution ($x = \dfrac{\pi}{4} + n\pi$)

This question is telling us that that there are more than two roots of this equation by stating "Find all the roots..." So we should be prepared to look for the general solution.

$2\sin^2 x + \sin x - 1 = 0$

$(2\sin x - 1)(\sin x + 1) = 0$ Factorising

Case 1

$2\sin x - 1 = 0$

$\Rightarrow 2\sin x = 1$

$\Rightarrow \sin x = \dfrac{1}{2}$

$\Rightarrow x = \dfrac{\pi}{6}$ First solution in the first quadrant

G.S is $x = \dfrac{\pi}{6} + 2n\pi$ Period is 2π. Hence G.S. is $\dfrac{\pi}{6} + 2n\pi$

and $x = \dfrac{5\pi}{6}$ Second solution in the second quadrant

G.S is $x = \dfrac{5\pi}{6} + 2n\pi$ Period is 2π. Hence G.S. is $\dfrac{5\pi}{6} + 2n\pi$

Case 2

$\sin x + 1 = 0$

$\Rightarrow \sin x = -1$

$\Rightarrow x = -\dfrac{\pi}{2}$

$= \dfrac{3\pi}{2}$

G.S is $x = \dfrac{3\pi}{2} + 2n\pi$ Period is 2π. Hence G.S. is $\dfrac{3\pi}{2} + 2n\pi$

Once again, this can be recognised as a quadratic equation in $\tan\theta$. The domain or interval has not been stated so we must assume that the general solution is required. A first check reveals that this cannot be factorised, hence we must used the standard formula for solving a quadratic equation $x = \dfrac{-b \pm \sqrt{b^2 - 4ac}}{2a}$

In this case $a = 1$, $b = 4$, $c = -3$

Hence we can write:

$$\tan \theta = \frac{-4 \pm \sqrt{16 + 12}}{2}$$

$$= \frac{-4 \pm \sqrt{28}}{2}$$

$$= \frac{-4 \pm 5.2915}{2}$$

$$= 0.64575 \ or \ -4.64575$$

$$\theta = \tan^{-1} 0.64575 = 0.5734$$

or $\theta = \tan^{-1} -4.64575 = -1.35878 = 4.9245$

Hence the General Solution is

$\theta = 0.5734 + 2n\pi$

$\theta = 4.9245 + 2n\pi$

20. Find the general solution of the equation $2\cos\theta - \sin\theta = 1$ (1)

This equation can be solved in more than one way so we will look at two methods to illustrate there is often more than one way to solve trigonometric equations.

Method (i)

In this case we recognise we have an equation of the form $a\cos\theta + b\sin\theta = c$, which has the solution:

$$\cos(\theta - \gamma) = \frac{c}{\sqrt{a^2 + b^2}}$$

$$\cos(\theta - \gamma) = \frac{1}{\sqrt{2^2 + (-1)^2}} = \frac{1}{\sqrt{5}} \quad (2)$$

$a = 2$, $b = -1$, $c = 1$

From the diagram we have $\cos\gamma = \frac{2}{\sqrt{5}}$, $\sin\gamma = \frac{-1}{\sqrt{5}}$ and $\tan\gamma = \frac{-1}{2}$

$\therefore \tan^{-1}\left(-\frac{1}{2}\right) = -0.4636$

$\Rightarrow \gamma = -0.4636$

Hence we can now write (2) as:

$$\cos(\theta + 0.4367) = 0.4472$$

Hence $\theta + 0.4636 = \cos^{-1} 0.4472$

$\Rightarrow \qquad \theta + 0.4636 = 1.1072$

$\cos\theta = c$ has general solution $2n\pi \pm \theta$

So $\qquad \theta = -0.4636 + 1.1072 = 0.6436$ $\qquad$ (3)

or $\qquad \theta = -1.1071 - 0.4636 = -1.5707 = -\dfrac{\pi}{2}$ $\qquad$ (4)

Therefore the general solutions are

$\qquad \theta = 2n\pi + 0.6436$ from (3)

and $\theta = 2n\pi - \dfrac{\pi}{2}$ from (4)

Method (ii)

Using the half angle formula

$\sin\theta = \dfrac{2t}{1+t^2}$

$\cos\theta = \dfrac{1-t^2}{1+t^2}$

Where $t = \tan\dfrac{1}{2}\theta$

We can write (1) as:

$\qquad 2\dfrac{1-t^2}{1+t^2} - \dfrac{2t}{1+t^2} = 1$

$\qquad 2(1-t^2) - 2t = 1 + t^2$ $\qquad$ Multiplying throughout by $1 + t^2$

$\qquad 2 - 2t^2 - 2t - 1 - t^2 = 0$ $\qquad$ Expanding bracket and rearranging

$\qquad 1 - 2t - 3t^2 = 0$ $\qquad$ Simplifying

or $\qquad 3t^2 + 2t - 1 = 0$ $\qquad$ Rearranging

so $\qquad (3t-1)(t+1) = 0$ $\qquad$ Factorising

Hence $\qquad t = \dfrac{1}{3}$ or -1

Giving $\qquad \dfrac{\theta}{2} = \tan^{-1}\dfrac{1}{3}$ or -1

Case 1

$\dfrac{\theta}{2} = \tan^{-1}\dfrac{1}{3}$

$\quad = 0.3218$

$\quad \therefore \theta = 0.6436$

Case 2

$\dfrac{\theta}{2} = \tan^{-1} -1$

$\quad = -0.7854$

$\therefore \theta = -1.5708$

These are the same results as we obtained previously in (3) and (4), leading to the same general solutions as before.

The graph below shows the solutions to $2\cos\theta - \sin\theta = 1$ on the interval $-10 \le \theta \le 10$
The red lines indicate the solutions of $\theta = 2n\pi + 0.6436$ and the green lines the

solutions of $\theta = 2n\pi - \dfrac{\pi}{2}$ on this interval and $n = -1, 0, 1$

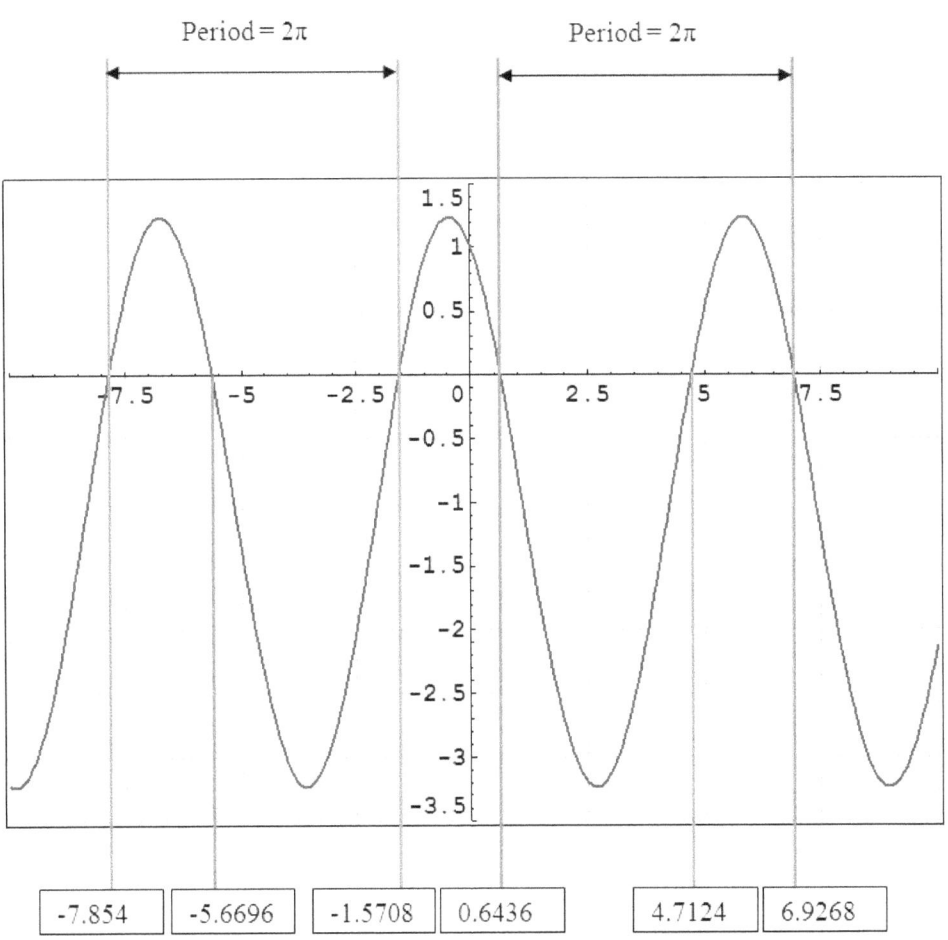

$$2\cos\theta - \sin\theta = 1$$

www.ingramcontent.com/pod-product-compliance
Lightning Source LLC
Chambersburg PA
CBHW081434170526
45166CB00008B/2202